Plastic Injection Molding

Robert A. Malloy

Plastic Injection Molding

A Practical Guide

HANSER

Print-ISBN: 978-1-56990-879-2
E-Book-ISBN: 978-1-56990-880-8

Bibliographic information of the German National Library:
The German National Library lists this publication in the German National Bibliography; detailed bibliographic data are available on the Internet at http://dnb.d-nb.de.

© 2026 Carl Hanser Verlag GmbH & Co. KG, Munich
Vilshofener Straße 10 | 81679 Munich | info@hanser.de
www.hanserpublications.com
www.hanser-fachbuch.de
Editor: Dr. Mark Smith
Production Management: Der Buchmacher, Arthur Lenner, Windach
Cover concept: Marc Müller-Bremer, www.rebranding.de, Munich
Cover design: Max Kostopoulos
Cover picture: © Robert A. Malloy
Typesetting: Eberl & Koesel Studio, Kempten

Dedicated to my friends and mentors Professor Stephen Driscoll and Professor Stephen Orroth, who have both shared their vast plastics engineering knowledge with so many, without ever seeking recognition for their efforts.

Stephen A. Orroth

Stephen B. Driscoll

The Author

Robert Malloy is a Professor Emeritus of Plastics Engineering at the University of Massachusetts Lowell. He served as a faculty member at UMass Lowell for more than 30 years, the last 12 of which he was the department chair prior to his retirement from the university in 2016. He has also served as a faculty member at the Algerian Petroleum Institute Plastics Program in Annaba, Algeria before joining the faculty at Lowell. Over the years, he has taught hundreds of courses on the subjects of plastic materials, plastic processing, plastic part design, injection mold design, and engineering fundamentals. He has also served as advisor to many plastics engineering, manufacturing engineering, and biomedical engineering graduate students.

Professor Malloy is now an active researcher and engineering consultant in the areas of thermoplastic processing, plastic part design, and thermoplastic recycling. He is the author of many book chapters, patents, and publications including books on the subjects of "Plastic Part Design for Injection Molding" and "Injection Molding". Over the years he has been active in a number of professional engineering societies. He is a past chairman of the Society of Plastics Engineers (SPE) Injection Molding Division, a recipient of SPE's International Education Award, and a Fellow of the SPE. Professor Malloy is also a member of the Plastics Hall of Fame being inducted in 2012.

Preface

I have spent the majority of my professional engineering career as an educator and researcher in the areas of plastic materials, plastic part design and plastic processing. From the beginning, I have been fascinated by the injection molding process, due in large part to the versatility of this plastics manufacturing process. The injection molding process can be used to manufacture plastic parts that range from micro-size, complex geometry parts weighing only milligrams, to large automobile and truck body panels weighing many kilograms. Unfortunately, it is this incredible versatility that makes writing such a "how to do" book difficult, because there are few (if any) rules or guidelines that apply to all injection molding applications. Fortunately, there are many excellent books dedicated to injection molding technology. While this is true, I felt that it would be beneficial for me to share some of my experiences and the various plastics processing philosophies that I have developed over the years working with injection molding process. While plastic molding professionals with many years of experience may derive only limited benefit from this introductory text, it is hoped that those new to thermoplastic processing and injection molding will benefit more from reading or referencing the material in this book. For this reason, the text includes a large number of concept illustrations that should help the reader gain a better understanding of the topics being discussed.

The material presented in this book is limited to discussion of the thermoplastic injection molding process. Even with this limitation, there are literally thousands of commercial injection molding grades of thermoplastic available on the market today. Having an understanding of the processing requirements for these materials is a prerequisite to successful part production. For example, a molder with decades of experience molding an amorphous material, such as polystyrene, may encounter completely different processing issues when molding a semi-crystalline, glass fiber reinforced polypropylene. These materials can both be injection molded, but their processing characteristics are very different. As a result, the first chapter of this text is dedicated to discussing the fundamental issues associated with thermoplastic materials and the additives they contain. It is my strong opinion that injection molding process technicians and engineers should have a very good understanding of the molding material properties before starting up a new injection molding process. Chapter 1 has been written in the hope that it will assist injection molding professionals with material supplier and customer communications.

The second chapter of the book is dedicated to the discussion of injection molding machinery and auxiliary equipment. Ideally, the molding machinery and auxiliary equipment used in manufacturing have been designed and purchased specifically for a specific molding application. While this is possible in some dedicated manufacturing operations, such as dedicated high-volume packaging applications, this is not always the case. Much of the injection molding industry is based on custom molding. A key with custom molding is to insure that all of the molding machinery used is closely matched to the application. Machinery considerations range from shot and clamp capacity to plasticating screw and nozzle geometry. The key variables associated with both molding and auxiliary machinery are given in this second chapter.

Chapter 3 of this book focusses on setting the process variables for the conventional injection molding process. The injection molding process is complex, and involves a large number of both sequential and concurrent process variables. Many different philosophies have been developed to assist molding professionals in setting these process variables, many of which exhibit complex interactions. These variable set points will influence both production rate and molded part quality. These two considerations are sometimes in conflict, making some set point decisions difficult. In this chapter, the effect that each primary process variable has on both production rate and part quality are discussed in an effort to provide some rationale or guidance on which process technicians or engineers can base their process set point decisions.

The fourth and last chapter provides a very basic overview of "other" non-standard injection molding processes. There are some molding applications where conventional injection molding is not the optimum process. For example, the low-pressure structural foam molding process may be a more appropriate process option when molding relatively thick, large thermoplastic parts. While this chapter does provide a basic description of these modified injection molding processes, the reader is directed to other sources if a more in-depth coverage of the specific modified molding process is required.

Lastly, I would like to sincerely thank the editorial staff at Hanser Publications, especially Rebecca Wehrmann and Mark Smith, for their encouragement and assistance with this project. I would also like to acknowledge and thank the following individuals from Wittmann USA, Inc. for their willingness to provide technical information that will hopefully benefit the injection molding industry in a broad way. Thanks to David Preusse, Steve Mussmann, Richard Heckbert, Chris Turnberg, and Crystal Brocious for hosting my visit to the Wittmann USA facility in Torrington, CT. I also thank my colleagues at UMass Lowell for the many discussions that have helped me grow personally and professionally.

July 2025

Robert Malloy

Contents

1

Injection Molding: Thermoplastic Material Considerations

1.1 Basic Injection Molding Process Overview

Injection molding is one of many plastic manufacturing processes that can be used to produce thermoplastic parts. The injection molding process has many variants but almost all of these injection molding processes have a number of commonalities. Fundamentally, the injection molding process involves a number of sequential process steps beginning with the melting or softening of a thermoplastic material formulation as shown in Figure 1.1. The charge or shot of plastic melt is then injected into a relatively cool closed mold, through the mold's flow passages (runners) and into the mold cavity(ies) where the molded plastic part shape is formed and cooled. The relatively cool mold will cause the runners and plastic part(s) to cool and solidify to a point where the mold can be opened, and the parts can be "ejected" or de-molded using a mechanical ejector system. The mold then closes and the process repeats itself over and over again for the length of the production run. Once ejected from the mold, the plastic parts will continue to cool and shrink until they reach the ambient temperature. The injection molding process description given above is overly simplistic. The thermoplastic injection molding process and the molding process machinery will be examined in more detail in subsequent sections of this text.

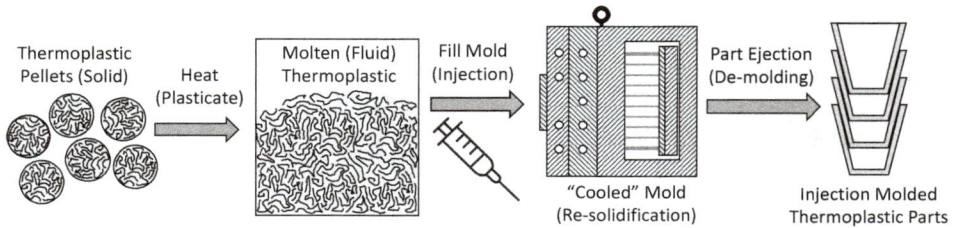

Figure 1.1 Fundamentally, the injection molding process involves a number of sequential steps, beginning with the melting or softening of a thermoplastic material, and ending with the ejection or de-molding of the cooled and solidified parts.

Figure 1.2 shows a simplified cross-sectional schematic of a typical reciprocating screw injection molding machine. Injection molding machines are typically described as having two main sections, the "injection unit" and the "clamping unit". Each unit must carry out a number of thermal and mechanical actions or functions. The injection unit is responsible for (i) melting and mixing the thermoplastic material, (ii) building up a charge of molten material (the shot), (iii) injecting the plastic melt into the mold, and (iv) applying pressure as the melt solidifies. The molding machine's clamping unit (i) holds the mold halves in place, (ii) opens and closes the mold, (iii) generates clamp force (to counteract the applied melt pressure), and (iv) energizes the mold's mechanical ejector system to eject the solidified parts at the end of the injection molding cycle. It is very important to note that the injection molding process can be a dangerous process since it involves high temperatures, high pressures and mechanical actions. As a result, injection molding machines incorporate a variety of safety features and guards designed to minimize the potential for injury. A safety first mindset is essential when injection molding plastic parts.

Molding machines also have very sophisticated control systems that regulate the injection molding process variables, of which there are many. The molding machine's mechanical functions can be driven either by hydraulic systems or directly by electronic servomotors or in some cases by both (in the case of hybrid injection molding machines). Injection molding processes also involve the use of various types of auxiliary equipment. This auxiliary equipment can include material pre-dryers, gravimetric feeders, mold temperature circulators, automation robots, granulators and the like.

Injection molding machines come in a very wide variety of shapes and sizes, which is one of the main reasons why this widely utilized process is difficult to describe in a universal way. Large molded plastic parts, such as automobile bumpers or facia, require very large injection molding machines featuring injection units with large shot capacities, and large, high force clamp units. The injection molding process is also used to produce tiny plastic parts, also known as micro molded plastic parts. Some plastic parts are so small that a single plastic pellet can produce multiple plastic parts.

The technical issues a molder may face with an auto bumper are very different than those for micro-molded parts. Most injection molded parts have a physical size that is somewhere between these two extremes.

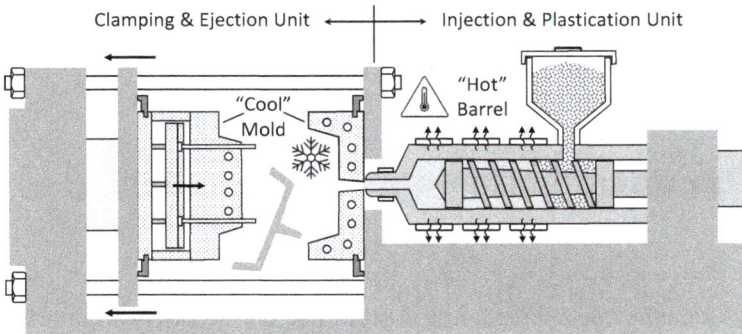

Figure 1.2 Injection molding machines are typically described as having an injection unit and a clamping unit, along with machine controls and auxiliary equipment. Each unit carries out a number of different functions. This simplified schematic shows a reciprocating screw injection molding machine at the point in the process where the solidified part is being ejected from the injection mold.

The injection molding process is commonly used for the production of plastic parts because it offers a large number of advantages and very few limitations. Some of the key advantages of the injection molding process are listed below.

- Most thermoplastics are available as injection moldable grades.

- The injection molding process can produce parts with a very complex geometry.

- Injection molded parts can have extremely tight dimensional tolerances.

- The process is economical, particularly for high production applications.

The most significant "barrier to entry" for the injection molding process is the injection mold or tool cost, which is why injection molding is generally limited to higher production applications. The tool cost must be amortized based on the number of parts that the mold will produce. The molding machinery itself is more cost effective because the injection molding machinery can be used for multiple injection molds and multiple materials. While some injection molding operations, particularly high-volume applications, utilize dedicated injection molding machinery, there are also many "custom" injection molding manufacturing businesses. Using custom molders for part production eliminates the lead times and costs of establishing a dedicated in-house injection molding facility. Whether molding is "dedicated" or "custom", the tool cost is associated <u>only</u> with the part to be molded. There are ways to reduce the cost of an injection mold for applications where part production volumes are relatively low and the tooling budget is limited. However, tooling short-cuts will generally lead to trade-offs with respect to the tool's durability or ability to maintain tight dimensional tolerances (for these low production applications). Note that while tool design is not the main focus of this text, the reader is directed to a number of helpful texts on the subject of injection mold design [1–4]. Certain aspects of injection mold tool design are discussed here, particularly where they involve mold features that have a significant influence on injection molding process variables.

1.2 Material Considerations for the Injection Molding Process

The focus of this text is on the injection molding process, beginning with raw material selection and ending with the production of an in-specification injection molded part. The overall concept associated with establishing a new injection molding process is depicted in Figure 1.3. New consumer products that incorporate plastic components are being developed every day. Many of these products are assemblies made up of multiple components, some of which may be injection molded plastic parts. The development process for new plastic parts can be a very complex task; it was discussed in an earlier text entitled "Plastics Part Design for Injection Molding" [5]. This plastic part development process involves defining the part's end use requirements, engineering design, material selection, prototyping, tool design and construction, and ultimately injection molding (i. e., manufacturing).

The methodology of new product development can be managed in a variety of ways, including the *"over the wall"* approach to development or by *"concurrent enginee-ring"*. While the silo-like *over the wall* method of development is known to be inefficient, it is still relatively common. *Concurrent engineering* will generally reduce development time, and lead to higher quality products because all groups within the product development team have an opportunity to weigh-in on important design decisions. Consulting with the injection molder is considered to be a very good practice when developing a new plastic part that will ultimately be produced using the injection molding process. The molder can provide a perspective on molding issues that could lead to part performance or cost concerns that may be overlooked by the part designer.

Regardless of which methodology is used for the development of a new plastic part, part production by injection molding will begin at some point. Process engineers will need to develop an economical injection molding process that is capable of producing the required quantity of high-quality plastic parts. Various aspects of this process engineering task are discussed in this text. It is assumed that at this point of the new part development process that the process engineer has a number of fixed and variable inputs. The fixed inputs are (i) the injection mold that has been built, (ii) the plastic material formulation that has been selected, and (iii) the molded part specifications (e. g., dimensional requirements), whichhave all been defined.

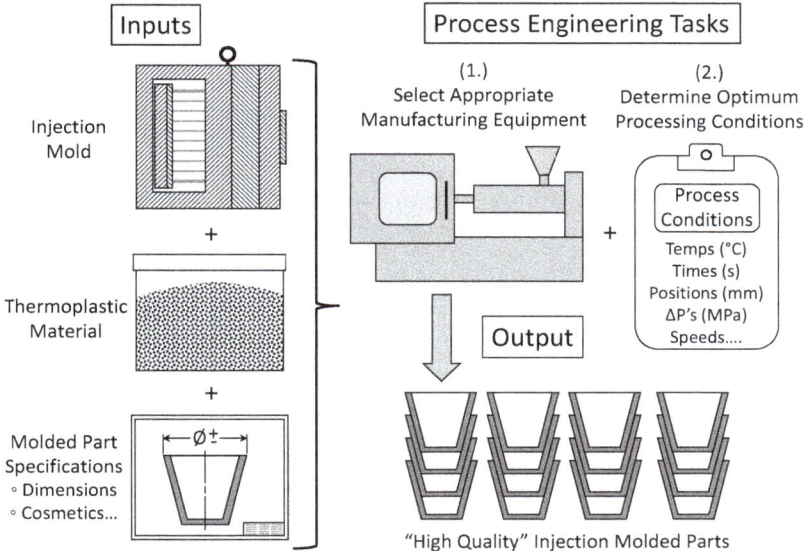

Figure 1.3 Process engineers are routinely given the task establishing a production injection molding process for a new material and new mold. They must first determine what type of molding machinery is most suitable for the new application and then develop a set of molding process parameters that will result in injection molded parts of acceptable quality.

Given these fixed inputs, the process engineer must then:

- Determine what type of injection molding machinery and auxiliary equipment is most suitable for production with the specific material formulation and injection mold to be used. Injection molding machinery is the focus of Chapter 2.

- The process engineer must then procure, acquire or locate suitable production machinery and schedule time for qualification trials and eventual production runs. Ideally, new, perfectly matched injection molding equipment would be specified and purchased for every new molding application. However, in most cases, process engineers must select from existing and available injection molding machinery. The key here is to select molding machinery that is the most appropriate for this new application. Some amount of machinery modification is typically required when new materials and injection molds are put into production (e. g., molding machine nozzle tips, end of arm tooling)

- The process engineer must also establish a production process for the new application. Every injection molding process parameter must be established. The goal is to establish process conditions that result in the production of high-quality molded parts (i. e., parts that meet the end use specifications) while keeping the molding cycle time as short as possible. The methods and concepts used to determine molding process parameters are discussed in Chapter 3 of this text.

There are many factors to consider when the process engineer is given the task outlined above. It is my opinion that learning as much as possible about the plastic material to be molded is the logical starting point. There are so many things about the plastic to be molded that will influence both the machinery selection and the injection molding process parameter settings. For example, if the plastic material is hygroscopic, it will need to be pre-dried. If the plastic material formulation contains abrasive or corrosive additives, appropriate injection molding machine metallurgy will be required. Different types of plastic may require different plasticizing screw geometries or nozzle tip designs. The list of possible material-machine interactions goes on and on. The sections that follow in this chapter are all plastic material related; they have been included because they all have the potential to influence some aspects of the injection molding process or molding equipment requirements. Having a complete understanding of a plastic's processing characteristics and material behavior will help guide a process engineer with decisions when establishing a new injection molding process.

1.3 Thermoplastic Material Nomenclature

Historically, the term "plastic" was used only as an adjective. The term is derived from the Greek word "plastikos" meaning *something (a material) that is capable of being shaped or molded.* As an adjective, the term plastic is essentially the opposite of elastic in the context of material behavior. Elastic deformation is described as being fully and instantly recoverable when an applied stress is removed from a deformed object, while plastic deformation is <u>NOT</u> recoverable. It was not until 1927 that the term plastic first appeared as a noun in that year's version of the *Century Dictionary and Cyclopedia.* In that reference, plastic, as a noun, was defined as *"A natural or synthetic (normally) organic polymer that can be formed into a desired shape when heated and then set into a solid or a more rigid state."*

Plastic: This noun was and is still used to describe these versatile materials that are very widely used to manufacture all kinds of consumer products that we use in our daily lives. Plastics have been described as the most versatile materials on earth. They can be flexible or rigid. They can be opaque, translucent or transparent. Most plastics have excellent electrical insulating properties complimenting the electrical conductivity of metals for the manufacture of electronic devices (that require both electrically conductive and insulating components). Plastics are available in a variety of forms including liquids for coatings, paints and adhesives, or as solids for the injection molded products that are the focus of this text. This text also focuses on the injection molding of thermoplastics, which are materials that soften when heated and solidify upon cooling. This text does <u>not</u> focus on the molding of thermosetting plastics (or thermosets), such as a phenolic. Thermosetting plastics are based on polymers that solidify as a result of an irreversible crosslinking chemical reaction. Most thermoplastics are available as grades that can be processed by injection molding. Some thermosetting plastics can be injection molded, but most thermosetting materials are processed using other manufacturing processes, such as compression molding or transfer molding.

The thermoplastics that are used in the injection molding industry are sometimes sub-categorized as being either:

- Commodity thermoplastics

- Engineering thermoplastics

- High-performance thermoplastics

Commodity thermoplastics, such as polystyrene (PS), are high volume, lower cost materials with general purpose properties. Engineering thermoplastics, like polycarbonate (PC), are medium volume, higher cost materials that exhibit improved mechanical performance and temperature capabilities compared to commodity thermoplastics. High performance thermoplastics, such as polyether imide (PEI), are lower volume, higher cost special-purpose materials that offer even better performance. High performance thermoplastics are sometimes referred to as "super engineering thermoplastics".

While the term "plastic" is very widely used in general materials engineering practice, there are a number of other material descriptors, phrases or terms that are sometimes used instead. These terms may have different meanings to different people, but they are listed and described here based on my interpretation and understanding of these terms.

Plastic material: The more descriptive term "plastic material" is also commonly used as synonym for the noun plastic.

Plastics: It is common to see an "s" added to the noun plastic. Some consider the terms plastic and plastics to be completely interchangeable. Dictionaries state that the term plastics is the plural of the noun plastic. ABS is a plastic (noun), while ABS and HDPE are both plastics (plural noun).

Polymer: The term polymer is also very common in material science. A polymer (or high polymer) is a large, usually chain-like molecule made up of many (poly) parts (mers). Fundamentally, all plastics are polymers, but not all polymers are plastics. For example, DNA, proteins, (unmodified) natural rubber, and cellulose are all polymers, but are not considered to be "plastics" since they *canNOT be formed into a desired shape when heated and then set into a solid or a more rigid state.* Polyethylene is both a polymer and a plastic. Polyethylene is a polymer because it is a long chain-like molecule. Polyethylene is considered to be a plastic (noun) since it can be melted, molded and resolidified. The phrase "polymeric material" is synonymous with "polymer".

Natural polymers: Polymers that exist as a naturally occurring materials are typically described as natural polymers. For example, cellulose (from cotton) is a natural polymer, but it is not moldable and is therefore <u>not</u> considered to be a plastic or plastic material. However, this natural polymer can be "chemically modified" and fortified with additives to produce a moldable plastic material formulation, such as plasticized cellulose nitrate (*celluloid*) or cellulose acetate. Celluloid is cellulose nitrate (i. e., a chemically modified cellulose) that contains camphor as a plasticizing additive. Such chemically modified natural polymers are sometimes described as "semi-synthetic" plastics.

Fully synthetic polymers: The vast majority of commercial plastic materials are "fully synthetic" polymers. The first commercially successful fully synthetic plastic, phenol formaldehyde, was commercialized in 1908. Most of these polymers do not exist in nature and are synthesized by chemically reacting low molecular weight raw materials, known as monomers or co-monomers. The monomers themselves, such as the ethylene glycol co-monomer used to produce PET, can be derived from petrochemicals or from more renewable resources, such as agricultural products. While there are only so many naturally occurring polymers, there are no limits as to the number of fully synthetic polymer structures that can be produced. In some cases, synthetic versions of naturally occurring polymers are produced to increase the material supply, as in the case of synthetic rubber, since the natural supply is limited.

Organic vs. inorganic plastics: Almost all commercial plastics are organic materials, having a carbon-based polymer backbone and other elements that determine the material's chemical structure and performance properties. Organic plastics may also contain elements that include; hydrogen, nitrogen, oxygen, sulfur, chorine, bromine and fluorine. Plastics that do <u>not</u> have a carbon-based backbone, such as silicone-based plastic, are inorganic plastics.

Resin: The term resin is often used as a synonym for the noun plastic. For example, the phrase "resin supplier" would be synonymous with "plastic material supplier" in the context of the plastics industry. Dictionaries generally define resins as *"natural organic substances that are usually transparent or translucent and yellowish to brown, that are formed especially in plant secretions"* (which is not a definition of plastic material). In some respects, the term resin as a synonym for plastic seems to be an older (perhaps outdated) term that may trace its roots back to the invention of liquid phenolics or other liquid plastics or plastic precursors, such as epoxy resins or unsaturated polyester resins. That said, the term resin is still very widely used as a synonym for the noun plastic.

Rubbers and Elastomers: Rubbers and elastomers (or plastomers) are polymers that exhibit good flexibility, extensibility and recovery. They can be naturally occurring or, more commonly, fully synthetic. They can be thermosetting, as with vulcanized natural rubber, or thermoplastic, as in the case of a thermoplastic elastomer (TPE). There seems to be general agreement that all rubbers and elastomers are flexible polymers. These materials are considered by some to be flexible plastics, while others consider that rubbers or elastomers are a separate category of polymers (i. e., they are thermoplastic or thermosetting elastomers, not plastics).

Composites: Dictionary definitions state that a composite material is a *"combination of two materials with different physical and chemical properties."* This is a very generic term that can be used for many material combinations. In the plastics industry, the term composite is commonly used to describe a material that consists of a polymer or plastic matrix material strengthened by fibrous or fabric reinforcements. Composites that are produced using a thermosetting plastics matrix material are often described as advanced composites. Injection moldable composites, that are produced using a thermoplastic matrix material and a reinforcement, such as glass fiber-reinforced nylon 66, are more appropriately described as either short fiber, or long fiber-reinforced thermoplastic composites.

1.4 The Four Plastic Material Pillars

The idea of using "Four Pillars" to describe the underlying dynamics of a subject area or discipline has been seen in many instances. There are the *Four Pillars of Life,* the *Four Pillars of Happiness,* the *Four Pillars of Sustainability.* This list goes on and on. This four-pillar model also provides for a convenient and systematic method of describing four key attributes that are associated with every plastic material in existence. The four underlying pillars that are associated with all thermoplastic materials are:

- Average chain length (i. e., average weight and molecular weight distribution).

- Repeating unit or chain link chemical structure.

- Morphology or chain arrangements.

- Additives that are incorporated into the thermoplastic matrix material.

Once a plastic material has been selected for a particular application, such as for an injection molded plastic part, it is important that process engineers and molding technicians gain a thorough and fundamental understanding of how each of these pillars affects the properties and processability of the plastic material formulation to be molded. Each of these four pillars will have a significant effect on both the processability and performance of the plastic material. The recommendation here is to research or learn as much as you possibly can about the selected plastic material prior to injection molding, especially if the engineers and technicians do not have previous experience of molding the plastic material. This type of information is available from a variety of sources, including plastic material textbooks [6-8], but it is typically the material supplier that can provide the most valuable plastic material information. The four underlying pillars that influence the properties and processability of thermoplastic materials are discussed in more detail in the following sections.

1.4.1 Plastic Material Pillar #1: Average Polymer Chain Length

All thermoplastics are chain-like molecules having a far higher molecular weight than non-polymeric materials. Most thermoplastics have linear, chain-like structures, although some do have a branched structure with side chains that are distributed over the length of the main backbone chain. The average length of these polymeric chains is one of the four variables that will influence the polymer's flow and physical properties. While it can be difficult to generalize when discussing plastics, it is generally true that higher molecular weight thermoplastics exhibit superior performance (in terms of, say, mechanical properties or chemical resistance) than their lower molecular weight counterparts (all other factors being equal).

Higher molecular weight materials exhibit improved performance for at least two main reasons. First, longer molecular chain molecules have a tendency to exhibit a greater degree of "entanglement". A material having more entanglement tends to offer greater resistance to an applied stress. Secondly, the intermolecular forces (the forces between the chains) will also dictate a plastic's performance. These electrostatic interactions, known as secondary chemical bonds (e. g., Van der Waals forces) ensure resistance to applied stresses. The sum of these forces will be greater when the polymer chains are longer, also helping to improve the material's performance.

As an example, consider the linear polyethylene thermoplastic material depicted in Figure 1.4. Material suppliers that produce polyethylene can create different polyethylene grades that have different chain lengths and branching types by controlling the polymerization process variables and catalysts used to make the polyethylene. It should also be noted here that, while material suppliers have a target chain length (or molecular weight) for each material grade they produce, the result of any polymerization process will be a thermoplastic having a statistical distribution of different chain lengths. This is why it is common to describe a material's molecular weight as a weight average molecular weight or number average molecular weight value.

Figure 1.4 shows some examples of polyethylene applications for polyethylene grades with different molecular weights. The candle is produced from a very short chain, low molecular weight polyethylene. The candle material can be melted and cooled and could be injection molded, but candle "wax", as it is called, has very limited mechanical properties due to its very low molecular weight. Wax is essentially a very low molecular weight polyethylene (plastic) but its very limited mechanical properties limit its use as an injection molding material. Most injection molded polyethylene parts, such as recycling bins, are produced using a much higher molecular weight grade of polyethylene. These higher molecular weight grades exhibit superior mechanical performance (compared to wax) as a result of their higher degree of entanglement. The extrusion blow-molded milk bottle shown in Figure 1.4 is produced from an even higher molecular weight polyethylene.

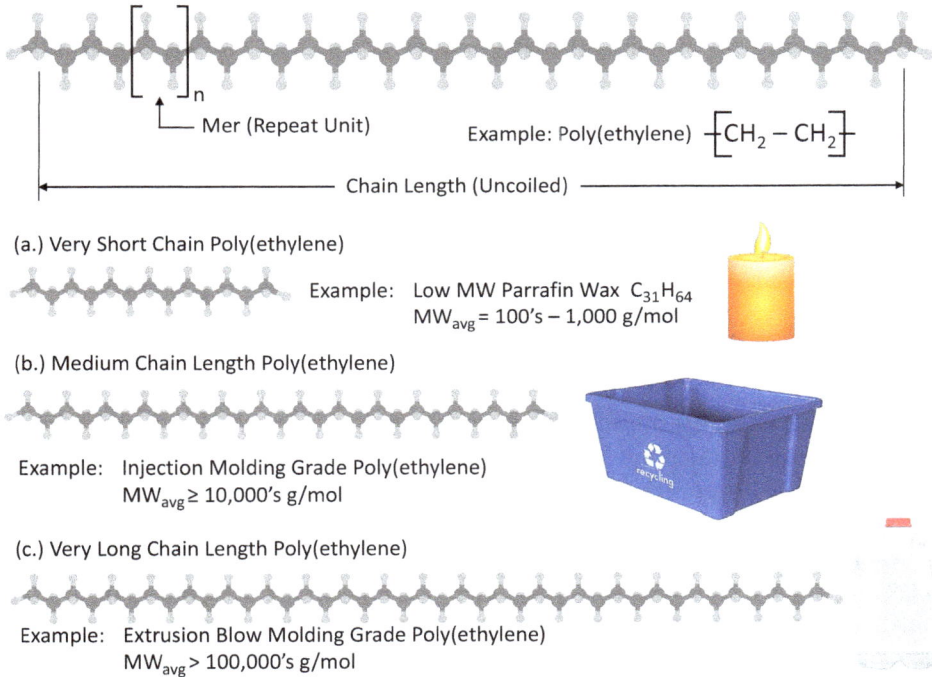

Mer (Repeat Unit)

Example: Poly(ethylene) $\left[CH_2 - CH_2 \right]$

Chain Length (Uncoiled)

(a.) Very Short Chain Poly(ethylene)

Example: Low MW Parrafin Wax $C_{31}H_{64}$
$MW_{avg} = 100's - 1,000$ g/mol

(b.) Medium Chain Length Poly(ethylene)

Example: Injection Molding Grade Poly(ethylene)
$MW_{avg} \geq 10,000's$ g/mol

(c.) Very Long Chain Length Poly(ethylene)

Example: Extrusion Blow Molding Grade Poly(ethylene)
$MW_{avg} > 100,000's$ g/mol

Figure 1.4 The average molecular weight (or chain length) of a plastic has a significant effect on both the material's processability and performance properties.

Higher molecular weight grades of a given plastic material will tend to exhibit improved mechanical performance, but they also have a higher melt viscosity. While it is true that increasing the chain length (i. e., the molecular weight or degree of polymerization) will generally improve the material's performance (for a given type of plastic), this effect is not linear. It is also true that increasing a polymer's chain length will also increase its melt viscosity (or resistance to flow). This is a critical point, since the injection molding process is best suited for thermoplastics that have a relatively low melt viscosity. The injection molding process, especially the mold filling phase of the process, is well suited for lower viscosity materials since it is a completely closed process. Lower viscosity materials require lower mold filling pressures, allow for greater flow lengths, and better surface replication. Consider the extreme case of ultra-high molecular weight polyethylene (UHMWPE). This material has "outstanding" performance properties, superior abrasion resistance in particular, making it suitable for applications like a hip or knee replacements. However, some UHMWPE grades are so viscous, due to their super high molecular weight, that they cannot be melted or plasticated by conventional means. These materials are typically processed as a compression molded, sintered powder.

This is one reason why selecting the material for injection molding applications is not a trivial exercise. On one hand, you may lean towards a higher molecular weight grade of a given material type in order to maximize the material's performance. On the other, a lower molecular weight grade of the same material type would offer improved flow behavior, and this is also important for the injection molding process. Ultimately, the material molecular weight (material grade) that is probably best for a given application is the one that offers the best balance of <u>performance</u> and <u>processability</u>. This balance is not as important for some other plastics manufacturing techniques, such as the extrusion blow molding process used to produce the bottle shown in Figure 1.4. Unlike injection molding, the extrusion blow molding process is well suited for very viscous thermoplastics.

1.4.1.1 Melt Flow Rate (MFR)

While a plastic's average molecular weight is the first of the plastic material pillars, this value is rarely stated on the material data sheets or in material databases provided by material suppliers. Most, but not all, data sheets report on a flow property that is known to correlate (inversely) with average molecular weight. This value, known as "Melt Flow Rate" (MFR) is a qualitative indicator of the average molecular weight for a given material type. A simplified schematic of the MFR test apparatus, known as an *extrusion plastometer*, is shown in Figure 1.5. Specifications for the MFR test equipment and testing procedures, including all variants, are given in test standards that include ASTM D1238 and ISO 1133. The flow test results are usually reported based on mass but can also be reported based on volume (convertible via the plastic material density). The MFR is essentially a heated funnel type test, in which melted plastic is extruded from the heated barrel through a standard geometry die. The specific test temperature and load given in the test standards are based on the type of plastic material being tested. Note that some materials can be tested at different temperatures and loads, which can complicate the comparison of MFR test results. The MFR results for different material grades can only be compared if the tests are run at the same temperature and with the same load conditions.

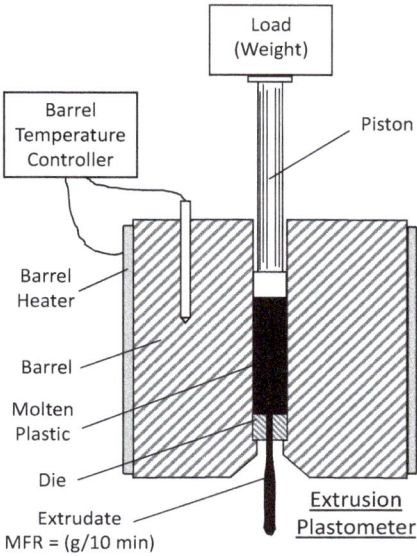

Figure 1.5
While the average molecular weight of a plastic material is a fundamentally important material variable, it is rarely reported as a property in the material supplier's data sheet. In practice, a flow property, such as the melt flow rate (which correlates inversely with a material's average molecular weight), is typically reported on material supplier data sheets or in their electronic databases. Material grades with higher melt flow rates (for a given material type) have lower melt viscosities due to their lower average molecular weight.

A material's MFR value is the weight of material (in grams) that flows through the standardized extrusion plastometer orifice, normalized to a time period of 10 minutes. It is normally reported to two significant digits. The units of the MFR are then grams/10 minutes. Plastic material grades that have a low average molecular weight will have a high MFR value. Plastic materials that have a high average molecular weight will have a low MFR (or a higher melt viscosity). Again, MFR values can only be compared if the comparative plastic material grades have been tested using the same temperature and load values. It should also be noted here that the extrusion plastometer test is used for most thermoplastics, and that the test result is normally described as the material's MFR value. However, in the case of the polyethylene materials (i. e., LDPE, LLDPE, HDPE), the extrusion plastometer flow test result is known as the polyethylene's Melt Index (MI) value. While the extrusion plastometer test results are reported in most material data sheets, this is not universal. Other more quantitative flow property tests, such as a solution viscosity test, are used as an indicator of molecular weight and flow behavior for some thermoplastics. Solution viscosity testing is widely used as an indicator of average molecular weight for grades of polyethylene terephthalate (PET).

The MFR test is simple and repeatable, although results can be influenced by operator error. The test is often criticized as being too simplistic, and as being a very low shear rate flow test that has little correlation with real-life injection mold filling process conditions. The flow properties of most thermoplastics tend to be highly dependent on the shear rate. The extrusion plastometer test is in fact a very low shear rate test. MFR values are given in grams/10 minutes. However, mold fill times for the injection molding process can range from a fraction of a second to several seconds (but not minutes). Possibly providing the best correlation with injection molding flow behavior, the extrusion plastometer test has been a deeply rooted industry standard for more than eighty (80) years. The simplicity of having a single number test result also makes the test result easily and instinctively understandable.

	Melt Flow Rate (gram/10 min.) @ 300°C, 1.2 kg.	Notched Izod Impact Resistance (ft. lbs. / in.)	
Increasing Melt Flow Rate	3.0	18	Increasing Melt Flow Rate
	7.0	17	
	10	15	
	17	14	
	25	12	Reduction in Material Performance: Due to Lower MW$_{avg}$
Lower Melt Viscosity Due to Lower MW$_{avg}$	32	12	
	60	11	
	80	5.0	

Figure 1.6 Melt flow rate (MFR) and Izod impact resistance data for several grades of neat, injection molding grade polycarbonates are shown. At one extreme, the polycarbonate with an MFR of 3.0 g/10 minutes is a relatively viscous injection molding grade. At the other extreme, the polycarbonate with a MFR of 80 g/10 minutes has a "very" low melt viscosity (and low average molecular weight). This very fluid PC grade is typically used in applications where superior surface feature replication is required (such a music CD or DVD), while the more viscous grades would typically be used in thicker-walled applications where greater mechanical performance is required [9].

The data shown in Figure 1.6 can be used to illustrate how the MFR can be a useful property value. In this case, the material supplier has provided property data for eight different injection molding grade neat polycarbonates (PC). The values reported here are the notched Izod impact resistance (ASTM D256), and the Melt Flow Rate for each grade of polycarbonate. One can draw conclusions regarding relative molecular weight in this case since all of the polycarbonates in Figure 1.6 are described as neat polycarbonates. The additives used to impart special properties to a plastic material formulation can also influence MFR comparisons. A neat polycarbonate having an MFR of 3.0 grams/10 minutes would be a very viscous, high molecular weight grade of polycarbonate. The polycarbonate with the melt flow rate of 80 grams/10 minutes would be a "very" fluid, low molecular weight grade of polycarbonate. As discussed earlier, both the melt viscosity and mechanical performance of a plastic material are influenced by molecular weight. The higher molecular weight grade is the most viscous but has the highest impact resistance or toughness. Figure 1.6 shows that the lower molecular weight grades of polycarbonate have a lower viscosity, but they also exhibit a significant reduction in impact performance. Each material grade has a specific MFR value which is typically used as a key material descriptor for material selection. Material engineers would use such data to screen or select the most appropriate polycarbonate grade for a particular application. For example, a lower MFR grade would typically be selected for a relatively thick-walled, easy to fill, injection molded part that requires good toughness. A higher MFR grade of polycarbonate might be selected for a thinner wall, harder to fill part geometries. The higher MFR grades have better flow properties, but reduced impact performance. To sum up, the key point here is to consider both the material's processability requirements and end use performance requirements when selecting the most appropriate material grade for a new application. Higher molecular weight grades (or those with lower MFR values) are inherently better in terms of mechanical performance, but their higher melt viscosity can lead to mold filling difficulties in injection molding applications. The use of lower molecular weight grades helps to minimize surface replication and knit line issues.

1.4.1.2 Molecular Weight Distribution

As stated previously, all commercial plastics have an average molecular weight value when they are synthesized by the material supplier. The result of most polymerization processes is a plastic material having a distribution of molecular weights where some molecules are above the average and others below it. Both the average molecular weight, and the breadth of this molecular weight distribution (MWD) will influence both the melt flow and performance properties of the plastic material. The exact shape of a MWD curve for a given material grade is not always disclosed by the material supplier and can be difficult to quantify without relatively expensive size exclusion chromatography test equipment. However, it is important for molding engineers to recognize (i) that in practice, such a molecular weight distribution does exist for thermoplastics, and (ii) that it can be difficult for material suppliers to (precisely) replicate the shape of the MWD curve each time they produce a new lot or batch of thermoplastic material. The processing characteristics of a new material lot can be different from a previous material lot if the shape of the MWD curve changes. In fact, a change in the shape of the MWD curve (such as that shown in Figure 1.7) is the most likely cause of what is known as lot to lot variation, which is sometimes encountered in plastics manufacturing.

It would be ideal for molders to screen incoming material lots for MWD curve changes, but this is not practical in most cases since molders would rarely have access to the relatively expensive size exclusion chromatography test equipment required to determine the shape of the MWD curve. This is also one of the main reasons why it is important for process engineers to develop molding processes with a wide process window (especially in terms of mold filling variables) so that molded parts of acceptable quality can be produced, even if there is a limited shift in the shape of the MWD curve for a new material lot.

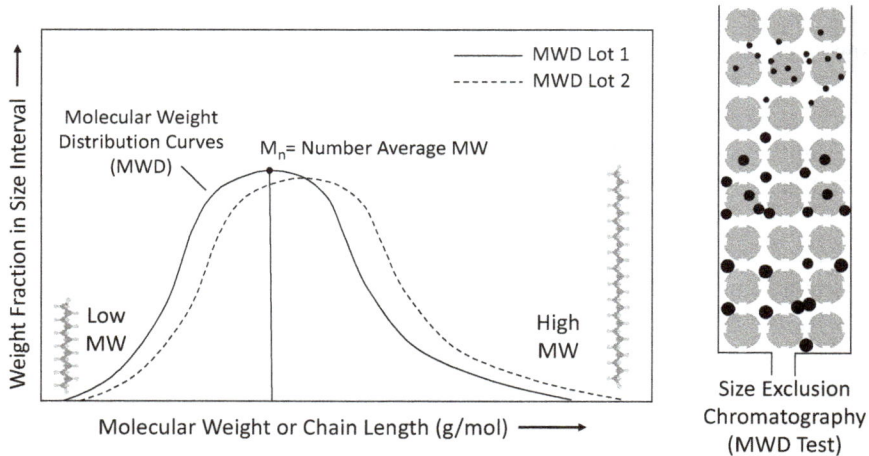

Figure 1.7 All commercial thermoplastic material grades have a target molecular weight distribution (MWD). In practice, different plastic material lots can have different MWDs, which will have some influence on the material's physical properties, especially on its melt flow behavior. The exact shape of a MWD curve can only be determined using relatively expensive size exclusion chromatography test equipment.

While most molders may not have the ability or equipment required to generate a MWD curve, there are other lower cost tests that could provide hints as to whether or not there is, a difference in the MWD curve for a new material lot. The most common incoming material test conducted for this purpose is the melt flow rate (MFR) test discussed above in Section 1.4.1.1. The MFR test is commonly used as an incoming material quality screening test since the test is relatively simple, and the test equipment relatively inexpensive. Normally, a routine MFR test is run to determine whether the MFR of the incoming material lot is within an acceptable tolerance of the material's published MFR value. The definition of acceptable tolerance is variable within the industry, but can be as much as +/- 20% according to some material suppliers. However, even if the incoming material lot MFR test results indicate that the material MFR value is within specification, the material may still have a different MWD, and the flow behavior at other melt temperatures and higher flow rates could be different.

Molders can gain additional information about a material's flow behavior by conducting extrusion plastometer tests using multiple test load values. Standard (or non-standard) MFR test procedures, such as the High Load Melt Index (HLMI) or the Melt Flow Index Ratio (MFIR) are used to gain a better understanding of a material's flow behavior. Multiple load MFR tests are typically run using the normal load indicated in the MFR test standard, and a significantly heavier load (typically ten times the normal load). A comparison or a ratio of these two MFR values can provide insights into the changes in the shear thinning flow behavior of the incoming material lot [10].

1.4.2 Plastic Material Pillar #2: Repeating Unit (Chain Link) Chemical Structure

A second variable that will determine both the performance and processability of a plastic material is the chemical structure of the chain link, or the repeating units, that make up the polymer chain. While the chain length (or average molecular weight) is an important plastic material variable as discussed above, two chains having the same length, but different chain link structures, will also exhibit different properties. Fundamentally, the chemical structure of the repeating chain link will affect both (i) the intramolecular strength of the chain (i. e., the axial strength of the chain itself) and (ii) the intermolecular strength of the chain (i. e., the electrostatic attractions between chains). Repeating units having a more polar chemical structures tend to result in a polymer that has greater strength, stiffness and temperature resistance when compared to materials having a more non-polar repeating unit structure. Non-polar repeating units have less electrostatic attraction between chains, leading to greater material flexibility and lower process softening temperatures. Polar polymers have more electrostatic attractions (i. e., secondary chemical bonds) between chains. These electrostatic attractions include Van der Waals forces, dipole interactions and hydrogen bonding. For example, polycarbonate is significantly stiffer, stronger and more temperature resistant than polyethylene because polycarbonate's polar, atomic structure offers both greater chain stiffness and greater intermolecular bonding in comparison to polyethylene. The processing temperature for polycarbonate is also significantly greater than that of polyethylene since additional energy is required to overcome the stronger electrostatic attractions associated with polycarbonate's polar structure.

1.4.2.1 Polymer Structure Nomenclature

The generic chemical name of a plastic is typically based on the chemical structure of the repeating chain link or the monomer from which the chain link is derived. For example, ethylene is the monomer that is used to produce polyethylene. This is also written as poly(ethylene) by one nomenclature convention where the repeating unit or monomer name is given in parentheses. Styrene is the monomer used to produce polystyrene or poly(styrene). All plastics do have a poly(xxxxxx) name, but the name can be complicated, especially for non-chemists. For example, poly(hexamethylene adipamide) is the chemical name for a plastic material more commonly known as nylon 66 in normal engineering practice. Poly(ethylene terephthalate) is more commonly described as PET for brevity. This type of non-standard nomenclature can commonly lead to confusion. This confusing nomenclature situation is further complicated by the fact that most material suppliers use registered tradenames to describe the materials they supply.

Thermoplastic materials are further characterized as being either homopolymers or copolymers.

Homopolymers: Many commercial thermoplastics are described as homopolymers, because they are produced or polymerized using a single monomer type. In some cases, this material descriptor is also used in conjunction with the material's name. For example, poly(propylene) homopolymer is a common material descriptor for grades of polypropylene produced from a single monomer type (i. e., propylene monomer).

Copolymers: Many other commercial thermoplastics are described as being copolymers. Copolymers are polymers that are synthesized using two monomers. The use of multiple monomers allows material suppliers to produce plastics that have a far wider range of properties by controlling (i) the comonomer chemical structures, (ii) the relative comonomer concentrations, and (iii) the distribution of the comonomers along the chain length. The plastic material property variations that can be produced by copolymerization are virtually limitless. Thermoplastics that are produced using three different monomers are typically described as "terpolymers".

For example, poly(acrylonitrile butadiene styrene), more commonly known as ABS, is a terpolymer. There are literally hundreds (or more) commercial grades of ABS available on the market. Material suppliers can control the ratio of each monomer type as well as their distribution along the main chain or side chains. An ABS having a high concentration of B (butadiene) would likely have superior impact resistance. An ABS having a high A (acrylonitrile) concentration would exhibit improved chemical resistance and stiffness. An ABS having a high S (styrene) concentration will tend to be less expensive. This ability to control the copolymer's "chemical structure" (as well as average molecular weight as described in Section 1.4.1) allows material suppliers to offer ABS grades that have an incredibly wide range of properties and processability. Many plastics, such as polypropylene or high-density polyethylene, are commercially available as both homopolymer and copolymer grades.

It should be noted here that there can be some confusion with the nomenclature associated with copolymerized thermoplastics and thermoplastic polymer blends. Copolymers have multiple links within the same polymer chain. Polymer blends are produced by physically mixing two different polymer chains (each polymerized separately). Fundamentally, a polymer blend consists of one primary polymer (the higher concentration polymer or matrix material) and a second, lower concentration polymer that is distributed throughout the primary polymer matrix material (essentially as an additive). For example, SAN is a copolymer produced using both styrene and acrylonitrile comonomers. A thermoplastic polymer blend could also be produced by physically mixing a polystyrene (PS) homopolymer with a polyacrylonitrile (PAN) homopolymer. The SAN copolymer and the PS/PAN blend would have very different material properties.

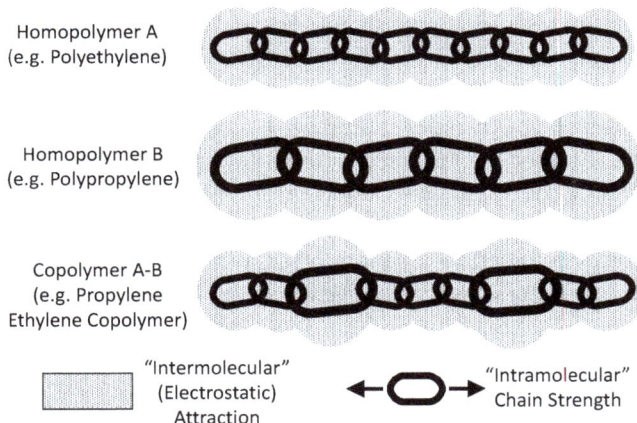

Figure 1.8 All thermoplastics are relatively long, chain-like polymeric molecules. The chemical structure of the individual chain links or repeating units is one factor that will determine both the performance and processing characteristics of the polymer. The chemical structure of the link will determine both the intermolecular attractions (the attractions between chains) and the intramolecular strength and stiffness of the polymer chain. Homopolymers have one type of chain link, while co-polymers or terpolymers, have multiple link structures within the same polymer chain.

1.4.2.2 Polarity and Hygroscopicity

Hygroscopicity is the tendency of a solid substance, such as a plastic material, to absorb moisture from the surrounding atmosphere. Since water molecules are polar, the tendency for a polymer to absorb moisture from the atmosphere is primarily dependent on the polarity of the polymer's repeating unit. Very polar polymers, such as a nylon 66, can absorb several weight percent of moisture (e. g., about 2% moisture) depending on the environmental conditions, while inherently non-polar polymers, such as polyethylene or polypropylene have little or no tendency to absorb moisture from the surrounding atmosphere. Non-polar polymers typically have an absorbed moisture content in the magnitude of hundredths of a percent. The moisture content can be higher if the polymer formulation includes a polar additive, such as a polar filler. In general, non-polar plastics, like neat polypropylene, do not require pre-drying, unless material storage or handling conditions lead to surface moisture or condensation on the plastic pellets.

Any hygroscopic plastic material formulation should be pre-dried prior to injection molding, unless it has been pre-dried and packaged in moisture resistant barrier packaging. Pre-dried plastic materials are used for small volume applications, but care must be taken to ensure the moisture resistant barrier packaging has not been compromised. Molding a plastic that contains even a small concentration of water can present a variety of problems if wet plastic material is injection molded. It is best to check with the material supplier to (i) obtain recommendations for pre-drying process conditions and drying equipment requirements, and (ii) to obtain a recommended value for the maximum acceptable moisture concentration or (iii) ideally both. Molders working with hygroscopic plastics should have moisture analysis test equipment, such as a gravimetric moisture analyzer, so that they can determine the percentage of moisture in dried material prior to injection molding. The moisture analyzer is used to confirm that the moisture concentration in the dried plastic material is below the upper limit recommended by the material supplier. This instrument is essentially a sensitive balance contained within a small oven. This method has some limitations but is the most widely used general purpose method for moisture concentration analysis due to both its simplicity and the relatively low equipment cost.

If moisture analysis equipment is not available to a molder, it is extremely important to ensure that the material supplier's recommendations regarding material pre-drying conditions are precisely followed. Drying variables include the type of pre-drying equipment, dryer capacity, drying times, drying air temperatures, air flow rates, and drying air dew point. Drying conditions that are outside of the drying window recommended by the material supplier can lead to a variety of issues. Under-drying the plastic material will result in an unacceptably high residual moisture concentration. Over-drying will result in an acceptably (low) moisture content but can cause other material issues such as oxidative degradation of the base plastic material or an additive.

Proper maintenance of the drying equipment is also critical. It is also important that there is no way for moisture to be re-absorbed as the pre-dried material is transferred from the dryer to the feed hopper of the injection molding machine. It should also be noted here that, historically, vented barrel injection molding machines have also been used to eliminate the need for pre-drying of some hygroscopic plastics (ABS being a popular example in the past). The use of vented barrel injection molding machines does not seem to be as popular today as it was in years past.

Molding a wet (or incorrectly dried) plastic material can lead to a variety of potential issues with the injection molding process or for the molded plastic part. The most common issue, which is likely to occur for any incorrectly dried thermoplastic is the cosmetic surface defect known as splay. The splay phenomenon as theorized is depicted in Figure 1.9 (bottom) and will occur when several criteria are met. First, the water within the plastic will volatilize and turn to steam at the high material temperatures associated with melt processing of a plastic material. Typical thermoplastic melt temperatures are significantly greater than the 100°C boiling temperature for water. Some of these gas bubbles will tend to break though the melt front during the mold filling phase of the molding process since the mold cavity is vented to the atmosphere. The melt front also exhibits fountain flow behavior [11] causing some of these gas bubbles to become trapped between the flowing melt and the mold cavity wall. The result in streak-like defects along the flow direction, generally described as splay or silver streaking. The corrective action to minimize or eliminate this surface defect is to ensure that the material to be molded is dried to a moisture concentration that is below the material supplier's recommended maximum moisture concentration. In most cases, splay will not occur with non-polar or correctly dried polar plastics as depicted in Figure 1.9 (top), unless volatiles (other than water, such as trapped air) are somehow present. Volatiles can also be generated during melt processing if the plastic material is exposed to excessive heat or shear stress. Undried moisture can also lead to a variety of other issues including; internal voids, weaker knit lines, mold vent clogging, and nozzle tip or hot runner drool [12].

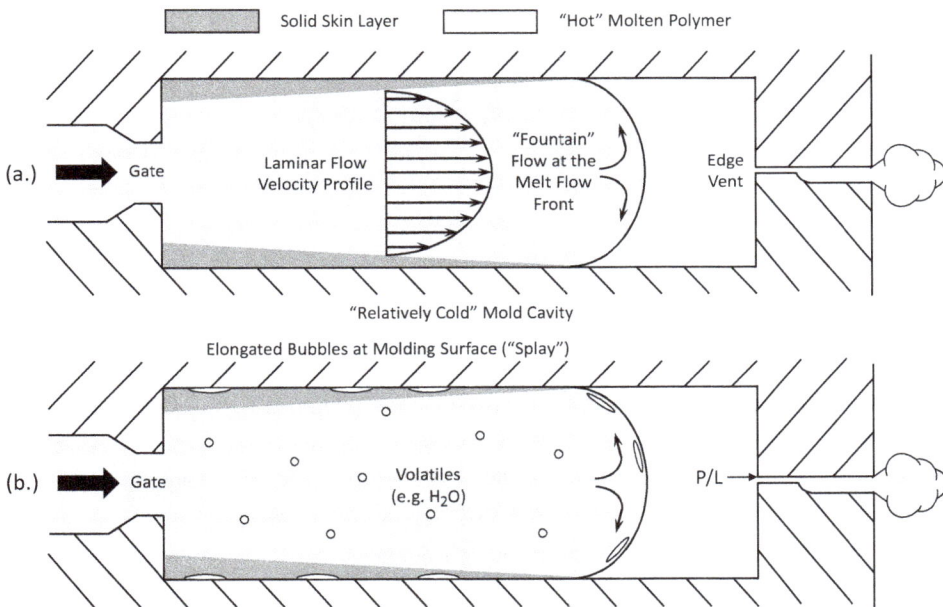

Figure 1.9 Polar or hygroscopic plastics, such as PMMA (acrylic) or nylon 66, have a tendency to absorb moisture from the surrounding atmosphere to a degree that is dependent on the material's polarity and the environmental conditions associated with plastic material storage. Such hygroscopic plastics need to be pre-dried prior to injection molding in order to bring the moisture level down to an acceptable concentration prior to melt processing. Undried (residual) moisture will vaporize at injection molding temperatures, which can lead to a number of molding or molded part issues, including nozzle drool and surface splay.

1.4.2.3 Hydrolytic Degradation for Condensation Polymers

While undried moisture can lead to splay or nozzle drool issues for most hygroscopic plastics, undried moisture in some plastic materials can result in significant material performance issues, caused by hydrolytic degradation. Most plastics are described as either (i) addition polymers, or (ii) condensation polymers, named for the type of poly-merization process used to synthesize the plastic material. Addition polymers, such as an acrylic plastic, are produced by self-reacting a monomer (or two monomers in the case of a copolymer) to produce a polymer chain. Generally, an addition polymeriza-tion reaction produces only the target polymer without any by-products. Molding a wet addition polymer will result in splay, but the water will generally not chemically react with the addition polymer. The effects of the water for addition polymers are purely physical, not chemical.

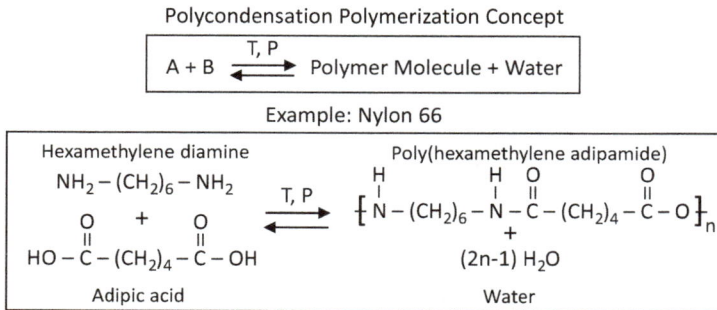

Figure 1.10 Correct pre-drying is particularly important for condensation polymers, such as polyesters or nylon polymers. When condensation polymers are synthesized, a low molecular weight by-product, such as water, is also produced. If a condensation polymer containing water is introduced into the heated barrel of an injection molding machine, some degree of de-polymerization or molecular degradation or chain scission can occur. This reaction reduces the average molecular weight of the polymer and therefore impacts on the mechanical performance of the molded part.

This is not the case for plastic materials that are produced by a condensation polym-erization reaction. Condensation polymers, such as a nylon 66, are far more chemi-cally sensitive to water, especially at the high temperatures the wet material will en-counter in the injection molding machine's barrel or hot runner system. Figure 1.10 depicts the condensation polymerization reaction for nylon 66. Like most condensa-tion polymers, nylon 66 is made by reacting monomeric constituents. The result of the polymerization process is a polymer (nylon 66 in this case) and a low molecular weight by-product. The low molecular weight polymerization by-product in this case is water. The opposing arrows in Figure 1.10 show that the condensation polymeriza-tion process is reversible. The right arrow shows how the polymer is made. The left arrow shows that the polymerization process can be reversed. Heating a condensa-tion polymer in the presence of moisture, can lead to a partial depolymerization reac-tion. This is exactly what can happen when a wet (i. e., incorrectly pre-dried) conden-sation polymer is heated in the barrel of an injection molding machine. The heated barrel is essentially a chemical reactor when it is used to process a wet condensation polymer. The net result of this partial depolymerization is a molded part having (i) a lower average molecular weight (compared to the feed pellets) and (ii) a cosmetic splay defect. The importance of a plastic's molecular weight was discussed in Section 1.4.1. The properties and performance of a plastic are very dependent on the material's average molecular weight. If the plastic material molecular weight is somehow degraded (due to hydrolytic degradation in this case), then the material performance, especially in terms mechanical properties, will be compromised. It is for this reason that pre-drying of condensation polymers is far more critical than it is for plastics produced by the addition polymerization process.

There is nothing in a polymer's name to indicate whether it has been produced by an addition or condensation polymerization process. Condensation plastic materials include; nylons, polyesters, polycarbonates, and polyurethanes. Because these materials can be so sensitive to moisture, it is important to initiate quality control procedures to ensure these materials are pre-dried to an acceptable moisture content before injection molding. In addition, many molders will perform tests on the molded parts to make sure that they exhibit the expected mechanical performance. It can also be helpful to evaluate the melt flow behavior of molded parts that have been granulated and pre-dried for comparison with the flow behavior of the virgin pre-dried plastic pellets. The MFR or flow properties of a plastic do have a correlation with the average molecular weight of the polymer as discussed in Section 1.4.1.1. This test is typically involves measuring the melt flow rate (MFR) of the dried virgin pellets and the MFR of the dried, reground parts. Ideally these two MFR values should be exactly the same if there is no chemical degradation associated with the injection molding process. However, in many cases the MFR of the reground parts will be slightly higher than that of the pre-dried virgin hygroscopic material pellets since trace levels of moisture maybe present. There can also be other possible reasons for molecular degradation (e. g., purely thermal, oxidative or shear degradation). The slight increase in MFR is inevitable and is acceptable to some degree. However, there is much debate as to how great an increase in the MFR is acceptable [13-15]. There is no simple answer as this value is material and application dependent. Various sources suggest that an MFR increase (part vs. pellet) as high as 10–40% (maximum) for unfilled plastics is acceptable for general purpose injection molding applications. The best molded part performance will be obtained when the MFR parts to pellet delta approaches 0%. Anything molders can do (in terms of pre-drying conditions, other processing conditions or machine barrel size selection) to minimize this MFR increase is desirable. The magnitude of an acceptable process related MFR increase for a filled or reinforced plastic is more difficult to rationalize or interpret since factors such as fiber length degradation also have a significant effect on a material formulation's MFR value.

1.4.3 Plastic Material Pillar #3: Morphology or Chain Arrangements

Almost all commercial thermoplastics have either an amorphous morphology, or a semi-crystalline morphology. These morphological categories indicate how the polymer chains are arranged within the solidified material. Both morphologies are shown in Figure 1.11 in the context of the injection molding process. Each morphology will be discussed in more detail in the section below. Unfortunately, there is nothing in the name of most plastics to indicate whether they are amorphous or semi-crystalline polymeric materials. There are exceptions in terms of nomenclature, such as amorphous nylon. Most nylon plastics have a semi-crystalline morphology (e. g., nylon 66, nylon 6), but some nylons, and specifically nylon copolymers, are amorphous because the co-monomer disrupts chain regularity, and this in turn interferes with the copolymer's ability to crystallize. Both amorphous and semi-crystalline plastics are very widely used commercially.

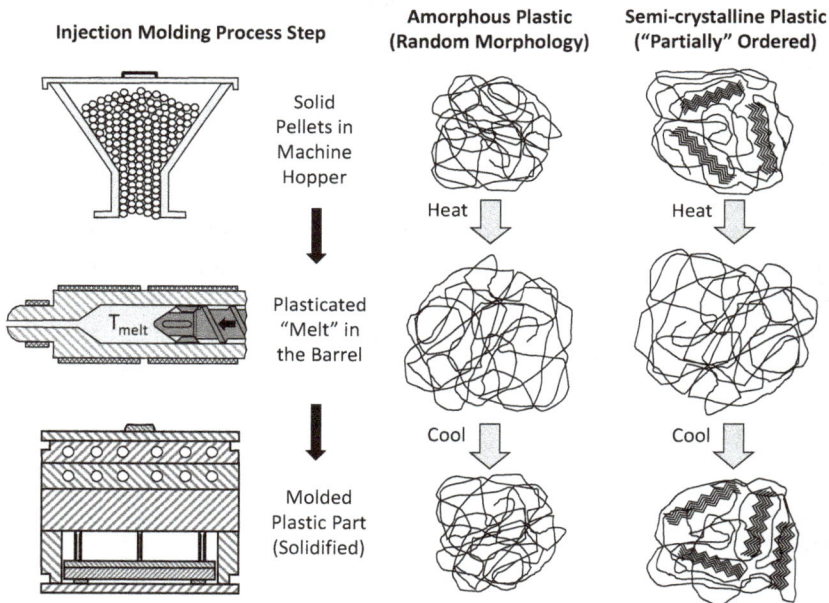

Figure 1.11 A plastic's morphology relates to the geometric form or arrangements taken by the long polymer chains. Many polymers, such as polystyrene or polycarbonate, are described as having an amorphous (random) morphology. Other polymers, such as polypropylene or nylon 66, are described as having a semi-crystalline morphology, meaning that they have both (ordered and tightly packed) crystalline regions and amorphous regions. Both amorphous and semi-crystalline plastics exhibit an amorphous morphology when they are in the molten state.

1.4.3.1 Amorphous Morphology

When describing the morphology of an injection-moldable thermoplastic, it is best to consider the material's morphology (i) before molding, (ii) during molding, and (iii) after molding, as depicted in Figure 1.11. Amorphous plastics, like ABS or polystyrene, have a completely random chain configuration in both the molten and solid state. People often use the analogy of a cooked bowl of spaghetti to help visualize the random configuration exhibited by amorphous plastics. Figure 1.11 shows that amorphous plastics have a random configuration before, during and after molding. The free volume for an amorphous plastic (i. e., the spacing between polymer chains) will be greater when the material is in the molten state. This is because almost all plastics, including almost all amorphous plastics, have a positive coefficient of thermal expansion (CTE), which means that the plastic expands when heated, and contracts when cooled. During injection molding, amorphous plastics must be heated to a temperature where they are fluid enough to allow for the shot plasticizing and subsequent mold filling process. At the processing temperature (i. e., temperature of the melt), the increase in free volume caused by thermal expansion (i. e., chain spacing) reduces intermolecular attraction to the point where the material will flow and enable injection mold filling. Once the relatively cold injection mold is filled, the amorphous material will cool, thereby decreasing free volume, which in turn causes greater intermolecular attraction and re-solidification. The material must be cooled to the point where it becomes rigid enough to resist the forces associated with part ejection and distortion after ejection.

All amorphous plastics exhibit certain common characteristics, but they can also exhibit widely varying behavior for other properties. For example, all amorphous plastics exhibit a relatively low mold shrinkage value, compared to semi-crystalline material. The mold shrinkage of an amorphous plastic is due only to thermal contraction. Material suppliers will generally state a range of shrinkage expected for an amorphous material because mold shrinkage is also affected by factors such as cavity pressure history and frozen-in molecular orientation (which causes directional effects). A typical mold shrinkage value for a neat amorphous plastic is about 0.005–0.007 m/m (i. e., 0.5–0.7%) but this shrinkage factor range is material grade specific. Material supplier data sheets normally include shrinkage factor recommendations. In general, amorphous plastics are relatively easy to work with for injection molding applications, due to both their wide softening temperature range or window, and their relatively low mold shrinkage factors (compared to semi-crystalline plastics). The latter allows for predictable part dimensions, low part warpage, minimal sink marks, and generally predictable and achievable, tight dimensional tolerances. Filled amorphous plastics exhibit even lower mold shrinkage values than neat amorphous plastics.

Amorphous plastics can have a very wide variety of mechanical properties. They can be flexible thermoplastic elastomers (TPE's) or they can be rigid and stiff plastics such as polystyrene. They can be inherently tough, like polycarbonate, or brittle, like general purpose polystyrene. Amorphous plastics can also be very transparent. Amorphous plastics, such as polycarbonate, polystyrene, and acrylics are transparent because their morphology is relatively homogeneous in terms of free volume. There are no crystalline regions (having a different density or refractive index) to diffuse light. It is interesting to note here that general purpose polystyrene is sometimes referred to as crystal polystyrene, but not because it is semi-crystalline. It is not. Polystyrene is amorphous, but the name derives from the fact that it is described as being crystal clear. Amorphous plastics can have visible light transmission values that exceed that of window glass. However, not all amorphous plastics are transparent. Multi-phase amorphous plastics or those containing pigment additives will be translucent or opaque. One disadvantage associated with most (but not all) amorphous plastics is their limited chemical resistance, especially their limited resistance to organic chemicals or solvents compared to semi-crystalline plastic materials.

1.4.3.2 Semi-Crystalline Morphology

Plastics with a semi-crystalline morphology have a far more complex molecular arrangement than amorphous plastics. As solids, semi-crystalline plastic materials have regions that are amorphous (random) and regions that are crystalline (ordered, more tightly packed regions). They are partially amorphous and partially crystalline. No thermoplastic material is 100% (or fully) crystalline, but the sum of the amorphous regions and crystalline regions must be equal to 100%. The performance of a semi-crystalline plastic will ultimately be determined by the relative ratio of each phase or what is commonly described as the material's or molded part's degree of crystallinity.

Figure 1.11 shows how the morphology of a semi-crystalline plastic changes during the injection molding process. A semi-crystalline material is introduced in the form of pellets with a semi-crystalline morphology. The initial degree of crystallinity is both material dependent, and dependent on the pellet manufacturing process conditions. Pre-drying conditions can also influence the percentage of crystallinity exhibited by the pellets. When a semi-crystalline material is injection molded, the material is first melted or plasticated within the heated molding machine barrel. The material is heated by conduction and shear forces (viscous heating), and its free volume increases. This heat causes the amorphous regions to soften due to thermal expansion. The crystalline regions must also be heated to the point where they melt and become amorphous. The crystalline regions of a semi-crystalline plastic have a true melting temperature where they transition from tightly packed crystals, to a more loosely packed, fluid amorphous melt. Once fully melted, the now amorphous melt shot (or charge) can be injected into the mold cavity. Once the melt has been injected into the mold cavity, the melting process is reversed as the melt cools as shown in Figure 1.11. The net result is a molded plastic part that has both amorphous and crystalline regions, each contributing to the performance of the molded part in its own way. The more tightly packed crystalline regions are loosely analogous to reinforcements as they impart greater stiffness and strength.

The final degree of crystallinity in the molded part is not necessarily equivalent to the degree of crystallinity of the pellets entering the process (which is actually irrelevant in itself). Any condition that promotes slower cooling will tend to promote a greater degree of crystallinity for a semi-crystalline material because crystals take time to form. Hotter melt and mold temperatures, and thicker part wall sections, will generally result in a greater degree of crystallinity. On the other hand, semi-crystalline materials molded at a low mold temperature or in a thin wall will cool faster, resulting in a lesser degree of crystallinity. This cooling rate phenomenon introduces a number of molding issues for semi-crystalline materials that do not exist when molding amorphous thermoplastics.

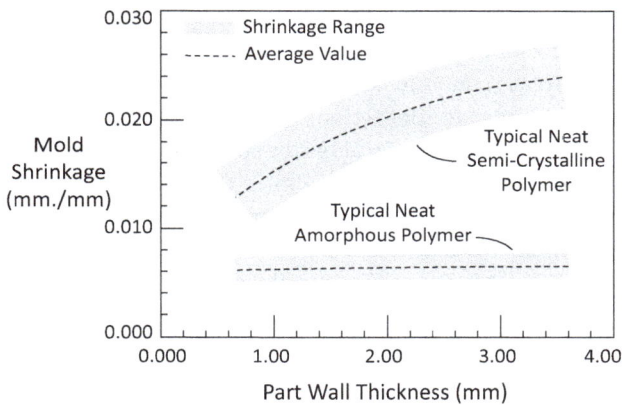

Figure 1.12 Neat semi-crystalline plastics exhibit far greater mold shrinkage factors than neat amorphous plastics because the crystalline regions of a semi-crystalline polymer are packed more tightly as they cool. In addition, the mold shrinkage of a neat semi-crystalline polymer is generally more variable, especially if the mold temperature and part wall thickness are non-uniform. In general, factors that promote slower cooling rates (e. g., thick part wall sections, warmer mold temperatures) will tend to promote a higher degree of crystallinity and therefore greater mold shrinkage for a semi-crystalline plastic material.

The degree of crystallinity associated with the molded part will influence virtually all of the part's performance properties, as well as its mold shrinkage behavior. The crystalline regions of a molded part shrink more than the amorphous regions because the crystalline regions are more tightly packed. A higher degree of crystallinity will result in greater mold shrinkage. A neat semi-crystalline plastic can shrink 2–5 times more than a neat amorphous material. The mold shrinkage value for a semi-crystalline plastic material is both greater and more variable than that of an amorphous plastic. This makes tight-tolerance injection molding more difficult for semi-crystalline materials compared with an amorphous plastic material. Unlike an amorphous plastic, where mold shrinkage values are published as a single number (or a narrow shrinkage range), the mold shrinkage data for a semi-crystalline plastic is often provided as a graph, such as that depicted in Figure 1.12. This figure shows that the mold shrinkage value (as a percent) for a typical semi-crystalline plastic material increases as part wall thickness increases. Thicker wall sections require longer cooling times, resulting in a greater degree of crystallinity and therefore a greater mold shrinkage value.

Plastic parts that have a variable wall thickness or a non-uniform mold temperature will tend to exhibit locally different mold shrinkage values when molded using a semi-crystalline plastic. This can lead to dimensional variations that can cause shrinkage stresses, warpage and difficulties in tight-tolerance molding applications. It is for this reason that one of the more common plastic part design rules is to maintain a uniform wall thickness whenever possible. On the other hand, there are other part design rules that promote the use of a non-uniform wall thickness. For example, flow leaders (or locally thick wall sections) are sometimes used to control mold filling pressure drop and promote flow to the harder-to-fill sections of a mold cavity. Another common plastic part design rule suggests that reinforcing ribs should be thinner than the nominal wall from which they extend, in order to minimize the depth of the cosmetic sink mark opposite the rib, like the one depicted in Figure 1.13. While the thinner rib wall will improve the depth of this cosmetic sink defect, it will also tend to cause warpage when working with a semi-crystalline plastic, since the thinner rib cools faster and shrinks less than the thicker nominal wall from which it projects. Designers and molders must balance these often conflicting design recommendations for semi-crystalline material applications.

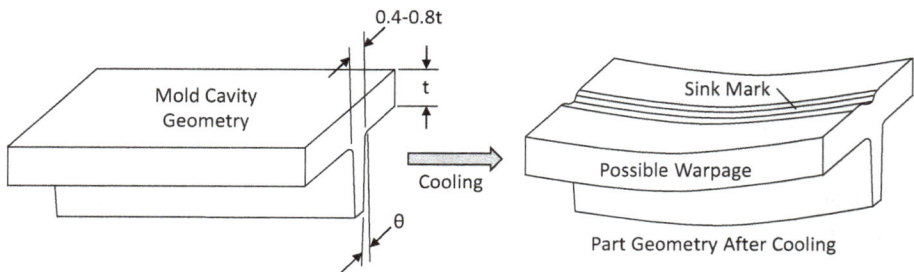

Figure 1.13 The variable shrinkage of semi-crystalline plastics can make it difficult to mold these materials to very tight tolerances in terms of overall dimensions, flatness, and possible sink marks. For example, it is common for designers to design molded parts that have some relatively thinner wall sections or features (such as ribs) in order to minimize the size of the sink marks on the part wall opposite the feature. This is a fundamental problem when using semi-crystalline materials since the thinner sections tend to shrink less due to their locally faster cooling rate, hence lower percentage crystallinity and therefore lower mold shrinkage. Variable mold shrinkage can either result in warpage of more flexible polymers or internal shrinkage stress of more rigid or stiff semi-crystalline polymers.

While semi-crystalline plastics can be more difficult to work with than amorphous plastic materials, especially for very tight tolerance injection molding applications, they do offer a number of very significant performance advantages, most notably chemical resistance. Semi-crystalline plastics tend to have significantly greater chemical resistance than amorphous plastic material, particularly when exposure to organic chemicals is a concern in the end-use application. Semi-crystalline plastics, such as HDPE, polypropylene, nylon 66, and polyphenylene sulfide (PPS), are commonly used for under-the-hood automotive applications where organic fluids or vapors, including gasoline, are present. Many semi-crystalline plastics are inherently lubricious (possessing a low coefficient of friction); this makes materials such as nylon, polybutylene terephthalate (PBT) or acetal (POM) common for gearing or bearing applications. Semi-crystalline plastics also offer a very wide mechanical performance range. Most neat semi-crystalline plastics tend to be opaque or translucent as they diffuse visible light, due to the amorphous and crystalline regions having different refraction index values. The optical transparency for some semi-crystalline plastics, such as a polypropylene, can be improved using nucleating additives or co-monomers that help control both the degree of crystallinity and the physical size of the crystalline regions.

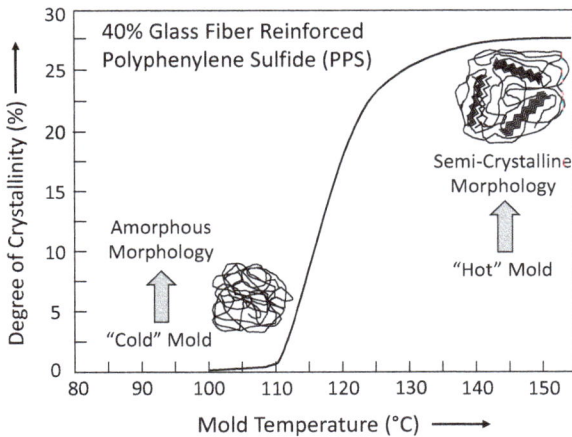

Figure 1.14 Mold temperature is a critical variable for many semi-crystalline plastic materials. Semi-crystalline plastics, such as polyphenylene sulfide (PPS), depend on the development of a high degree of crystallinity for optimum material performance. This requires very hot mold temperatures. Oil heated molds or molds heated with high-pressure water mold temperature controllers are typically utilized for molding PPS formulations to ensure that an optimum degree of crystallinity (and therefore material performance) is achieved [16].

It should also be noted here that proper crystal structure development is critical for semi-crystalline plastics. Most amorphous and semi-crystalline plastics have a relatively wide process window in terms of molding variables, such as acceptable melt or mold temperature ranges. It is key not to venture outside of the material supplier's recommended melt or mold temperature range. As an example, consider the effect that mold temperature has on the degree of crystallinity for a 40% glass fiber reinforced polyphenylene sulfide (PPS), as shown in Figure 1.14. PPS is a high performance semi-crystalline plastic material used in a variety of engineering applications. Like most semi-crystalline plastics, the performance properties of this material are dependent on achieving a high degree of crystallinity in the molded parts. The material supplier indicates that this can only be accomplished if the material is molded at mold temperatures greater than 120–140°C as shown in Figure 1.14. While it is technically possible to mold PPS at lower mold temperatures, the parts will not exhibit the expected or achievable performance. Materials like PPS that require very high mold temperatures are typically molded using a high-pressure water mold temperature controller or hot oil mold temperature controller. In some cases, injection molded semi-crystalline parts are annealed or heated after molding to further develop their crystalline structure.

1.4.3.3 Post-molding Dimension Changes

One final point regarding the mold shrinkage of both amorphous and semi-crystalline plastics is that of post-molding dimensional changes. A thermoplastic's mold shrinkage value is essentially a scaling factor that determines the magnitude of the dimensional differences in the molded part and injection mold cavity. Most mold shrinkage occurs while the part is cooling in the injection mold (some of this is constrained until part ejection). A lesser amount of mold shrinkage occurs as the part continues to cool down to room temperature after part ejection. In most amorphous plastic applications, the part's mold shrinkage is considered to be complete once the part reaches room temperature. However, for some plastic materials, especially some semi-crystalline plastics, part dimension changes can still occur even after the part has reached room temperature. The dimensional changes are usually small, but can be important, particularly for tight tolerance moldings. These dimensional changes can be due to (i) post mold crystallization, or (ii) moisture re-absorption as depicted in Figure 1.15. Amorphous plastic materials that do not absorb moisture, such as polystyrene, reach stable dimensions quickly, as there are no confounding dimensional or shrinkage variables. Plastic parts molded from a hygroscopic amorphous plastic (that does absorb moisture from the atmosphere), such as polycarbonate, are demolded in a dry state but will slowly re-absorb moisture from the atmosphere after molded part ejection. This effect is not very significant for most hygroscopic thermoplastics as most of them absorb only very small concentrations of water. This effect is more significant for a material like nylon 66, which absorbs a higher-than-average concentration of water compared to most other plastics.

Parts molded from a semi-crystalline plastic, especially those with a low glass transition temperature (see Section 1.8.2) can continue to crystallize even after the parts have reached room temperature. Post-mold crystallization is an asymptotic phenomenon that occurs primarily within hours (or possibly even days) after molding. Post molding dimensional changes may or may not be important depending upon the specific material grade and the application's dimensional tolerance requirements. However, is a safer practice to wait a certain period of time before part dimensions are measured during a dimensional qualification molding trial, as inconvenient as that can be.

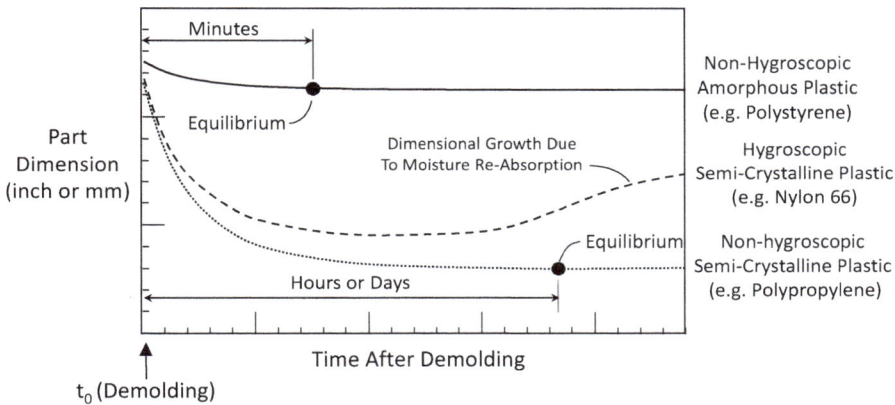

Figure 1.15 In most applications, mold shrinkage (and therefore the molded part dimensions) are considered to be stable once the de-molded (ejected) plastic part reaches room temperature. This is true for most amorphous plastics, especially non-hygroscopic amorphous plastics. Some semi-crystalline molded plastic parts, particularly those produced from a semi-crystalline plastic having a low glass transition temperature, can continue to shrink (for hours, even days) after the molded part reaches room temperature. This is known as post-mold crystallization, which is the cause of the post-mold shrinkage. This situation is further complicated for hygroscopic plastics, since dry (as molded) plastic parts will re-absorb moisture from the surrounding atmosphere over time, which can cause swelling or dimensional growth. Post-mold expansion due to moisture effects is only significant for those polymers that have a relatively high equilibrium moisture content.

1.4.4 Plastic Material Pillar #4: Additives

The fourth plastic material pillar involves the addition of other materials to the thermoplastic material. These other materials are generally described as additives and they are used to enhance certain physical properties that the base thermoplastic material may be lacking. Virtually all commercial plastics have one or more additives that are normally melt-compounded into the base polymer by the material supplier [8]. Details regarding the type and concentration of the additives in a plastic may or may not be disclosed by the plastic material supplier as this information is sometimes considered proprietary. Virtually any material in existence could be used as an additive in a plastic material formulation. Additives can be liquids, powders, fibers, other polymers, or even gases as in the case of foamed plastic material. Additive concentrations can vary from parts per million to as much as 40 percent or more by weight. A list of common plastic additives is given in Figure 1.16.

Common "Additives" Used in Thermoplastics
Concentration Range: [ppm to 40% Typical]

- Pigments
- U.V. Stabilizers
- Anti-oxidants
- Flame Retardants
- Internal Lubricants

- Particulate Fillers
- Fibrous Reinforcements
- Anti-static Agents
- Anti-microbial Agents
- Fragrances

- External Lubricants
- Foaming Agents
- Other Plastics (i.e polyblends)
- Impact modifiers
- Plasticizers (etc.......)

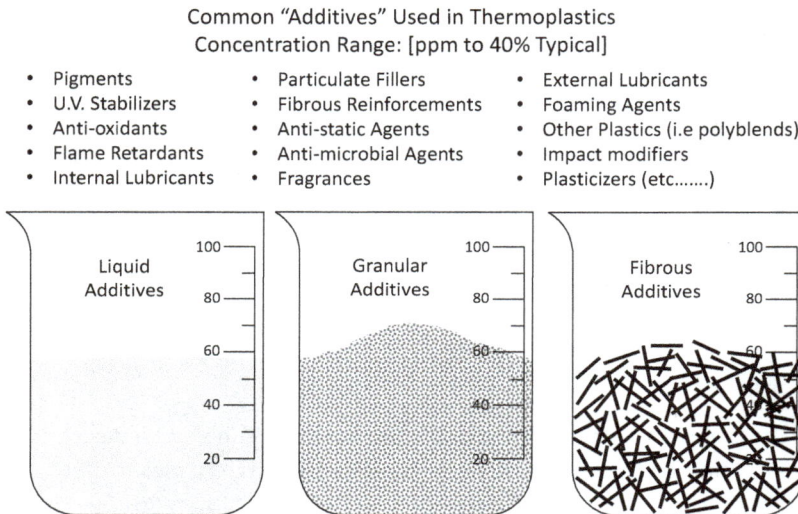

Figure 1.16 Virtually all commercial plastic materials contain one or more additives. Additives are used to enhance certain physical properties that the base polymer lacks. Additives can be used to enhance either the performance of the molded part or the material's processability or both. Additives can be liquids, granules (powders), pellets, or fibers. They are added at concentrations that range from parts per million to more than forty weight percent.

1.4.4.1 Additives: Method of Incorporation

While additives are usually incorporated into the base polymer matrix material by the material supplier (i. e., the material grade or formulation is sold with the additives already incorporated), additives can also be incorporated in other ways. In some cases, the base plastic material and the additives are purchased separately. The raw materials can then be shipped to a compounder who will use an intensive melt mixer, such as an intermeshing twin screw extruder, to create the desired material formulation. Many compounders offer standard product lines, but they can also offer custom material grades containing additives specified by the customer. Large material suppliers tend to focus their efforts on high volume formulations, while custom compounders fill a niche for lower volume, custom color, or other specialty applications. It should be noted that custom melt compounded formulations will have an additional heat history that is associated with the melt compounding and re-pelletization process.

In some cases, additives are incorporated into the base plastic material using the injection molding machine to do the melt compounding. This concept is economically attractive, but the reality is that most injection molding machines are not very good mixers. Almost all injection molding machine plasticating units are based on single screw extrusion type plasticating technologies (with very rare exceptions). Injection molding barrels normally have only one feed port and they have a relatively short L/D ratio compared to other single screw processes (such as extrusion). While injection molding machines do have some mixing capabilities, it is possible to incorporate some additives during the injection molding process with good results. Acceptable additive mixing results can often be achieved with injection molding plasticating units if the additives do not require extensive mixing, especially if they do not require extensive dispersive mixing. Additive concentrations can be controlled by pre-blending the materials or using loss-in-weight additive feeders. Additives may also require pre-drying, sometimes under different conditions than the plastic being molded, which can complicate the material handling process. It is also very helpful if the materials to be mixed have similar melt viscosities and chemical compatibility. Pelleted additives, such as masterbatch pigments (or other masterbatch additives) are the most common type of additive incorporated at the injection molding machine. When masterbatch additives are used, it is important that the masterbatch carrier polymer is either equivalent to or at least compatible with the base thermoplastic being molded. For example, using a masterbatch that has a polyolefinic carrier (like low density polyethylene) would be inappropriate when molding a styrenic plastic. These two polymer types are generally not compatible, which leads to both mixing difficulties and the potential for delamination defects in the molded part. Ideally, the masterbatch additive used when injection molding a styrenic plastic should itself have a styrenic carrier polymer.

Liquid additives, such as a liquid colorant, are also gaining popularity for injection molding applications as it is relatively easy to pump a fluid into the machine's feed throat or through a liquid injection port. Liquid additives are convenient but can cause an increase in screw recovery time in some cases. If additives are to be incorporated at the molding machine, it would make sense to modify the molding machine to enhance its mixing capabilities. These modifications could include the use of a plasticating screw modified with a suitable mixing element or the use of a static nozzle mixer (see Chapter 2). The use of additional back pressure during plastication is also recommended in applications where enhanced melt mixing is important.

1.4.4.2 Additives: Negative Side Effects

Thermoplastic materials would not be as versatile or useful without the use of additives. For example, an additive such as glass fiber can greatly improve the stiffness, strength, creep resistance and overall dimensional stability of a thermoplastic material. These property improvements can be so significant that glass fiber reinforced plastics can often be used in very demanding metal replacement applications. While the use of glass fibers is very helpful in terms of many performance property improvements, their use can also introduce a number of negative side effects. Like most medications, additives do have these negative side effects, or unintended consequences associated with a given additive, which should not be overlooked as they can have a very significant impact on both the molding process and molded part performance.

While the use of a glass fiber additive is very desirable in terms of structural stiffness, its use can introduce a number of other processing and molded part performance issues. The list of negative side effects can be quite long. While the use of a fibrous glass additive will generally impart higher strength, stiffness and creep resistance, the glass fiber will usually reduce the ultimate elongation or break strain of the compounded material (because the glass fiber itself has a very low break strain). Injection molded glass fiber reinforced plastic parts also exhibit a reduction in surface finish or surface gloss when compared to the neat base plastic. Knit line (or knit plane) strength can also be a significant concern when working with glass fiber reinforced plastics in comparison to unreinforced plastics [5]. The use of glass fibers as a reinforcement can also result in differential (directional) mechanical properties and mold shrinkage due to flow-induced fiber orientation. Unlike oriented polymer molecules, oriented glass fibers have no ability to re-randomize even if warm mold temperatures are used. Glass fibers also increase the plastic material's viscosity and its thermal conductivity, making mold filling more difficult, especially for thin wall molding applications.

Glass fiber reinforced plastics also tend to be "very" abrasive compared to their unreinforced counterparts. Glass has a very high Mohs hardness value compared to most materials. The Mohs scale of mineral hardness is a qualitative ordinal scale, from 1 to 10, which is used to characterize the scratch resistance of minerals and other materials. A diamond has a Mohs hardness of 10, while glass has a relatively high Mohs hardness that ranges from about 5.5–7.0. Other mineral fillers or reinforcements, such as talc, are far softer than glass, having Mohs harness values from 1.0–2.0. Therefore, mineral fillers like talc tend to be much less abrasive than glass fiber reinforcement.

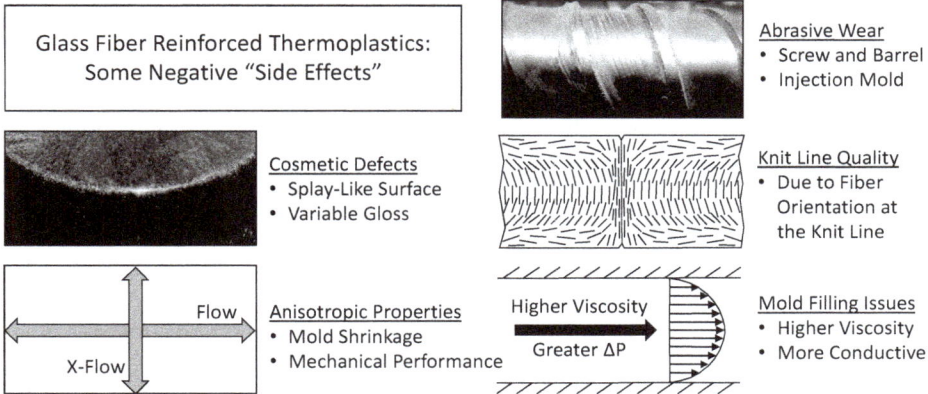

Figure 1.17 Additives are used to enhance certain properties that the base polymer lacks. However, like medication, additives can have negative side effects that should be taken into account during designers and manufacturing engineers. For example, glass fiber is an additive that is commonly used to improve a plastic's stiffness and creep resistance, and to reduce its coefficient of thermal expansion. On the other hand, glass fibers can also have many negative side effects that should be considered, including; knit/weld line performance issues, molding machine and mold steel abrasion, differential mold shrinkage, achievable molded part surface finish and higher melt viscosity.

The high hardness of glass can also lead to significantly greater than normal abrasive wear for the injection molding machine components, such as the plasticating screw, the barrel, the non-return valve and the nozzle assembly. Likewise, the injection mold itself must be built with higher hardness and more wear resistant steels for mold components that are in contact with the material flow (i. e., the runner bar, the cavity and core inserts etc.). This abrasion is attributed largely to the glass fiber ends. The degree of abrasion will increase as the glass fiber concentration increases. High hardness mold and machine components can be expensive, resulting in an increase in molding cost compared to a non-reinforced plastic.

Similar metallurgical issues arise when the plastic formulation contains an inherently corrosive base polymer or a corrosive additive, such as a halogenated flame retardant. Corrosion-resistant molding machine components and mold steels are required when the material formulation contains a corrosive additive. While the idea of using property enhancing additives is obvious and apparent to most people, the negative side effects are often overlooked and may not be considered when a plastic material is selected. This can result in unanticipated molded part defects or processing issues.

1.4.4.3 Additives: Short and Long Fiber Reinforced Plastics

Glass fiber reinforced plastics are available as either a short glass fiber reinforced plastic (SGFRP) or as a long glass fiber reinforced plastic (LGFRP). The more traditional SGFRPs are more commonly used for most injection molding applications, but LGFRPs can offer superior performance and are widely used for structural or metal replacement applications. Figure 1.18 depicts a cut-away section of both a short fiber and a long fiber reinforced plastic.

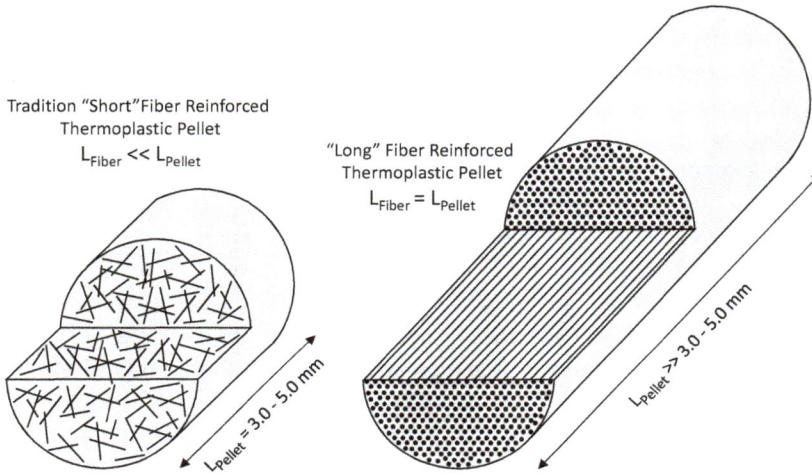

Tradition "Short" Fiber Reinforced Thermoplastic Pellet
$L_{Fiber} << L_{Pellet}$

"Long" Fiber Reinforced Thermoplastic Pellet
$L_{Fiber} = L_{Pellet}$

$L_{Pellet} = 3.0 - 5.0\ mm$

$L_{Pellet} >> 3.0 - 5.0\ mm$

Figure 1.18 Glass fiber reinforced thermoplastics can be purchased as either short fiber or long fiber reinforced pellets. Short glass fiber reinforced plastics are more traditional and common, while long glass fiber reinforced plastics can offer superior performance, as long as fiber length degradation (damage) is minimized during the plastication and injection phases of the injection molding process. Lighter weight carbon or graphite fibers are also used as reinforcing additives for thermoplastics in some higher performance injection molding applications.

Short glass fiber reinforced plastics (or composites) are made using a traditional melt compounding operation, where short glass fiber bundles are melt mixed with the base thermoplastic material at a specific concentration (e. g., 30% typical) in a compounding extruder. The extruder pumps the compounded melt through a multi-strand die, to form strands that are cooled and then cut into conventional cylindrical pellets using a strand pelletizer. The pellets typically have an overall length of 3.0–5.0 mm, while the dispersed glass fibers within the pellets have a far shorter length, as shown in Figure 1.18. While the glass fibers within the pellet are relatively short, they are very effective at imparting greater stiffness, strength and creep resistance, provided the glass fiber length is maintained and not significantly reduced or degraded further during the injection molding process.

Long glass fiber reinforced plastics are made by a different process. They are made by a melt coating or a thermoplastic pultrusion type of process, where continuous rovings (or yarns) of glass fiber are drawn through a melt extrusion coating or impregnation die. The impregnated plastic/glass fiber strands are then cooled and cut to produce reinforced plastic pellets that are far longer than conventional pellets. Long glass fiber pellets typically have a pellet length of 12 mm or more. It is notable that the glass fibers in a long pellet all have the same length as the pellet itself. These very long fibers provide greater reinforcement, as long as the amount of fiber length damage or degradation is minimized during the injection molding process.

Maintaining fiber length is key to property retention for both long and short fiber reinforced plastics. Fiber damage is most commonly associated with the plastication and mold filling phases of the injection molding process. The use of regrind is also a concern when molding glass-reinforced thermoplastics as the regrinding process will also result in fiber length degradation (see Section 1.7). In terms of plastication variables, the use of a low compression ratio plasticating screw, slow screw rotational speeds, low back pressure, and free flow non-return valves all help minimize glass fiber length degradation. Fiber damage can also occur during mold filling. Any variable that decreases the shear forces acting on the plastic material formulation will help minimize fiber damage. Large diameter nozzle tips, large diameter sprue bushings, large diameter full round runner systems, and large gate cross sections (e. g., fan gates) are all beneficial in terms of minimizing the glass fiber damage that can occur before the melt enters the mold cavity.

When molding reinforced plastics, it is also important to confirm that the glass fibers are uniformly dispersed within the molded part. There are times where the glass fibers, particularly long glass fibers, may not flow into thin sections of a molded part. The part may fill completely, but the concentration of glass in the thin sections can be less than it should be. Ideally, part wall thicknesses for glass fiber reinforced plastic parts are greater than those used for their neat plastic counterparts, in order to provide for additional geometric stiffness and lower resistance to flow, thereby reducing the potential for fiber hang-up or fiber degradation. Generous corner radii and gradual wall thickness transitions are also recommended. Tests, such as a burn out test (or ash test) are can be conducted on glass fiber reinforced molded parts to ensure (i) that the fiber concentration within the molded part is uniform, and (ii) to determine the amount of fiber length degradation. Non-destructive CT scanners are also utilized as tools to quantify the fiber length distribution and fiber concentration for molded fiber reinforced plastic parts.

1.4.4.3.1 Additives: Mold Shrinkage for Glass Fiber Reinforced Thermoplastics

The addition of glass (or other) fiber reinforcements to a thermoplastic can also lead to differential mold shrinkage. If the glass fibers in the polymer melt are exposed to the laminar flow velocity profile that occurs during the mold filling phase of the injection molding process, they will tend to preferentially align (or orient) in the flow direction. Unlike molecular orientation, which can be recoverable under favorable injection molding conditions, glass fiber orientation is irrecoverable once the injection mold cavity is filled. As a result, the rigidly oriented glass fibers (which themselves do not shrink appreciably), tend to disproportionally restrict the mold shrinkage of the plastic formulation along the flow direction. Since volumetric shrinkage must occur as the part cools, the less restricted, transverse (or cross flow) mold shrinkage for a glass fiber reinforced plastic tends to be greater than expected. This anisotropic mold shrinkage behavior can be quite complex and difficult to quantify in advance of molding, especially for glass fiber reinforced semi-crystalline plastics like polybutylene terephthalate (PBT). The mold shrinkage behavior for a 30% glass fiber reinforced PBT is depicted graphically in Figure 1.19. The mold shrinkage is both directional and thickness dependent. The directional effects are attributed to glass fiber orientation, while the thickness effects are associated with a greater degree of crystallinity for the thicker sections due to their longer cooling time. It should also be noted here that, in addition to mold shrinkage, glass fiber orientation will also have an effect on almost all other physical and performance properties for a glass fiber reinforced injection molded part. For example, a fiber reinforced thermoplastic composite's strength in the flow direction is generally greater than that in the cross-flow direction.

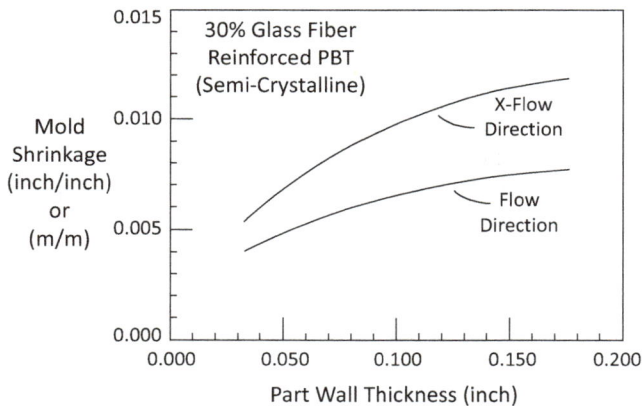

Figure 1.19 Additives, such as glass fibers, can also have a significant effect on a material's mold shrinkage behavior. The figure shows that the mold shrinkage behavior of a 30% glass fiber reinforced, semi-crystalline, polybutylene terephthalate (PBT) is quite complex. Directional shrinkage differences can be attributed to glass fiber orientation, while the increase in shrinkage with wall thickness can be attributed to a greater degree of crystallinity for thicker wall section [5].

1.4.4.3.2 Additives: Hybrid Reinforced Thermoplastics

A number of techniques have been studied in an effort to minimize differential (directional) mold shrinkage for fiber reinforced plastics. One of the more common approaches is to use a thermoplastic composite that contains <u>both</u> a fibrous reinforcement, and a particulate mineral filler or flake-like reinforcing filler, as shown in Figure 1.20 (b). These hybrid materials offer a unique balance in terms of mechanical performance and a more isotropic mold shrinkage (compared to conventional fiber reinforced grades). Some reduction in mechanical performance is expected with a hybrid composite material (compared to a purely fiber reinforced thermoplastic) although the hybrid material's mechanical performance should also be more isotropic. As a result, these hybrid materials are well suited for tighter tolerance applications that necessitate the use of a reinforced thermoplastic.

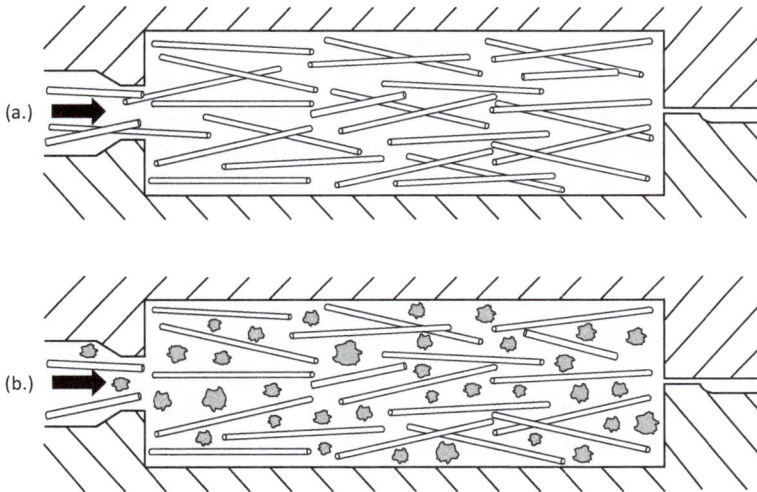

Figure 1.20 Hybrid thermoplastic composites that incorporate both a fibrous reinforcement and a particulate filler (or flake-like filler) can exhibit a good balance of mechanical performance and a more isotropic mold shrinkage compared to conventional fiber reinforced plastics (the fiber and filler geometries are not to scale).

1.4.4.4 Additives: Other Polymers

Other thermoplastics are also used as additives for injection molding applications. When two thermoplastics are mixed together, the resulting material formulation is described as a polymer blend, or polyblend, or just a blend. Polymer blends are sometimes referred to as polymer alloys, although strictly speaking, this alloy term is limited to compatible polymer blends. A polymer blend will generally have a balance of physical properties that reflect the behavior of each of the polyblend component materials based on their relative concentrations. One polymer is usually considered as the primary (higher concentration) constituent, while the second, lower concentration polymer is considered to be the additive. Polyblends are a convenient way for material suppliers to fine tune or alter the properties of the primary polymer. Polyblends are in many ways analogous to co-polymers, but they differ in one major respect. Co-polymers are produced by reacting co-monomers, resulting in a polymer chain containing more than one type of chain link as shown in Figure 1.8. Polyblends are produced by mixing two different polymers, each of which has been polymerized separately. The net result of these two approaches could be similar in some cases. For example, a polyblend produced by melt-mixing general purpose polystyrene (PS) with polyacrylonitrile (PAN), could have very similar properties to a co-polymer produced from styrene and acrylonitrile monomers, resulting in a co-polymer known as poly(styrene acrylonitrile) commonly designated as SAN.

Polyblends can offer a number of advantages over the use of a single polymer. They may be formulated to create a desired performance property profile or to improve the processability of the primary polymer. The polyblend may also offer an economic advantage by using a lower cost polymeric additive. For example, commercial blends of polyphenylene ether (PPE) and polystyrene (PS) have a great balance of performance properties, and exhibit improved processability, compared to pure PPE. Polystyrene is also significantly less expensive than PPE. Blends of polycarbonate (PC) as the primary polymer and polyacrylonitrile butadiene styrene (ABS) as the additive are also very common. These blends are commonly used for the production of molded electronic enclosures. The blends offer a great balance of properties, improved processability, and a cost advantages over pure polycarbonate.

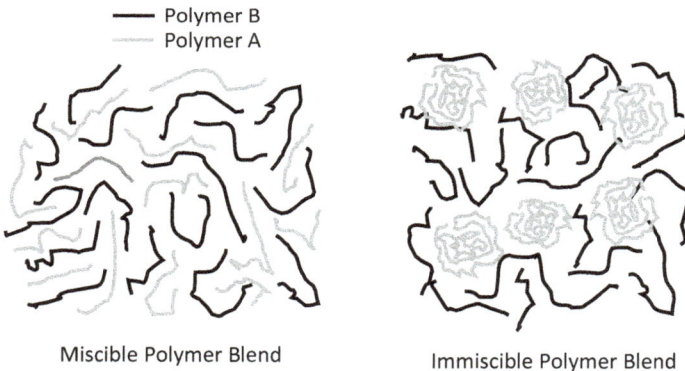

Miscible Polymer Blend   Immiscible Polymer Blend

Figure 1.21 Other polymers are also used as plastic material additives, as in the case of a polymer blend or polyblend. Some polymer blends, produced using chemically compatible polymers, are described as miscible blends or polymer alloys, while others, especially impact modified blends, are described as immiscible polymer blends. The degree of mixing and dispersion is also important for polyblend applications.

Some polymer blends are described as being miscible or compatible, while other blends are immiscible or incompatible. A blend of PPE and PS is an example of a compatible polyblend, while other polymer blends that contain an elastomeric or rubber polymeric additive are incompatible. Both compatible and incompatible polyblends can exhibit useful properties as shown in Figure 1.21. For example, a blend containing an immiscible elastomer or rubber, will form macroscopic, sphere-like domains, which improve the primary polymer's toughness by acting as crack arresters that stop crack propagation.

The degree of mixing (or intensity of mixing) is important for polyblends as it is for most applications involving an additive. Commercial polymer blends produced by material suppliers or compounders have typically been produced using intensive melt compounding lines having very good mixing capabilities. Even so, the use of injection molding screws with enhance mixing capability (compared with a general-purpose injection molding screw) is advantageous when molding a polymer blend.

In some cases, the injection molding machine's plasticating unit itself is used to produce a polymer blend. The two polymers are either pre-blended at a specific ratio, or they are introduced using separate gravimetric feeders. The later method is preferred and easier to implement for materials having different pre-drying requirements (i. e., drying temperatures etc.). The use of a plasticating screw with enhanced mixing capability is recommended to enhance mixing. The use of a static nozzle mixer can also improve the degree of mixing. Processors have also been known to blend polymers having the same chemical structure, but different melt flow rates (or average molecular weights), in an effort to modify the material's flow behavior. While this may seem to be an acceptable practice, it can lead to a variety of performance issues, particularly when working with semi-crystalline plastics. Even though two polymers may be chemically similar, their crystalline structures may not be equivalent or compatible.

Molded parts produced with polymer blends that are incompatible or incorrectly mixed may show defects such as delamination caused by phase separation, which can occur during mold filling. While it may not seem obvious, material formulations that contain a base thermoplastic, and a masterbatch additive could be considered to be polyblends, since the masterbatch additive (such as a color concentrate) has a carrier polymer. While the concentration of this masterbatch polymer in the molded part is relatively low, it does exist. The properties of some polymers, like polycarbonate, can be impacted in a negative way by even a small concentration of contamination or incompatible additives. It is for this reason that the carrier polymer used for the production of masterbatch additives should be equivalent or at least compatible with the primary thermoplastic material being molded.

1.5 Physical Forms of Plastic Raw Material

Most thermoplastics are solid materials supplied as granules having some type of regular geometry. These granules could be in powder form or more commonly as bead-like geometries known as plastic pellets or pre-production plastic pellets. Pellets are the most widely used raw material geometry for injection molding applications for a number of reasons. First, the free-flowing nature of pellets makes them relatively easy to handle, transport, pre-dry or convey, especially when using pneumatic or vacuum material handling systems. Pellets are also largely dustless which eliminates problems, such as dryer or loader filter clogging, which can occur when dealing with finer particle geometries. Pelleted geometries are also well suited for the single screw plasticating units found on most injection molding machines. While powdered thermoplastics may offer advantages for some plastic processes, such as rotational molding, powdered thermoplastics are generally more difficult to work with for injection molding applications.

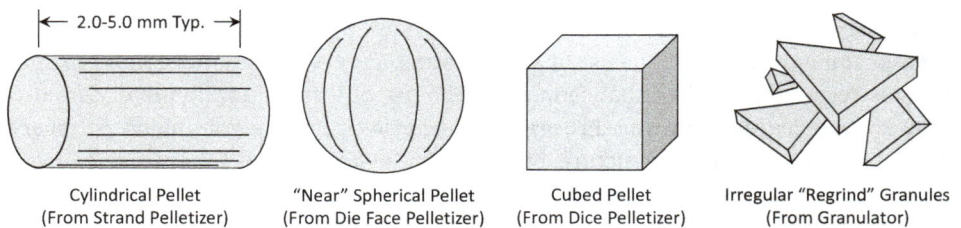

| Cylindrical Pellet | "Near" Spherical Pellet | Cubed Pellet | Irregular "Regrind" Granules |
| (From Strand Pelletizer) | (From Die Face Pelletizer) | (From Dice Pelletizer) | (From Granulator) |

Figure 1.22 Injection molding grade commercial plastics can be supplied in a variety of physical forms. Small pellets, having dimensions in the 2–5 mm range, are the most common due to their ease of handling and good conveying characteristics. Cylindrical pellets, made by strand pelletizing, and (near) spherical pellets, made by hot die face pelletizing, are the most common pellet forms. Cubed pellets, made by sheet dicing, are far less common. Powdered plastics are rarely utilized for injection molding. In some cases, reground molded parts or reground runner scrap may also be used at some concentration. In such a case, the shape of the reground granules will relate to both (i) the geometry of the parts and runners from which they are cut, and (ii) the granulator screen size.

The plastic pellets used for injection molding applications can have a variety of different shapes or sizes as shown in Figure 1.22. Plastic pellets with a cylindrical geometry are very common. These pellets are produced using a strand pelletizing process, where extruded strands are cooled and cut to a fixed length. The pellets typically have a length and diameter in the 2.0–5.0 mm range. "Sphere-like" pellets are also quite common for injection molding applications. Sphere-like pellets are usually produced using a die face pelletizing operation where extruded strands are cut immediately as they exit the extrusion strand die. The cut (hot) molten strands tend to naturally ball-up and then cool into a sphere-like shape. These sphere-like pellets can be more dust free than strand cut cylindrical pellets. Some plastics are supplied as cube-like pellets, which are dice cut from an extruded strip or sheet. Dice cut pellets are less common than cylindrical or near sphere pellets.

All of these pellet geometries are suitable for use in injection molding applications. The size and shape of the pellets have been shown to effect both the feeding and melting behavior of the plastic during injection molding. The pellet geometry should be consistent, having a narrow size distribution, as well as size consistency from lot to lot. Ideally, any fines (or small particles) will have been screened from the pellets by the material manufacturer before shipping, although fines can form anytime the plastic pellets are shipped or conveyed, especially if the pelleted plastic material is rigid and brittle. It is also worth noting here that hopper magnets should also be used when melt-processing plastic pellets to help ensure that any ferrous metal contamination (if present) is captured prior to plastication. Pellet count is sometimes used as one measure of incoming plastic material quality. Bulk density and angle of repose are also bulk material properties that influence the feed characteristics of any granular plastic material.

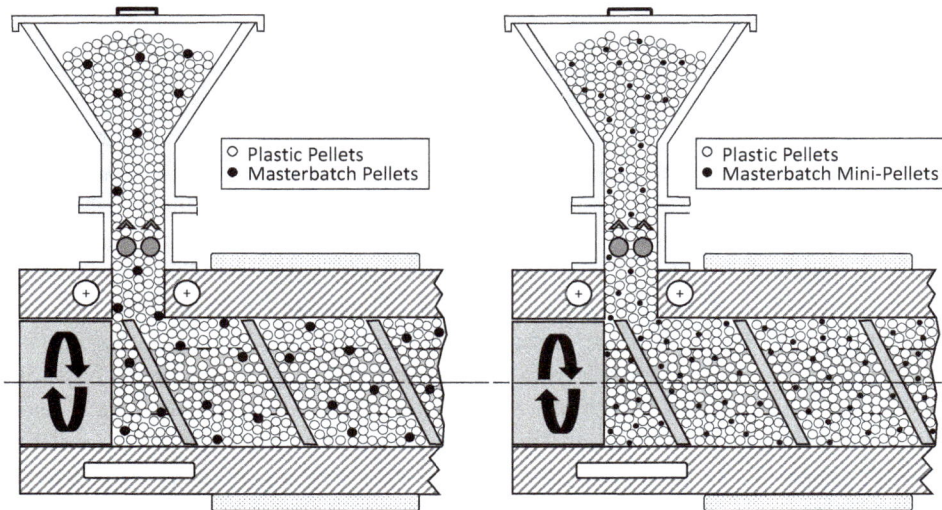

Figure 1.23 The physical size of plastic pellet is also a variable that can influence the plastication process and product quality. For example, the pellet size for a masterbatch additive, such as a color concentrate, is important. Similar pellet sizes minimize segregation or settling, while smaller pellets (e. g., micro-pellets) may be advantageous in terms of more uniform distributive mixing as long as they do not tend to settle out.

In many injection molding applications, the extruded pellets received from the material supplier contain both the base thermoplastic, and all of the additives associated with that particular formulation or material grade. However, in many applications, additives (or regrind) are introduced along with the base polymer as part of the injection molding process. This would be the case for applications where a color concentrate or masterbatch is used. This can be accomplished by pre-mixing the granular materials or using separate loss in weight or volumetric feeders. If multiple materials are introduced at the molding machine hopper, it is probably best if the individual material pellets have a very similar geometry, especially if they are pre-mixed (in order to avoid post-mixing segregation). On the other hand, smaller pellets (e. g., micro-pellets or mini-pellets) may be advantageous for a masterbatch additive in terms of distributive mixing. This is particularly true for applications where the additive concentration is very low (e. g., several weight percent) as shown in Figure 1.23. If pellet segregation or settling is an issue, separate feeders would be advantageous.

1.5.1 Housekeeping Considerations for Thermoplastic Pellets

Plastic pellets that are accidentally released into the manufacturing environment from spills and the like should be cleaned up immediately and discarded, since spilled pellets represent both a safety hazard to workers and to the environment. Spilled plastic pellets are also likely contaminated with other materials. From a safety perspective, pellets represent a slip hazard for the molding personnel (at a minimum). From an environmental perspective, spilled pellets may find their way into floor drains or other conduits that may lead to release into the external environment. The plastics industry, has developed and implemented an industry-led stewardship program designed to (i) improve housekeeping techniques and practices for pellets and granular materials, and (ii) to work towards achieving zero material loss. This effort, known as Operation Clean Sweep®, is Administered by the American Chemistry Council (ACC) and the Plastics Industry Association (PLASTICS) [17-18]. This environmentally responsible stewardship program has been implemented in thousands of plants around the world in an effort to protect the environment. Plastic pellets have been identified by the National Ocean and Atmospheric Administration [19] as one possible source of micro-plastic ocean pollution. In this context, the organization defines a micro-plastic as a plastic particle having a size ≤ 5.0 mm. This is typical for the size of a pre-production plastic pellet. In recent years, the term "nurdle" has been used as a synonym for plastic pellets, particularly for rogue plastic pellets found in the environment [20].

1.6 Commercial Plastic Designations

1.6.1 IUPAC Chemical Nomenclature

Plastic material nomenclature can be quite confusing due to the complex chemistry associated with these materials as discussed earlier in this chapter. While all polymers do have an internationally standardized chemical name, which is specified by the *International Union of Pure and Applied* Chemistry (IUPAC), the IUPAC chemical name is rarely used in common engineering material practice to describe commercial plastic materials for several reasons [21]. For one, these chemical names can be quite lengthy and complicated. More importantly, commercial plastics are almost always formulations, containing one (or more) polymers, along with other additives. Commercial plastic suppliers and users of plastics tend to describe plastics using other non-standard generic terminology, or more commonly trademarked nomenclature.

1.6.2 Trademarked Thermoplastic Nomenclature

As an example, consider polycarbonate, a widely used engineering thermoplastic. There are actually many grades of polycarbonate available commercially, from a large number of material suppliers. The commercially available polycarbonate grades differ from one another in that they can have different average molecular weights (or molecular weight distributions) and different additives. The most commonly used polycarbonates are produced by reacting bisphenol A (BPA) with a diacid chloride co-monomer (or co-reactant). Chemically, the resulting polymer is known as poly(bisphenol A carbonate). Commercial grades of polycarbonate (based) plastics are generally marketed using trademarked names and a grade-specific number designation (or alpha numeric code). While the grade numbers are associated with a specific polymer molecular weight distribution and very specific additives, supplier disclosures regarding these details are often very limited. As an example, *Lexan*™ is one material supplier's trademarked name for the polycarbonates they market [22]. There are many commercial grades of *Lexan*™ polycarbonate available on the market, all of which are somehow different from one another. Some of these grades are marketed specifically for injection molding applications.

One of these molding grades is *Lexan*™ 101 polycarbonate. The *Lexan*™ trademark indicates this material is a polycarbonate, while the 101 grade designation or code that is associated with <u>both</u> the material's molecular weight distribution and the additives it contains. Color is typically specified separately for pre-colored plastics. While the grade number for any plastic is associated with a certain molecular weight distribution, the material's melt flow rate (MFR) is usually provided on datasheets, in place of actual molecular weight data. This point was discussed in more detail earlier on this chapter. The data sheet for *Lexan*™ 101 describes this material grade as a BPA based polycarbonate, having a melt flow rate of 7.0 grams/10 minutes (@ 300°C/1.2 kgf), and a UL94 rating of HB. The material is described as being suitable for both injection molding and extrusion applications [22].

This material supplier also markets a grade of polycarbonate designated *Lexan*™ 123SRMV polycarbonate. The data sheet for this polycarbonate grade describes it as an easy flow, U.V. stabilized, impact modified grade of polycarbonate with enhanced mold release characteristics. It has an MFR of 18 g/10 minutes (@ 300°C/1.2 kgf), confirming the improved flow behavior compared to *Lexan*™ 101. Additives are most likely used to enhance the UV stability, impact behavior and release characteristics, although specific details regarding the chemical compositions or concentrations of the additives are not typically disclosed on the material's property data sheet. Material suppliers will usually disclose information about additives that are used at high concentrations (e. g., glass fibers, talc, plasticizers etc.) but details regarding lower concentration additives, such as stabilizers, antistatic agents, anti-oxidants and the like can are often held as proprietary. Gaining more additive information from the supplier through conversation can be helpful in practice, as many additives can have negative side effects that you should be aware of, as discussed in Section 1.4.4.2. *Material Safety Data Sheets* (MSDS sheets) or *Product Stewardship Documents* can sometimes be sources of additional information about the composition and concentration of the additives that are incorporated into a plastic material formulation.

Summarizing, it is imperative to have an understanding of the four plastic material pillars that determine both the performance and processability of the plastic material being injection molded. The materials trademarked name is linked to the material's chemical structure (Plastic Pillar 2). The grade number or grade code designation is linked to <u>both</u> the plastic's molecular weight distribution (Plastic Pillar 1), or its MFR in common practice, and the additives that have been melt-compounded into the plastic material (Plastic Pillar 4). While most injection molding grade thermoplastics have either an amorphous or a semi-crystalline morphology (Plastic Pillar 3) there is nothing associated with the plastic's chemical or tradename to indicate which morphology applies.

1.6.3 ASTM D4000 Nomenclature Standard

The *American Society for Testing and Materials* has established and publishes a standard classification system and subsequent line call-out (specification) that is intended to be a means of specifying or identifying plastic materials. The ASTM D4000 standard is entitled "Standard Classification System for Specifying Plastic Materials" [23]. The purpose of this classification system is to provide a method of adequately identifying plastic materials using a generic or non-trademarked system that can be used for any plastic material. It further provides a means for specifying these materials, including their additives and key physical properties, by the use of a line call-out designation. The standard facilitates the coding and decoding of the line call-out material designation. An example of an ASTM D4000 line call-out for a nylon formulation is shown below.

Example ASTM D4000 Line Call-out: PA0120G33A53380

- PA0120 = Nylon 66, heat stabilized
- G33 = Glass fiber reinforced, 33% ±2%
- A = Table A from ASTM D4066 for nylons
- 5 = Tensile Strength, ISO 527, 175 MPa minimum
- 3 = Flexural Modulus, ISO 178, 7500 MPa minimum
- 3 = Izod Impact Resistance, ISO 180/1A, 75 J/m minimum
- 8 = Deflection Temperature Under Load, ISO 75, 235° minimum
- 0 = Unspecified

Figure 1.24 ASTM has established and publishes a generic, standardized classification system and line call-out specification that is intended to be a means of specifying or identifying a plastic material. This is an example of an ASTM D4000 line call-out for a 33% glass fiber reinforced nylon 66.

1.7 Re-use of Manufacturing Scrap or "Regrind"

By definition, all thermoplastics are technically recyclable and can be reground and reprocessed in much the same way that metals can be recycled. The inherent recyclability of thermoplastics is an advantage in comparison to crosslinked plastics (or thermosetting plastics) that cannot be remelted once they have solidified. Since many thermoplastic injection molding processes do involve the generation of manufacturing waste, such as cold runner scrap, the potential for re-use of this manufacturing scrap should be discussed by the customer, molder and material supplier to ensure there is no mis-communication as to whether regrind use is acceptable for the application. The reuse of runner scrap (or out of specification part scrap) can offer a significant economic advantage, but can also lead to a very wide variety of manufacturing and performance issues for the molded parts. There are many factors to consider if regrind is to be used as discussed below. The discussion here is limited to the re-use of post-industrial manufacturing regrind, and not post-consumer plastic regrind.

The decisions associated with the use of regrind should be discussed by the product development team early in the part design process, long before manufacturing begins. In some critical applications, the use of regrind may be prohibited. Eliminating the use of regrind eliminates all of the technical issues associated with regrind use. While this practice is conservative or "safe", the molded part cost will generally be greater than if the use of regrind was allowed. In many applications, the use of regrind is allowed, particularly in more general purpose applications. If the use of re-grind is allowed, then the methods and practices associated with the use of the re-grind must be discussed, documented, and properly implemented.

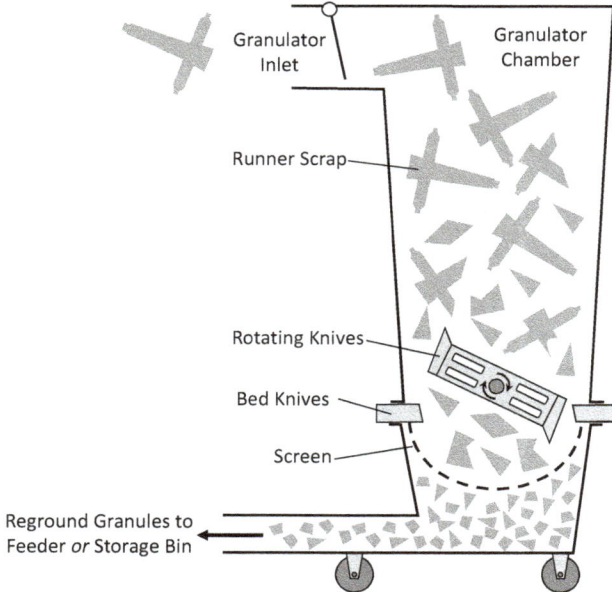

Figure 1.25
In theory, all thermoplastics are fully recyclable. Reground manufacturing scrap (such as reground cold runner systems) can be used if the application permits. The use of regrind (or recyclate) can be an economic advantage, but may also lead to a variety of manufacturing and part performance issues.

1.7.1 Source of the Regrind

The most common source of manufacturing waste for the injection molding process is runner system scrap produced by cold runner molds. Designers of cold runner molds often consider the potential for regrind use as one factor when designing the geometry of the mold runner system, in an effort to keep the regrind concentration to a specific value. Ideally, the cold runner volume represents a relatively low percentage of the overall shot volume since many applications allow the use of a limited concentration of regrind (e. g., 25% is a typical target). Regrind use becomes problematic for molding operations that produce very small plastic parts (or micro molded parts) that are produced in cold runner molds that have relatively large volume runner systems. In such cases, the volume of the runner can easily exceed the volume of the molded parts, resulting in molded parts with multiple heat histories if the runners are reground and re-used. Without question, the easiest way to eliminate all of the issues associated with the re-use of cold runner scrap is to eliminate the cold runners altogether by using a "hot runner mold". Hot runner molds do not generate runner scrap, but they do have a higher capital cost. The higher cost for a hot runner mold is most easily justified for higher production injection molding applications (e. g., plastic packaging, plastic closures etc.).

MATERIAL: G27 ABS
COLOR: S895 BLUE

REGRIND SPECIFICATIONS
• 25% BY WEIGHT
• RUNNERS ONLY
• DISCARD ALL PURGINGS
• SCREEN FINES
• RE-DRY IF STORED

72.2 Ø

76.2 Ø

A   A

2.0

1.5 R

25.4

ALL DIMENSIONS IN MILLIMETERS

SECTION AA

PART NAME: CUP
DATE: 6/10/2025
TOLERANCES: ± 0.127
SURFACE FINISH: A-3
SCALE: FULL

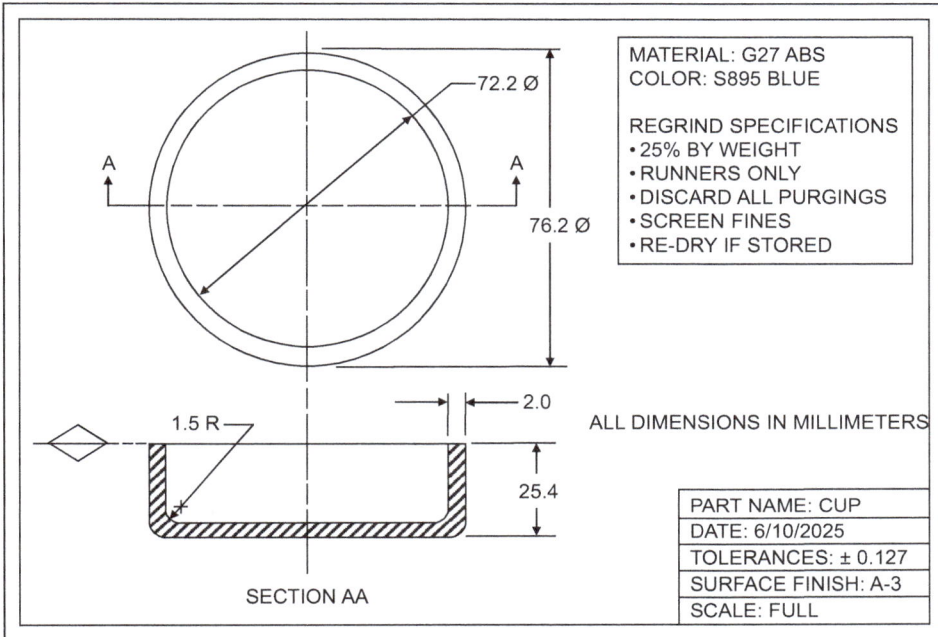

Figure 1.26 It is important that molders have a clear understanding as to whether the use of cold runner regrind is sanctioned or prohibited. If the use of regrind is sanctioned, the target regrind concentration and the procedures associated with regrind use should be well documented.

Cold runner scrap is the most common source of regrind. Other possible sources of regrind include (i.) off-specification molded parts and (ii) purgings (solidified air shots). It is best (mandatory in my opinion) to avoid the regrinding and re-use of purgings. The purgings produced during a material change are contaminated with at least trace quantities of another plastic. Even purgings that are not contaminated, such as cycle delay purgings, should not be used because hot purgings tend to oxidize and discolor as they cool down in the presence of oxygen.

Off-specification parts can also be used as a source of regrind, but their use should also be avoided, when possible, for several reasons. Most importantly, off-specification parts should not be produced if the manufacturing process is in control, except during process start-up. If larger quantities of off-specification molded parts are being produced, the molding process or mold should be improved (at least in the long term) such that this source of regrind material is eliminated. In addition, it becomes very difficult to control the uniformity of regrind concentration when rejected parts are the source of the regrind. That is not the case when cold runner scrap is used as the source. The concentration of regrind from runner scrap will be consistent from shot to shot, especially when a closed loop (beside the press) regrinding process is used.

1.7.2 Material Considerations for Regrind

While all thermoplastics are technically recyclable, some plastic materials exhibit better inherent recyclability than others. Ideally, the physical properties and process-ability of a reground (recycled) thermoplastic should be equivalent to its virgin counterpart. Unfortunately, processes like injection molding involve high temperatures and high shear rates that can lead to thermal, oxidative, and mechanically induced degradation. The heat and shear history the material encounters during injection molding can have an effect on both the base polymer and any additives in the material formulation. For example, some polymers, such as HDPE or ABS, have been shown to exhibit very good inherent recyclability when processed correctly. Other polymers, such as a glass fiber reinforced thermoplastic, will generally exhibit a property reduction when the fibrous composite material is recycled, due in part to the glass fiber length reduction caused by melt processing and regrinding. In other cases, regrind property reduction could be caused by oxidation of the base polymer or an additive such as a sacrificial anti-oxidant. Unfortunately, the inherent recyclability of a plastic material is very material dependent and material supplier information regarding a plastic material's inherent recyclability is generally very limited.

The issue of material pre-drying should also be addressed when considering the use of regrind. The molded parts and runners produced from a hygroscopic virgin plastic that has been properly pre-dried prior to molding will be ejected from the injection mold in a dry state. However, the dry, as molded parts will begin re-absorbing moisture from the surrounding environment as soon as they are ejected from the mold. The rate of moisture reabsorption is dependent upon many factors, such as ambient environmental conditions and the part surface area to volume ratio. Depending on the moisture sensitivity of the plastic material, it may be possible to use the regrind from these dry runners successfully, without the need for regrind drying, if the regrind is used immediately after part ejection. This is most easily accomplished by using a beside the press granulation process, where the runners are reground and re-used within minutes after the part is ejected from of the mold. If the use of regrind involves a significant time delay before re-use, as would be encountered with a central granulation process, then the regrind should be re-dried prior to molding in order to eliminate possible manufacturing defects, such as splay (for any hygroscopic plastic) or hydrolytic degradation (for condensation polymers).

1.7.3 Processing Considerations for Thermoplastic Regrind

The two most common methods used to generate and incorporate regrind are (i) a beside the press regrinding process, or (ii) a central granulation process. A beside the press granulation process offers significant advantages over a central regrinding process. Advantages of beside the press regrinding include the potential to eliminate regrind drying, automatic regrind concentration control (based on runner volume), automated materials handling and conveying, and limited potential for contamination. Beside the press granulation processes are more capital intensive and may not be as suitable for lower production applications with limited automation.

In either case, the potential for contamination of the regrind is yet another issue that deserves careful consideration in terms of granulation equipment and materials handling procedures. Contamination of the runner scrap can occur before, during, or after granulation. Care should be taken to ensure that the runner scrap does not come into contact with contaminants such as hydraulic oil, hand oils, mold release, water, dust or dirt. The granulator itself is sometimes a source of polymeric contamination (from a previously granulated plastic material). Granulators (and associated pneumatic conveying equipment) can be difficult to clean but this should be done thoroughly any time there is a material change. It is best if a granulator can be dedicated for specific material type, but if not, cleaning the granulator with high suction vacuum cleaners is preferable to cleaning with high pressure shop air which simply disperses the contamination. The use of hopper magnets is strongly recommended when processing regrind as the potential for ferrous metal contamination is ever present. Injection molding nozzle filters are also used in some applications where regrind is used, particularly for injection molds that have very small gates.

An optional process that is sometimes used in applications where regrind is utilized is that of fines separation. Unlike pelletizing processes, which produce relatively consistent and uniform pellet shapes, granulation or regrinding processes produce granules having a relatively wide range of geometries and a wide particle size distribution. Small particles, known as fines, are sometimes screened-out or removed from the granulated runner regrind. If the fines are not removed, they can cause a variety of molding issues. The wide particle size distribution for regrind can also lead to more variability for the plastication (melt preparation) process compared to molding virgin pellets alone. The variation in plastication rate can be related to both feeding difficulties during plastication, such as bridging or inconsistent feeding. Reground runners (or reground parts) tend to have lower bulk densities than the virgin pellets that normally feed the injection molding process. Lower bulk density materials will generally require a longer screw recovery time compared to that required for virgin pellets (for a given screw rpm). If regrind is used, it is good practice to monitor screw recovery time as an important process variable to ensure feeding inconsistencies are not significant. Plastication conditions should be adjusted to minimize any variation in screw recovery time to ensure the molding process and part quality are more consistent. It is good to plan for some expected variation in screw recovery time when regrind is used. The plastication conditions should be set such that the additional cooling timer will not time-out before shot preparation is complete.

1.8 Thermal Transitions

All thermoplastics have thermal transitions that (i) determine the useful temperature range for the material in terms of high and low temperature performance, and (ii) determine the material's melt processing temperature range. Understanding these thermal transitions is important since thermoplastics are far more sensitive to temperature changes than more traditional engineering materials such as metals or ceramic. The thermal transition temperatures for thermoplastics can be measured or determined using several different testing methods. Commonly used thermal transition test methods include; Differential Scanning Calorimetry (DCS), Thermomechanical Analysis (TMA), Dilatometry, and Dynamic Mechanical Analysis (DMA). Typical DMA test results for both neat amorphous and neat semi-crystalline plastics are given in the following section. With DMA testing, an injection molded test sample of the plastic material is subjected to an oscillating stress (or oscillating imposed deformation) and the material's response to this input is monitored as the testing environment (and test sample) temperature ramps up or down. The stress state for a DMA test may be compression, tension, flexure or torsion.

1.8.1 Thermal Transitions for Amorphous Thermoplastics

The results of a DMA test can be presented in a variety of ways, but it is very common to plot a graph that shows how the material's modulus (or rigidity) changes with temperature. Typical amorphous plastic modulus-temperature graphs are given in Figure 1.27 for both a rigid amorphous material and a flexible amorphous material (e. g., a thermoplastic elastomer). The overall shape of a modulus-temperature curve is very similar for most neat amorphous plastics. While the overall shape of the curves in Figure 1.27 is very similar, the curves for two amorphous materials shift along the x-axis from left to right, or right to left, based on the material's response to temperature.

The modulus–temperature behavior for typical rigid, neat amorphous plastic, polystyrene, is shown in Figure 1.27 (left). The material is rigid and stiff at room temperature. Its stiffness increases slightly at lower temperatures, and decreases slightly as temperature increases, along what is known as the glassy plateau. However, as temperature continues to increase, the material's modulus suddenly drops dramatically within a very narrow temperature range (a few degrees). The midpoint of this drastic drop in stiffness is known as the glass transition temperature or T_g. A material's T_g occurs at a temperature where a critical value for free volume is achieved (due to thermal expansion). This value is about 100°C for polystyrene. The material becomes very flexible or rubber-like at or above the T_g. Polystryrene is therefore useful as a rigid material in an end use application only at temperatures below its T_g. The material exhibits a second plateau, known as the rubbery plateau, at higher temperatures and eventually softens to the point where the material can be melt-processed or plasticated. The processing temperature range for most amorphous plastics is very wide due to the very gradual softening behavior. For polystyrene, this transition to the melt phase occurs at around 200 to 220°C. Polystyrene is considered to be in the melt state or melted at these temperatures, although strictly speaking, amorphous plastics do not have a melting temperature or melting point in the classical sense. Amorphous plastics just soften very gradually.

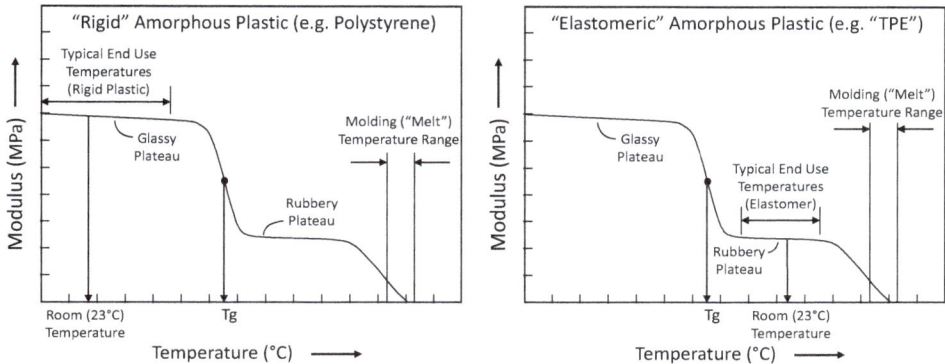

Figure 1.27 Amorphous thermoplastics can have a softening or "glass transition temperature" (T_g) that can be above or below room temperature. Rigid amorphous plastics, such as polystyrene, have a T_g that is above room temperature. Flexible or rubbery amorphous plastics, such as thermoplastic elastomers, have a T_g below room temperature. Amorphous thermoplastics will all become fluid and melt-processable over a fairly wide range of temperatures, but do not have a melting temperature or melt point (T_m) per se.

The modulus-temperature behavior for other rigid amorphous plastics would be similar to that of polystyrene, but the curve would shift along the x-axis. For example, the modulus-temperature behavior for a neat polycarbonate, which is also a rigid amorphous plastic, would shift to the right along the x-axis in comparison to polystyrene. Polycarbonate has improved high-temperature capabilities compared to polystyrene. The T_g for polycarbonate is about 147 °C, while its processing temperature range is 310–330 °C.

Other amorphous plastics, such as flexible thermoplastic elastomers, have modulus-temperature curves that shift to the left (relative to room temperature) as shown in Figure 1.27 (right). If the T_g of an amorphous plastic is below room temperature, it means that the material is in a rubbery state at room temperature. It behaves as a thermoplastic elastomer at room temperature. Thermoplastic elastomers are useful at various temperatures because they are above their T_g where they are flexible or rubbery. An amorphous thermoplastic elastomer will become as brittle as polystyrene at temperatures below its T_g.

1.8.2 Thermal Transitions for Semi-crystalline Thermoplastics

Semi-crystalline plastics have a somewhat more complex modulus-temperature behavior than single-phase neat amorphous plastics. That is because semi-crystalline thermoplastics have two distinct phases, namely the amorphous phase (amorphous regions) and the crystalline phase (crystalline regions) as depicted in Figure 1.11. Each of these two phases of a semi-crystalline thermoplastic has its own thermal transitions, leading to the characteristic semi-crystalline modulus-temperature behavior shown in Figure 1.28.

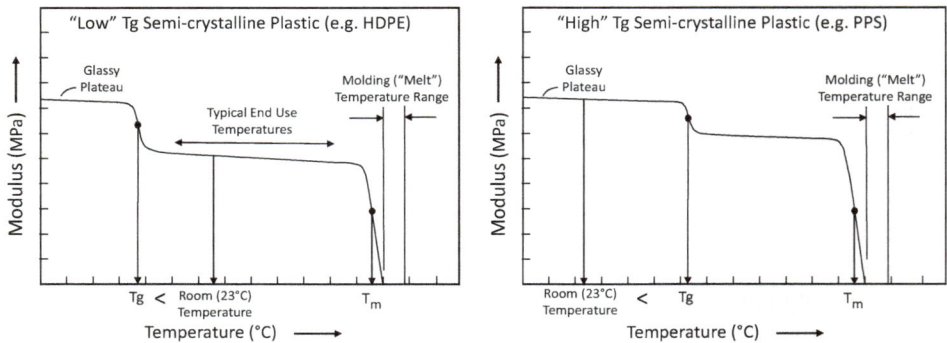

Figure 1.28 Semi-crystalline thermoplastics have both a T_g and a T_m. The T_g is associated with the amorphous regions of the material, while the T_m is associated with the melting of the crystalline regions of the semi-crystalline material. The T_g for a semi-crystalline thermoplastic can be above room temperature (in the case of a rigid semi-crystalline plastic like PPS) or below room temperature for a more ductile semi-crystalline plastic (such as HDPE). The recommended processing temperature range for a semi-crystalline material is somewhat higher than its T_m.

Both the amorphous and crystalline regions of a semi-crystalline plastic will influence the bulk material modulus-temperature behavior for a semi-crystalline material. The degree of crystallinity (or percent crystallinity) of a semi-crystalline plastic is therefore a factor influencing the material's performance. One way to rationalize or explain the modulus-temperature behavior of a semi-crystalline plastic is to imagine the material as a continuous amorphous matrix, that is reinforced with more tightly packed, highly ordered, crystalline regions. The crystalline regions tend to reinforce and stiffen the amorphous matrix. For example, consider the modulus-temperature behavior for a very ductile, semi-crystalline plastic, such as high-density polyethylene

(HDPE) as shown in Figure 1.28 (left). HDPE is a very tough material at room temperature because, at room temperature, the amorphous regions of the HDPE are in the rubbery state, due to the fact that the amorphous regions of HDPE have a T_g well below room temperature (of about −100 °C). The amorphous regions of HDPE are elastomeric at temperatures above its T_g (like room temperature), which gives the material its inherent ductility. At room temperature, the low T_g elastomeric amorphous phase gives the material its toughness, while the crystalline regions impart stiffness. HDPE will become as brittle as rigid polystyrene in a very cold environment (i. e., at temperatures below its T_g).

HDPE also has crystalline regions. The crystalline regions of a semi-crystalline plastic contribute to the material's stiffness, strength and chemical resistance. Figure 1.28 also shows that, as the temperature continues to increase, the material continues to gradually soften, until there is a sudden loss of modulus once the material's crystalline regions melt. This crystalline melting transition typically occurs over a fairly narrow temperature range (compared to the gradual softening temperature range for amorphous regions). The crystalline regions of a semi-crystalline thermoplastic <u>do</u> have a true melting transition or melting point T_m. Therefore, semi-crystalline plastics have both a T_g that is associated with the amorphous regions, and a T_m that is associated with the crystalline regions.

Figure 1.28 (right) also shows the modulus-temperature behavior of a more rigid semi-crystalline plastic, such as neat polyphenylene sulfide (PPS), that has a T_g above room temperature. At room temperature, this semi-crystalline material is relatively rigid (compared to the more ductile HDPE) because the amorphous regions of the PPS are below T_g at room temperature. Highly crystalline semi-crystalline plastics having a high T_g (i. e., above room temperature) tend to have relatively high upper use temperature ranges as their behavior is dominated by the crystalline regions. It is also important to note that semi-crystalline plastics typically have a very complicated degree of crystallinity distribution within an injection molded part. Sections of the part that cool quickly will tend to be less crystalline, while sections of the part that cool slowly tend to be more crystalline (as described in Section 1.4.3). The net result is that plastic parts molded from semi-crystalline plastic will tend to have a variable degree of crystallinity through their thickness. Parts molded from a semi-crystalline plastic tend to have at least three distinct layers, having two skins and a more highly crystalline core (i. e., a skin-core-skin type of morphology).

1.8.3 Other Thermal Properties: DTUL and Vicat Softening Temperature

1.8.3.1 Deflection Temperature Under Load (DTUL)

While DMA tests are typically used to determine the mechanical properties of a thermoplastic over a wide temperature range, there are several other standardized short-term or screening tests that are commonly used to provide a rough indication of the upper use (high-) temperature capabilities of a plastic material. The test that is most commonly utilized for this purpose is the *Deflection Temperature Under Load* (DTUL) test. This test is also referred to as the *Heat Deflection Temperature* (HDT) test. Both ASTM D648 and ISO 75 outline the test equipment and test procedures for DTUL or HDT testing. The DTUL test provides a measure of the elevated temperature at which a molded beam-like test specimen that is subjected to heating and a three-point bending load, deflects a fixed distance under a specified load (i. e., an outer fiber stress of 1.82 MPa or 264 psi). The test sample is immersed in an oil bath that is heated at a rate of 2 °C per minute. The sample temperature is assumed to be equivalent to that of the oil as the heating rate is very slow. The load and deflection associated with this test are arbitrary, but they are standardized. Unlike DMA testing, the DTUL test does not provide any information regarding the shape of a material's modulus-temperature behavior as it is a short-term, single-point test. As such, the DTUL test is most suitable for initial material selection screening.

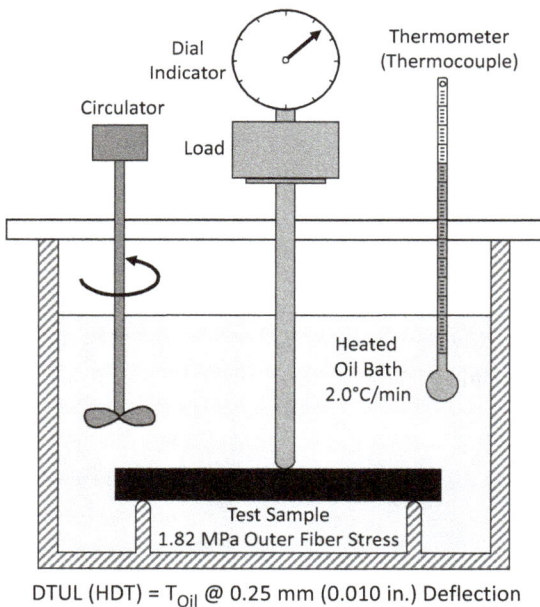

DTUL (HDT) = T_{Oil} @ 0.25 mm (0.010 in.) Deflection

Figure 1.29
A material's Heat Deflection Temperature (HDT) and Deflection Temperature Under Load (DTUL) are equivalent terms that indicate the elevated temperature at which a molded thermoplastic test sample subjected to a three-point bending load deflects a fixed distance. This temperature is a rough indication of a material's upper use temperature limit.

The DTUL test is most appropriate for relatively rigid amorphous plastics and relatively rigid semi-crystalline plastics. The test is not suitable for very flexible plastics or thermoplastic elastomers as they are unable to support the test load, even at room temperature. While a test load of 1.82 MPa (264 psi) is most commonly used for DTUL testing, there is a second variation of this test that is used for testing semi-rigid plastics, such as low density polyethylene (LDPE). This lower load DTUL test uses a load that creates an outer fiber stress of only 0.46 MPa (or 66 psi) to allow more flexible materials to be tested. However, there really is no way to compare or correlate DTUL results for tests that have been run at the different test load values.

1.8.3.2 Vicat Softening Temperature

A variant of the DTUL test is the Vicat Softening Temperature Test shown in Figure 1.30. These two tests utilize the same basic test apparatus, but different sample fixtures. Unlike the three-point bending test configuration associated with the DTUL test, the Vicat test provides a measure of the temperature at which a lightly loaded flat pin penetrates a fixed distance (1.0 mm) into a molded test specimen that has been immersed in oil that is gradually heated. Both ASTM D1525 and ISO 306 outline the test equipment and test procedures for the Vicat softening temperature test. The Vicat test apparatus is shown in Figure 1.30. The objective of the test is to provide a relative indication of the ability of a material to withstand short-term contact with a heated object. A material's Vicat softening temperature is also commonly used for process design purposes (e. g., molding simulations) as a measure of the maximum temperature at which an injection molded plastic part can be ejected from a mold (without distortion of damage from the mold's ejector pins). This Vicat temperature is often used as the theorical ejection temperature for cooling time estimations or calculations. This temperature is assumed to be an estimate of the temperature at which the material would be rigid enough to withstand the forces of ejection. Such a value is required for theoretical injection mold cooling time calculations or cooling time predictions. Both DTUL and Vicat temperature can be used for this purpose, but the Vicat temperature is generally preferred because it more closely simulates the effect an ejector pin or ejector sleeve would have on a plastic part during demolding. Some plastic material supplier data sheets report both Vicat and DTUL temperature values, although of these two tests, DTUL test results are more commonly reported. The DTUL and Vicat test results can also be used as a rough indication of the intrinsic resistance of a thermoplastic to distortion or warpage at elevated temperatures. The values are useful only as a guide because the tendency towards warpage is influenced by many factors including the degree of orientation, residual stress, external loads, and part geometry.

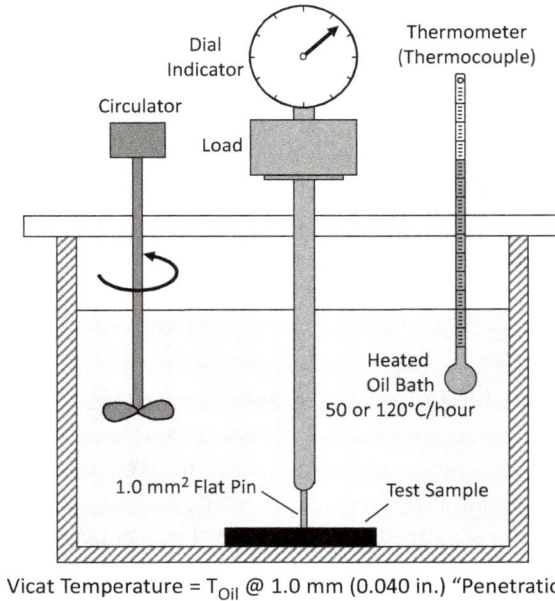

Figure 1.30
The apparatus used to measure a thermoplastic's Vicat softening semperature is very similar to that used for DTUL testing, except that the test fixture geometry differs. DTUL is an elevated temperature 3-point bending test, while the Vicat test is an elevated temperature penetration test conducted using a lightly loaded flat, circular pin, having a 1.0 mm² cross sectional area.

1.8.3.3 Correlation With Glass Transition Temperature

The Vicat and DTUL temperatures can show some correlation with glass transition temperature (T_g) for more rigid, neat amorphous plastics. For example, the T_g of general-purpose neat polystyrene is reported to be about 100 °C, while the DTUL of polystyrene (at 1.82 MPa) is reported to be about 85 °C. These two values are similar, but the DTUL temperature value or a rigid amorphous plastic is typically less than the material's T_g since the DTUL test involves a relatively high continuously applied stress. This DTUL to T_g correlation is further complicated for rigid amorphous materials that incorporate fillers or reinforcing additives that help support a load even though the matrix material may have softened. While thermal properties like DTUL (or Vicat) show a limited correlation with T_g for rigid, neat rigid amorphous thermoplastics, there is no correlation between DTUL and T_g for thermoplastic. In fact, most thermoplastic elastomers are too flexible for the DTUL test.

Properties like DTUL are a bulk material property, having a value that is influenced by all regions within the polymer. For example, high density polyethylene (HDPE) is a semi-crystalline plastic having both amorphous and semi-crystalline regions. As stated earlier, HDPE has a T_g of about −100 °C, yet the material has a DTUL of about +50 °C. The crystalline regions of a neat semi-crystalline thermoplastic help it support a load, even if the amorphous regions of the HDPE are rubbery or elastomeric at temperatures above its T_g. Therefore, there really is no correlation between T_g and DTUL for a semi-crystalline thermoplastic, especially for semi-crystalline plastics that have a T_g below room temperature.

1.9 Injection Molded Part Cooling Time Estimation

The overall production cycle time for an injection molded part is dictated by a number of part design, mold design, molding process variables, and plastic material variables. The injection molding cycle time used in actual molded part production is typically determined experimentally during initial molding or qualification trials. However, it can also be useful to predict or estimate the theoretical molding cycle time required for molding prior to part production, in an effort to optimize part design or estimate molded part cost. In most injection molding applications, the time required to cool the part accounts for the vast majority of the injection molding cycle time and therefore the production rate at which parts can be manufactured. Mold filling times and demolding times add to the overall cycle time as well, but the time required to cool the part is typically the dominant variable contributing to the required molding cycle time.

In injection molding, the initially molten plastic parts must be cooled down to a temperature that is low enough for the part not to become distorted due to the forces associated with part ejection. The distortion can be due to (i) the mechanical ejection forces imparted by the ejectors (e. g., ejector pins) or (ii) the relaxation of residual molding stresses that can cause the part to warp or distort after ejection. As stated earlier, there are many variables that influence the cooling time and therefore cycle time. For example, tooling variables include the materials used to produce the mold cavity and core, as well as the layout of the mold cooling channels. Important molding process variables include the material's melt temperature and the mold temperature. In fact, there are so many plastic material, molding process, mold design, and part design variables involved with cooling, that it is impossible to theoretically specify the exact cooling time that will be required for a given part design (in advance of actual manufacturing) with complete certainty. However, it is possible to obtain a theoretical estimate of the required cooling time if these variables can be estimated. Cooling time estimates can be very helpful to the designer of an injection molded plastic part, particularly when choosing a nominal wall thickness or when estimating part and manufacturing costs. The best estimates for the cooling time that will be required for a plastic part are obtained if a computer-aided process simulation can be performed. Cooling simulations can be used to determine cooling time, and they can be used to optimize the cooling system design for the injection mold. Cooling simulations are typically conducted in an effort to achieve balanced and efficient cooling for all areas of the injection molded part. Without a cooling analysis of this type, it is likely that certain sections of the injection mold will run hotter than others, leading to delays in the cycle and the potential for differential shrinkage and part warpage.

If a cooling simulation software is not readily available, approximate numerical solutions for unsteady state, conductive heat transfer (for regular geometries) can be used to provide an estimate for cooling time. For example, cooling times for plastic parts having a width and length that are both significantly greater than the part's nominal wall thickness can be estimated using cooling equations for a plate-like geometry. Likewise, conductive heat transfer solutions for a cylindrical geometry can be used to estimate cooling the cooling time for a full round runner. Numerical solutions for unsteady state conduction of plate-like and cylindrical geometries are given in Equation 1.1 and Equation 1.2 respectively [24].

$$t_c = \frac{h^2}{\alpha\,\pi^2}\ln\left[\frac{4}{\pi}\left(\frac{T_m - T_w}{T_e - T_w}\right)\right] \qquad (1.1)$$

$$t_c = 0.173\frac{R^2}{\alpha}\ln\left[1.6023\left(\frac{T_m - T_w}{T_e - T_w}\right)\right] \qquad (1.2)$$

Where:
t_c = time required for the centerline temperature to reach the ejection temperature (s)
h = nominal part wall thickness (m)
R = radius of a cylindrical molding or cylindrical runner (m)
T_m = melt temperature at the start of cooling (°C)
T_w = mold cavity/core wall temperature (°C)
T_e = the theoretical ejection temperature of the plastic material (°C)
α = thermal diffusivity of the plastic material = $k/\rho\,C_p$ (m²/s)
k = thermal conductivity of the plastic material (W/m K)
C_p = specific heat of the plastic material (J/kg°K)
π = constant pi (3.1416)
ρ = density of the plastic material (kg/m3)

Many assumptions have been used in order to derive these numerical heat transfer solutions. Most of these assumptions are not actually realized during the actual injection molding process. For example, one assumption is that the temperature of the melt in the mold cavity at time zero (i.e., the start of cooling) is uniform. This is unlikely in real life. Another assumption is that the mold cavity (and core) wall temperatures are both uniform, and constant over the course of the mold cooling cycle. Or in other words, it is assumed that the mold cooling system can remove heat from the mold faster than the melt can transfer heat to the mold steel. This is also unlikely in real life since completely uniform mold cooling is very difficult to achieve in practice. Yet another assumption is that there is complete interfacial contact between the mold cavity steel and the plastic part, which may not be the case, especially late in the cycle due to in-mold material shrinkage.

These equations can be used by designers or engineers to (i) estimate cooling time for a given application, and (ii) to determine how changes in key material, process, and design variables will impact cooling time. For example, note that terms for part wall thickness or runner radius are both squared in the cooling time equations. This means that a small change in thickness or radius has a very significant impact on cooling time. Doubling the part thickness would (theoretically) increase cooling time by a factor of four. The thermal properties of the plastic are also required for the cooling time calculation. Thermal conductivity (k), specific heat (C_p), and density (ρ) must be obtained for the material to be molded. Many material data sheets or databases may list these individual values, or they may list the material's thermal diffusivity or thermal diffusion coefficient (α). The thermal diffusion coefficient is a lump sum thermal constant. None of these thermal properties are really constant; they change with temperature, but average values of these thermal properties are typically used so that the cooling time can be estimated. These thermal properties are required for the cooling time calculations, but they also relate to the energy required for processing. For example, materials with a high specific heat will require more energy to heat and cool (all other factors being equal).

The cooling time equations also require process temperature inputs. Melt temperature (T_m), mold wall temperature (T_w) and part (or runner) ejection temperature (T_e) are all required for the calculation. Material supplier data sheets or processing guides will usually list the recommended ranges for the melt and mold temperatures. It is generally best not to venture outside of these recommended temperature ranges in practice to avoid molding or part performance issues.

The one cooling time equation variable that is the least intuitive or instinctive is the ejection temperature. The ejection temperature is essentially the end point for the cooling time calculations. The cooling time equations given above determine the amount of time it takes the midplane of the plate (or center of the cylinder) to cool from the T_m value, to the T_e value. The T_e value has no real physical meaning, other than to say it is a temperature where the material has cooled and solidified to a state where it is rigid enough to withstand the forces of ejection without any permanent deformation or distortion. While material supplier datasheets or databases do not list a T_e value as physical property, they do list other property values that could be used in place of T_e for the purpose of a cooling time calculation. Specifically, a material's deflection temperature under load (DTUL) or preferably its Vicat softening temperature are both used to represent T_e. These properties are normally used to provide a rough indication of a material's upper use temperature limits, or a temperature where the material starts to soften and deform under a load (see Section 1.8.3.1 and Section 1.8.3.2). However, in the case of a cooling time calculation, DTUL or the Vicat temperature can be used in reverse, thinking of them as a transition temperature where the material goes from being flexible or deformable, to more rigid. Both DTUL and Vicat are commonly used as an indication of a material's T_e. Both values are similar, but the Vicat temperature is thought to be the more appropriate value for T_e because the Vicat test uses a flat pin indenter, much like an injection mold ejector pin. In reality, the cooling time estimates given above may over-estimate the actual cooling time required in a real life situation, since moldings are often ejected before the center has fully cooled, as long as the solid skins are thick and rigid enough to prevent demolding or post-molding distortion. This is especially true for runner systems, since (i) full round runners are well drafted and easy to eject, (ii) and distortion (unlike part distortion) for a cold runner system is generally acceptable since the runners are either discarded or reground for reprocessing.

1.10 Thermoplastic Material Flow Behavior

The flow or rheological behavior of a thermoplastic material is important for all plastic processes, including injection molding. The flow property that is of greatest interest for the injection molding process is the material's shear viscosity behavior. A material's shear viscosity is a measure of the material resistance to flow, when the material is subjected to shear forces associated with laminar flow in the case of injection molding. For injection molding applications, a thermoplastic material's shear viscosity will primarily influence the material's plastication characteristics, and molding variables associated with the mold filling and packing phases of the molding process. The shear viscosity behavior of a plastic material is influenced by both the base polymer and the additives that have been incorporated into the formulation.

Like most materials, the shear viscosity of a plastic is influenced by both temperature, and hydrostatic pressure. Almost all plastics have a positive coefficient of thermal expansion which means they expand when heated and contact upon cooling. Increasing a plastic's temperature, increases its free volume and the distance between the polymer molecules. As this molecular separation distance increases, secondary chemical forces of attractions are reduced, thereby reducing the material's resistance to flow at higher temperatures. The magnitude of the effect that temperature has on a plastic material's viscosity is quite variable depending on the material type. For example, the viscosity of polystyrene melt is far more sensitive to temperature changes than that of a polyethylene melt. Hydrostatic pressure also has an effect on a plastic's shear viscosity since hydrostatic pressure also effects a plastic's free volume or molecular separation. The effect of pressure on the viscosity is often ignored, especially for lower pressure plastics processes. However, the effect of pressure can be very important for injection molding applications since peak melt pressures during mold filling can be as high as 200 MPa or more in some applications. Hydrostatic pressure has the opposite effect on a plastic's melt viscosity when compared to the effect of temperature. A higher hydrostatic pressure, reduces the free volume, resulting in closer molecular spacing, higher intermolecular attractive forces and more steric hindrance. Commercial injection molding process simulation software packages have the ability to consider the effects of hydrostatic pressure on shear viscosity. That said, most of the shear viscosity data published for thermoplastics, such as that shown in Figure 1.31, shows how temperature and shear rate effect a thermoplastic melt's viscosity, but pressure effects are not presented.

A plastic's flow behavior is also greatly affected by the type or nature of the deformation that occurs when the material is subjected to the forces that cause flow during plastication or mold filling. As fluids go, a typical thermoplastic melt is a relatively high viscosity fluid, which typically results in very low Reynolds numbers for plastic melts that are processed by injection molding. The plastic flow behavior for an injection molded plastic is almost always laminar flow. There is laminar drag flow during melt plastication and laminar pressure flow during melt injection. Laminar flow is essentially a layer-like flow, as depicted in Figure 1.31. Unlike turbulent flow, there is no radial or vertical mixing associated with laminar flow. There is also a velocity gradient through the thickness of a plastics flow channel (e. g., nozzle, sprue, runner, gate) where the velocity at the wall is theoretically zero, and at its maximum at the center of the flow channel. The relative speed difference between adjacent material layers is known as the apparent shear rate in the case of laminar drag or pressure flow. Faster screw rotational speeds, faster injection velocities, or thinner walls, result in higher apparent shear rates. Shear rates also tend to be greater near the flow channel walls, and are lower towards the midplane or center of the channel, where the non-Newtonian flow front tends to be blunted.

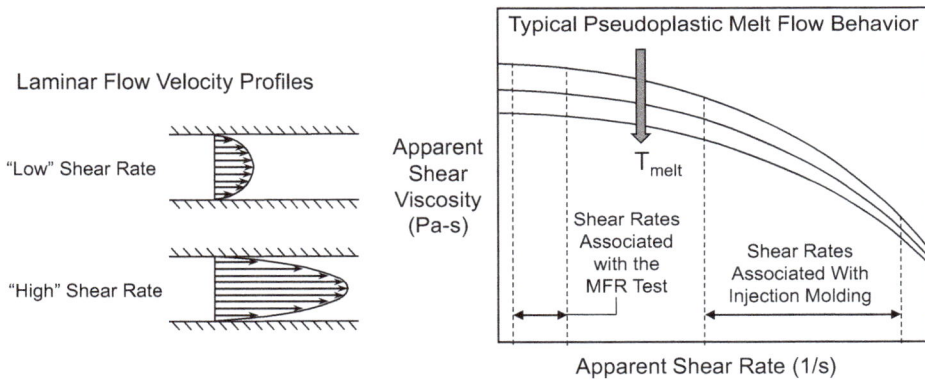

Figure 1.31 The flow behavior of a thermoplastic can be quite complicated. Most thermoplastic formulations exhibit pseudoplastic flow behavior where the thermoplastic melt's shear viscosity decreases with the increasing rate of deformation (or increasing shear rate) and increasing melt temperature. The effect of pressure on a melt's viscosity is often neglected in practice, even though it can be significant, especially at the high pressures that are common with the injection molding process.

Most of the plastics that are used for injection molding applications are described as pseudoplastic meaning that their apparent shear viscosity will decrease with increasing apparent shear rate. The term apparent is used for viscosity data that is determined assuming Newtonian flow behavior and equations, while the term "true" is typically used if the viscosity data has been corrected for non-Newtonian effects. Viscosity curves, such as that shown in Figure 1.31, are normally generated by, and available from, the plastic material supplier. These curves show how the apparent or true shear viscosity of the plastic melt is influenced by both temperature and shear rate. These viscosity graphs are typically run at several different melt temperatures (which are within the recommended melt temperature range for that material) and over a very wide range of apparent shear rates (typically 100–10,000 s^{-1}). A capillary rheometer is the instrument used to generate this viscosity data, but viscosity data of this type can also be generated using an instrumented laboratory extruder or an instrumented injection molding machine [25].

The pseudoplastic behavior exhibited by most plastics is caused by the molecular orientation that occurs during flow. The long-chain nature of a polymer causes the material to become aligned when it is exposed to a laminar velocity field or profile. It is the differential speed of the adjacent (infinitely thin) material layers that causes the alignment or orientation of the chains. The more the chains are aligned, the lower the material viscosity at that layer. At low flow rates, or low shear rates, the material can recover fully or partially as it is being deformed, and the material's melt viscosity is relatively high. Given the chance (time), oriented polymer chains will recover, because the always amorphous plastic melts "want to" have a random (completely amorphous) configuration when they are in the melt state. But, as the shear rate (or rate of deformation) increases, the material will eventually be deformed at a rate that is faster than the rate at which it can recover, and the material's molecular configuration (at the higher shear rate) will be at least partially aligned. Melts having an aligned molecular configuration will tend to exhibit a reduction in melt viscosity (all other factors being equal). The material's viscosity will continue to drop as shear rates continue to increase (due to even more molecular alignment) as shown in Figure 1.31. This graph shows how a typical plastic material's shear viscosity is influenced by temperature and by rate of deformation. The conceptual shear viscosity curve shown in Figure 1.31 is typical of that required for engineering and process design purposes. Process simulation software packages used for plasticating screw design, plastic part design, or injection mold design all require shear viscosity data at different temperatures and shear rates.

Tests of this type are sometimes used to evaluate the flow behavior of incoming production material lots as a quality control test. While shear viscosity curves generated using a capillary rheometer do provide a relatively complete description of a material's melt flow behavior, the Melt Flow Rate (MFR) test discussed in Section 1.4.1.1 is a simpler and more common incoming material lot screening test. It is true that the MFR test is very simple, but it is a very low shear rate test, and as such, the data it generates does not correlate very well with the higher shear rate injection molding process. A typical injection molding grade plastic material has a melt for rate of say 20 grams/10 minutes. The fill time for a typical injection molding process is measured in seconds or fractions of seconds, NOT in minutes. This means that the shear rates associated with the MFR test are orders of magnitude lower than those encountered during the injection molding process. The MFR test is NOT a waste of time as a material lot screening test, since a material having an out of tolerance MFR value will no doubt require different process conditions when it is injection molded. However, it is also possible that two material lots having the same MFR (i. e., within acceptable tolerance limits) may still process differently during injection molding, since the MFR test does not provide data on the material's flow behavior at the far higher shear rates encountered during injection molding. Material lot differences are more easily detected by conducting flow property tests over a far wider, and higher range of shear rates using the more sophisticated and expensive capillary rheometer. If access to a capillary rheometer is not available, extrusion plastometer tests run using several different weights can provide more flow property information than the single point melt flow rate test as described in Section 1.4.1.1.

1.11 Generic Identification and Marking Codes for Thermoplastic Parts

Many injection-molded plastic parts are marked in some way to provide a permanent indication of what plastic material a molded part is made from. Ideally, these markings should be molded into the plastic part and remain visible for the life of the part. Adding the material of construction marking to a plastic part can have a variety of purposes. It can be used to facilitate sorting for plastic recycling. In other cases, such as for a thermoplastics automotive body panel, it can also be used to facilitate painting and repair procedures. Not all injection molded parts require material markings, but many do. In such cases, a number of decisions regarding the markings must be made. These decisions include;

- Which marking protocol should be implemented.
- How the plastic part should be marked.
- Where the plastic part should be marked.
- Type of marking font and symbol sizes.

1.11.1 The SPI (now PLASTICS) Resin Identification Coding System

The earliest thermoplastic plastic material marking protocol was developed and implemented by the Society of the Plastics Industry, SPI (now PLASTICS). It was named the "Resin Identification Coding System" (RIC). This marking system was originally developed to facilitate and improve the recycling rates for rigid plastic packaging by simplifying the material identification and sortation processes. While the SPI Coding System was developed for packaging applications, it has also been used for many non-packaging applications. Administration of this coding system was eventually transferred to the American Society of Testing Materials (ASTM) and the system was revised in 2013 when a new graphic without chasing arrows was introduced as shown in Figure 1.31 [26]. The ASTM Coding System, designated ASTM D7611, is still based on only six generic families of plastics (#1–#6), and a number seven (#7) as a catch-all for other types of thermoplastic materials. While this system is relatively simple, it is clearly overly simplistic due to the very large number of material variables associated with plastics. For example, there are many grades of polypropylene available commercially. There are homopolymer PPs and copolymer PPs. PPs are available having a very wide range of molecular weights or melt flow rates. PPs can incorporate a wide variety of different additive packages. While all of these PPs are very different materials, they would all be marked as #5 based on this material coding system.

Revised Resin Identification Code

Figure 1.32 The original *Resin Identification Coding System* developed by SPI (now PLASTICS) in 1988 has only seven options (#1–#7), each having chasing arrows. Administration of this coding system was eventually transferred to ASTM and the system was revised in 2013 to introduce a new graphic without the chasing arrows.

1.11.2 Other Injection-Molded Plastic Part Marking Protocols

There are several other plastic marking protocols commonly used for injection-molded plastic parts [26-29]. Some part marking protocols are more generally applicable, while others are specific to certain industries. For example, both the American Society of Testing Materials and the International Standard Organization (ISO) both have general part marking designated protocols. Others, like the Society of Automotive Engineers (SAE), publish a plastic part marking standard designated specifically for molded plastic automobile parts [29]. The purpose of the SAE document is to provide information to facilitate (i) the selection of materials and procedures for repairing and repainting automotive plastic parts, and (ii) the collection and handling of automotive plastic parts for subsequent recycling. While all of these standards are different, they do have some commonalities and similarities.

These part marking standards are based on plastic material line call-out designations, in much the same way that the ASTM D4000 generic material naming system creates line call-out (as discussed in Section 1.6.3). The coded designations use uppercase letters to abbreviate the base polymer and additive types, and numbers to indicate additive concentrations. These coded characters are placed between greater than and less than chevrons. Several examples are given below.

A single neat thermoplastic	>ABS<
A polymer blend	>ABS+PC<
A single thermoplastic with 33% glass fiber reinforcement	>PP-GF33<
A single neat thermoplastic with 30% post-consumer recycle content	>ABS(REC30)<

These part marking codes are all overly simplistic, but they are far more rigorous than the #1–#7 Resin Identification Code given in ASTM D7611. For example, these part marking standards do not consider material factors such as a material's average molecular weight or additives that are used at a low concentration. It is also unfortunate that the plastics industry has not developed a single, comprehensive and universally applied part marking standard. It is my opinion that each grade of plastic should have one, and only one, part marking code. A blow-molded HDPE used to make a detergent bottle, should be marked in exactly the same way as a blow-molded HDPE windshield washer reservoir used for an automobile (if it was made from the same material grade). Unfortunately, different industries use different part marking standards. These multiple markings (for the same material) will undoubtedly lead to unnecessary confusion for both consumers and thermoplastic recycling industries.

1.12 Get to Know Your Plastic Material

One of the more challenging tasks that molders face is establishing a suitable molding process for applications involving a new injection mold and new plastic material. The focus of this section is on the new plastic material, meaning a material that the molder does not have previous experience processing. There are also certain attributes of a new injection mold that will cause a molder to select a specific type and size of the injection molding machinery for this new application. These will be discussed in more detail later in this text, but obvious requirements such as the physical size of the injection mold or the required shot volume will direct the molder towards the selection of suitable molding machinery. There are also many plastic material attributes that can have an influence on molding machinery and auxiliary equipment selection.

For example, abrasive or corrosive thermoplastics may require harder or more corrosion resistant metallurgy for the molding machine's injection and plasticating unit. The options associated with injection molding machinery are discussed in more detail in Chapter 3. Once the molding equipment selection is complete (based on the new material and new mold attributes), the molder will need to determine all of the injection molding process variables that are required to produce molded parts that meet the dimensional and end use performance requirements associated with the application. The methodology for establishing molding process variables is discussed in more detail in Chapter 4.

The key when working with a new plastic material for the very first time is to learn as much about the material from reliable sources. Other work colleagues may have had previous experience molding the thermoplastic material. Networking with other plastics professionals at technical conferences or local engineering society meetings opens up great opportunities. There are plastic material textbooks that provide information about generic plastic material families, but not necessarily information for the specific thermoplastic material grade that is to be molded. In practice, the best source of technical information for a particular thermoplastic grade (i. e., the new material in this case) is the plastic material supplier. Material suppliers have the most relevant and reliable information regarding the specific properties and processing characteristics of the material grades that they market. This information is sometimes obtained via conversations with the supplier technical support team, supplier web sites or from downloaded technical literature.

1.12.1 Plastic Material Data Sheets and Injection Molding Process Guides

All plastic material suppliers provide technical information for the specific material grades they market. While there is no standardized format associated with this technical information, most material suppliers do publish detailed material data sheets (MDSs) or databases that provide both a qualitative and quantitative description for the materials they supply. These material data sheets typically provide information regarding the thermoplastic material type, its additives and the material's physical properties. Many of these material data sheets also include recommendations for some injection molding process parameters, especially those molding parameters that are universally applicable. In some cases, material suppliers publish both material data sheets and injection molding guides as separate documents. While material suppliers are the go-to source for technical information for a specific material grade, there are third-party searchable databases that are compilations of material data sheets from multiple suppliers [30]. Some of these third-party material databases cover as many as 100,000 individual plastic material grades.

It is difficult to discuss the topic of material data sheets because there is no standardization as to exactly what technical information is presented in the data sheet. Most material suppliers do test their materials using standardized physical property tests, such as standardized American Society of Testing Materials (ASTM) tests or International Standard Organization (ISO) tests. Some suppliers test according to ASTM standards, while others test according to ISO standards. This makes comparing the data between suppliers difficult when the data is used for material selection. Some material suppliers test using both ASTM and ISO standards which is ideal. It is therefore best to have at least a working knowledge of both ISO and ASTM test standards. Perhaps a more significant issue is that there is little or no standardization regarding which specific set of material property tests is conducted and reported. Some material data sheets may provide impact resistance data based on Izod Impact Testing, while others may report Charpy Impact Resistance. This is a fundamental problem when attempting to use material data sheets for material selection, since an "apples to apples" comparison is not possible. Industry attempts to standardize material database content between suppliers are commendable, but have met with limited success to date. The most widely used standardized material database is CAMPUS, which is an acronym for Computer Aided Material Pre-selection by Uniform Standards [31]. Hopefully there will be more cooperation between material suppliers and more material data sheet standardization in the future.

Many of the physical properties given on material data sheets are used to describe material performance and its suitability for an end use application. For example, an electrical property such as volume resistivity, is important if the material will be used in an electronic application. However, this particular property is unrelated to the material behavior during injection molding, and as such is of no particular interest to a molder. However, other properties, such as Vicat Softening Temperature (see Section 1.8.3.2), may be important to both a part designer (or end user) and the molder. From a part design perspective, Vicat relates to the material's upper use temperature limit. From a molding perspective, a material's Vicat Temperature does provide a rough indication of the temperature at which a molded part can be ejected from the mold (at least theoretically). Some of the other material data sheet properties (or molding guide recommendations) that should be of interest to an injection molding engineer or technician are discussed in more detail below. A mock molding guide for a neat nylon 66 is given in Figure 1.33. This mock guide includes the type of information that is typically contained in a material supplier's material process guide. Guides of this type are extremely useful as a starting point when establishing a new molding process.

Material: Unreinforced (neat) nylon 66

$$\left[N - (CH_2)_6 - N - C - (CH_2)_4 - C \right]_n$$

with H on first N, H O on second N-C, and O on final C

Machine Related Recommendations

Screw and Barrel Recommendations:
  Barrel: Bi-metallic liner
  Screw: L/Ø ratio ≥ 20:1
          Compression ratio = 3:1
Nozzle Recommendation:
  Nozzle Type: One- or two-piece construction
  Nozzle Tip: Reverse taper nylon tip
Clamp Tonnage Recommendation:
  Tonnage = 2.0 to 3.0 (tons/in^2) x projected area (in^2)

A Reverse Taper (Nylon) Nozzle Tip is Recommended

Mold Design Recommendations

Mold Shrinkage Rate: 0.012 m/m (1.2%)

Recommended Vent Depth: 0.0005–0.0010 in.
                          (0.013–0.025 mm)

Molding Process Variable Recommendations

Pre-drying Recommendations:
        Maximum moisture content: 0.05–0.10 % (by weight)
        Properly sized desiccant dryer:
            Inlet air temperature: 180°F (82°C)
            Air dew point: ≤ -20°F (-30°C)
            Minimum drying time: 4.0 hrs.
            Maximum drying time: 6.0 hrs.
            Air flow: 1.0 ft^3/min. per pound/hr.

$H_2O \uparrow$

Recommended Melt Temperature: T_M = 470–540°F (243–282°C)

Recommended Mold Temperature: T_W = 110–220°F (43–104°C)

Regrind Use: Same material grade regrind "may be" used at
               concentrations in the range of 25–30% if the regrind
               material is not degraded or contaminated.

Back Pressure: "Moderate" back pressure is recommended to ensure
                proper melt homogeneity and shot size consistency.

ΔP_b

Injection Velocity: "Fast" injection speeds are generally recommended.   ← V_P

Screw Rotation: Screw recovery RPM should be set such that recovery time
                  represents 75–90% of the available time for plastication
                  (i.e. "additional" cooling time).

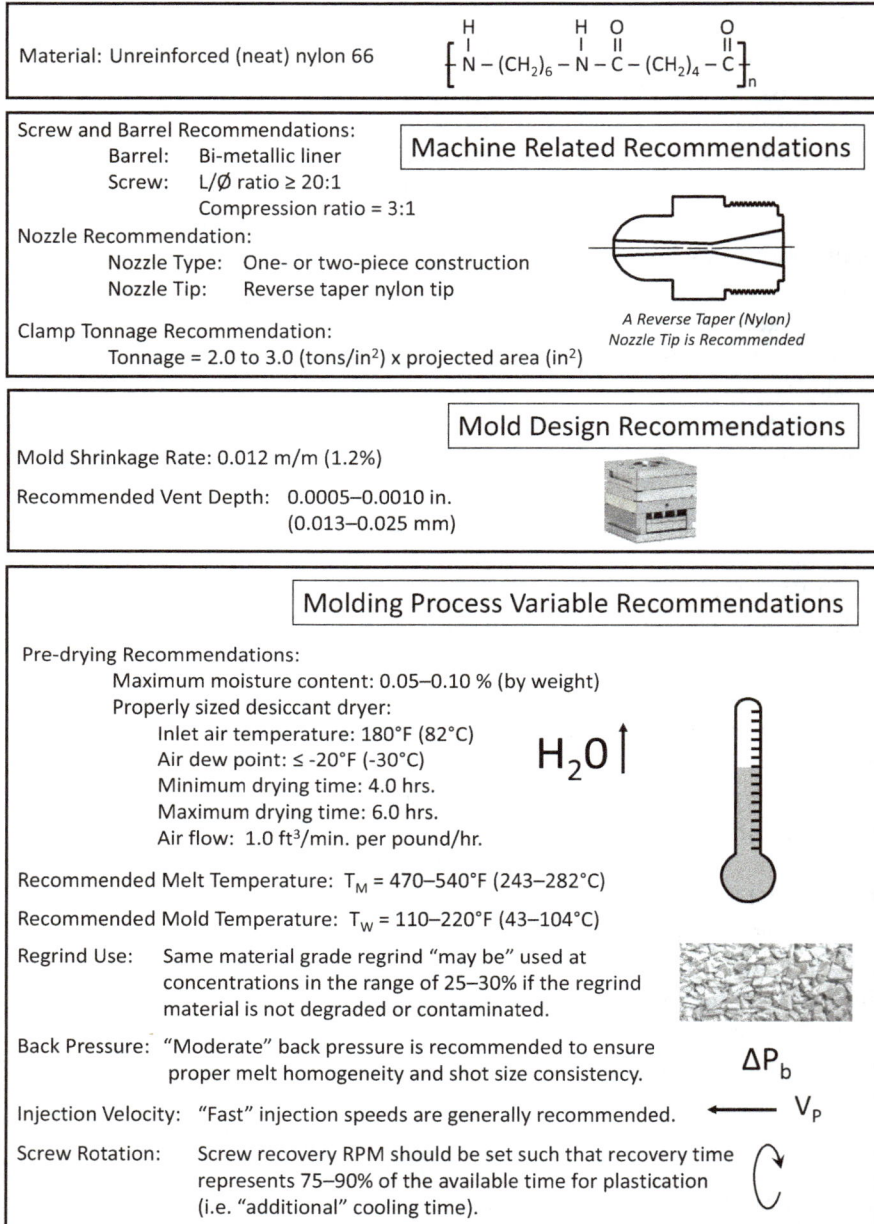

Figure 1.33 A mock but realistic material supplier process guide for the injection molding of a neat nylon 66 is shown. Molding process recommendation guidelines are not standardized in any way and the breadth of the process information varies greatly from supplier to supplier. The process information recommended by the material supplier is the best starting point when establishing a new injection molding process

Generic Material Description: Most material data sheets or molding guides do provide a generic description of the thermoplastic material and optionally information on the formulation's additives. Additive information is often limited since many material suppliers consider additive information proprietary. Ideally the material description would indicate if the base polymer or additives may be corrosive or abrasive, but these determinations may require further material research by the process engineer.

Melt Flow Rate: Most material data sheets list the material's melt flow rate (see Section 1.4.1.1). The material's MFR provides a rough indication of the material's flow behavior or melt viscosity. Materials having a low MFR value are more viscous and likely require more mold filling pressure. Higher melt flow rate materials have a lower melt viscosity and generally better processability for injection molding. Lower viscosity materials require less mold filling pressure, exhibit better mold surface replication, and have fewer knit line issues. On the other hand, higher MFR material grades may be more prone to parting line, ejector pin or vent flash.

Molding Machine Hardware: Material data sheets or molding guides may also provide guidelines related to molding machinery selection or molding screw design recommendations. While much of the injection molding industry is based on what are referred to as general purpose plasticating screws, it is ideal if the plasticating screw has a design that is appropriate for the material to be molded (especially the screw's compression ratio). Some material process guides also recommend the use of a specific type of nozzle tip, such as a reverse taper tip shown in Figure 1.33 that is commonly used when molding nylons.

Mold Design Variables: Most material data sheets or process guides also provide some information that is useful for mold design. The material's mold shrinkage value(s) are typically reported, as are recommended vent depths. These mold design variables are of greater interest to the mold designer or mold builder, since these mold variables are in no way adjustable by the molder once the mold has been built.

Pre-drying Conditions: Hygroscopic plastics must be pre-dried before they are injection molded (see Section 1.4.2.2). A material supplier's literature will almost always provide recommendations for pre-drying requirements if the thermoplastic grade is hygroscopic. This pre-drying information can be presented in various ways. In some cases, the material supplier will provide a recommendation that it deems acceptable for the material's maximum moisture content. This presumes that the molder has the ability to make such a moisture content measurement (ideally all molders should have this capability). In other cases, the material supplier may recommend pre-drying process conditions that should lead to an acceptable moisture content. They typically specify the pre-drying temperature, minimum and maximum pre-drying times, drying air dew point, and dryer air flow rates. It is difficult for a material supplier to specify pre-drying conditions with any certainty, since the drying equipment type and design (unknown to the supplier) are also factors. Even when the recommended pre-drying conditions are implemented in production, it is good practice to use a moisture analyzer to confirm that the moisture concentration after pre-drying the hygroscopic material is within acceptable limits.

Melt Temperature: Material suppliers will always provide guidance for a recommended range of melt temperature for the thermoplastic grade. This temperature range is typically selected so that the material will be (i) fully melted (or softened), and (ii) exhibits acceptable flow behavior and (iii) minimal thermal degradation. There are many factors to consider when selecting the most appropriate melt temperature, but it is best <u>never</u> to venture outside of the melt temperature range recommended by the material supplier. Melt temperatures towards the lower end of the range are attractive as they reduce cooling time and thermal degradation issues, while temperatures towards the higher end of the range may be required if the part is hard to fill. It should be noted here that in reality, melt temperature is <u>not</u> a machine variable that is simply set. Melt temperature is a by-product of many other variables including the plasticating screw type, plasticating conditions, and barrel temperature profile. In addition, melt temperature is very difficult to measure reliably, particularly in-line.

Mold Temperature: The temperature of an injection mold is typically managed by controlling the temperature of the coolant being circulated through the mold. Like melt temperature, material suppliers publish a recommended range for mold temperature. Mold temperatures above the range would result in an excessive cycle time, while those below the range lead to issues such as excessive frozen-in molecular orientation or difficulties filling thin part wall sections. Mold temperatures towards the higher side of the range offer a number of part quality advantages, while those on the lower end of the range reduce cycle time. It should also be noted here that mold temperature is normally not a machine variable that is set. Mold coolant temperatures are typically the controlling variable. Like melt temperature, mold temperatures should be confirmed and measured using a surface pyrometer.

Other Process Variables: Material suppliers can also make recommendations on other process variables, but with far less certainty than those discussed above. They include process variables such as back pressure, injection velocity, screw recovery speed, barrel temperature profiles and the like. They can make qualitative recommendations, like slow or fast, but cannot specify exact values since these variables are also machine- (or sometimes mold-) dependent. For example, the value selected for screw recovery speed (RPM) is dependent on many other variables including shot size, available screw recovery time, and screw diameter, none of which are known to the material supplier. The same is true for barrel temperatures, packing pressure variables, holding pressure variables and many other process variables.

The next two Chapters of this text provide an overview of injection molding machinery, and the methodologies associated with establishing optimum injection molding processing conditions.

References

[1] Kazmer, D., "Injection Mold Design Engineering", Hanser Publications, Munich (2022).

[2] Catoen, B., "Injection Mold Design Handbook", Hanser Publications, Munich (2021).

[3] Beaumont, J., "Runner and Gating Design Handbook", Hanser Publications, Munich (2019).

[4] Mennig, G., "Mold-Making Handbook", Hanser Publications, Munich (2013).

[5] Malloy, R., "Plastics Part Design for Injection Molding", Hanser Publications, Munich (2011).

[6] Campo, E., "Industrial Polymers", Hanser Publications, Munich (2007).

[7] Ehrenstein, G., "Polymeric Materials", Hanser Publications, Munich (2001).

[8] Zweifel, H., "Plastics Additives Handbook", Hanser Publications, Munich (2009).

[9] Technical Bulletin, "Molecular Weight and the Effects on Polymer Properties", AMCO Polymers, Orlando, FL., January (2018).

[10] Yan, D. et. al., "Effect of Long Chain Branching on Theological Properties of Metallocene Poly-ethylene", *Polymer*, **40**, p. 1737 (1999).

[11] Kulkarni, S., "Understanding Fountain Flow in Injection Molding", *Plastics Technology*, (2022).

[12] Kulkarni, S., "Robust Process Development and Scientific Molding, Hanser, Munich (2017).

[13] Sepe, M., "Melt Flow Rate Testing – Part 5", *Plastics Technology*, November (2013).

[14] Sepe, M., "Melt Flow Rate Testing – Part 6", *Plastics Technology*, December (2013).

[15] Rose, D., "Plastic Degradation During Injection Molding", American Injection Molding Institute, (2019).

[16] Greer, M. et. al., "The Importance of Mold Temperature on the Properties of PPS Parts", Solvay Specialty Polymers (2015).

[17] American Chemical Society, 1155 Sixteenth Street, NW Washington, DC 20036.

[18] Plastics Industry Association,1425 K St. NW, Suite 500 Washington, DC 20005.

[19] National Ocean Atmospheric Administration, *https://oceanservice.noaa.gov/facts/microplastics. html*

[20] *https://www.worldwidewords.org/topicalwords/tw-nur1.htm.*

[21] "Compendium of Polymer Terminology and Nomenclature", IUPAC Polymers Division, Research Triangle Park, NC 27709-3757, USA.

[22] *https://www.sabic.com/en/products/polymers/polycarbonate-pc/lexan-resin*

[23] American Society for Testing and Materials Standard ASTM D4000, "Standard Classification System for Specifying Plastic Materials", West Conshohocken, PA 19428.

[24] Technical Bulletin, "Estimating Cooling Times in Injection Moulding", BASF Plastics, *www.plastics portal.com.*

[25] Malloy, R., Chen, S. and Orroth, S., "Viscosity Measurements Using an Instrumented Injection Molding Nozzle", SPE Annual Technical Conference, 34, 279 May (1988).

[26] American Society of Testing Materials Standard ASTM D7611, "Standard Practice for Coding Plastic Manufactured Articles for Resin Identification", West Conshohocken, PA 19428.

[27] International Standards Organization, ISO/DIS 11469 "Plastics –Generic Identification and Marking of Plastic Products", 1214 Vernier, Geneva, Switzerland.

[28] Juster, Herwig, "Miniguide for Plastic Part Marking", *http://findoutaboutplastics.com.*

[29] Society of Automotive Engineers International Standard SAE J1344, "Marking of Plastic Parts", Warrendale, PA 15096 (2017).

[30] *https://www.ulprospector.com/en/na.*

[31] *http://www.campusplastics.com.*

2

Injection Molding Machinery and Auxiliary Equipment

2.1 Injection Molding Process Evolution

Injection molding is one of the most important thermoplastic manufacturing processes in the plastics industry. This incredibly versatile process can be used for the high volume, cost-effective production of high-quality, tight tolerance thermoplastic parts or components having almost any size and shape. Injection molded parts can be as small as a single grain of sand or larger than an automobile bumper. Most injection molded thermoplastic parts have a physical size that is somewhere between these two extremes. The versatility of this process extends beyond part size as virtually all thermoplastics are available as injection moldable grades. As a result of this versatility, injection molding is the go-to process for thermoplastic part manufacturing in industries that range from food packaging to the life saving medical device industry. The sections that follow in this chapter provide an overview of the basic injection molding process and the molding machinery required to produce injection molded thermoplastic parts. The section that follows reviews some of the historical process developments that have led to the evolution of current day injection molding technologies.

2.1.1 Compression Molding Process

Plastic parts can be manufactured using a variety of different plastic manufacturing techniques. The earliest, commercially successful plastic manufacturing processing technique was compression molding, which was initially used to produce parts from naturally occurring material formulations in the late 1800s. The compression molding process can be used with both thermosetting and thermoplastic materials, although compression molding is best suited for molding thermoset plastics. A basic compression molding process, for both a thermosetting plastic material and a thermoplastic material is depicted in Figure 2.1. In both cases, the compression molding technique utilizes a heated, two-plate compression mold that is first loaded with a pre-weighed charge of a plastic material formulation. The heated mold causes the plastic to soften by conductive heat transfer. After the plastic material softens, the mold is closed causing the softened material to replicate the mold cavity geometry. The mold is typically overfilled or overcharged to (i) ensure there are no incomplete parts (i. e., no short shots) and (ii) to help with internal cavity pressure generation. Overfilling of the mold will result in flash which will need to be removed after demolding.

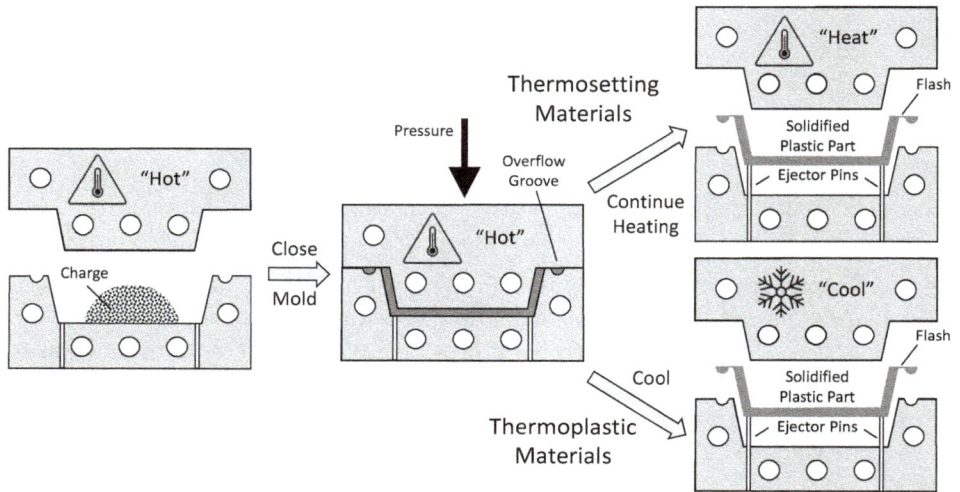

Figure 2.1 The compression molding process predates the injection molding process and has been used with both thermoplastic and thermosetting plastic materials. The compression mold is heated continuously when molding thermosetting materials. When compression molding thermoplastic materials, the compression mold temperature must be cycled from hot to cold each cycle. This can be time consuming and energy intensive.

When a thermosetting material is compression molded, the mold is heated continuously. While heat does cause thermosetting material to soften (initially), additional heat triggers a chemical cross-linking reaction that causes the thermosetting material to solidify. Historically, thermoset compression molding processes were used to mold crosslinked natural rubber as far back as1839 when Charles Goodyear discovered the vulcanization process for natural rubber.

Thermoplastics can also be compression molded, although the process differs from that used with a thermosetting plastic material, as shown in Figure 2.1. The compression molds used for molding a thermoplastic material are also heated in order to soften the pre-weighed material charge. The overfilled mold is then closed causing the material to replicate the cavity shape as it does with a thermosetting material.

However, once the thermoplastic part is formed, the compression mold must then be cooled in order for the thermoplastic material (part) to re-solidify. Once cool, the part is de-molded and the mold must be re-heated for the next molding cycle. This repeated heating and cooling of the compression mold is both time consuming and very energy intensive. Over the years, other more automated and energy efficient molding processes, like the injection molding process, were developed specifically for processing thermoplastic materials.

In the early days of the plastics industry, compression molding of both thermoplastics and thermosets was popular. The mirror frames shown in Figure 2.2 are among the earliest compression molded thermoplastic parts ever made. They were compression molded from a (natural) wood flour-reinforced shellac-based plastic molding compound known as *Florence Compound*. Its inventor, Alfred Critchlow, founded the Pro Molding Corporation, Florence, MA in 1847.

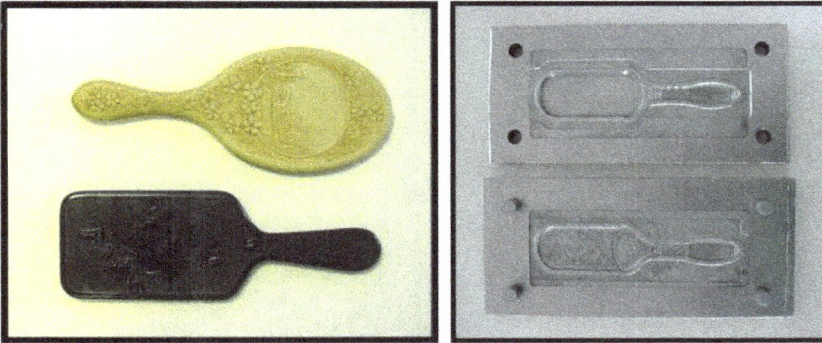

Figure 2.2 The compression-molded, wood flour-reinforced shellac mirror frames and the compression mold date back to the 1860s.

John Wesley Hyatt, the inventor of *celluloid*, a thermoplastic blend of nitrocellulose with sap from the laurel tree (camphor) also used compression molding to produce molded plastic items in the 1860s. Shortly thereafter, he and his brother Isaiah invented and patented their so-called "stuffing machine", which has been described as an early plunger injection molding machine. This novel molding machine included a device described as the predecessor to a plasticating torpedo used in later plunger injection molding machine designs [1].

While compression molding was not the ideal process for molding thermoplastics, it was and still is a very common process for processing thermosetting plastics. The first fully synthetic plastic, phenol formaldehyde (a.k.a. phenolic), was discovered in 1907 when a Belgian born chemist, Dr. Leo H. Baekeland, reacted phenol and formaldehyde under pressure using hexamethylenetetramine as a catalyst for the reaction. The resulting material was a thermosetting plastic he named *Bakelite*. Compared to other plastics available at the time, such as *celluloid*, Baekeland's thermosetting phenolic was more chemically and thermally stable. Once molded, this new material would not burn or soften when reheated. This greater performance made phenolic stand out from the limited number of plastics available on the market at that time. Baekeland's *Bakelite* was an instant commercial success. It was a moldable, electrically insulating, chemically stable, heat resistant, rigid, moisture- and weather-resistant material. Phenolic was very widely used for its electrical insulating capability. The radio housings shown in Figure 2.3 are compression molded from thermosetting phenolic. The compression molding process is still a very common processing method for thermosetting plastic material. The thermoset compression process has also evolved over the years. One variation, known as the transfer molding process, uses a separate chamber or pot, where the thermosetting material charge is softened, and then injected into a closed mold using a ram or plunger. Transfer molding has some similarities with early injection molding processes.

Figure 2.3 Most early thermoset plastic parts were produced by compression or transfer molding. These molded phenolic *EKCO radio* housings are being de-flashed after being compression molded (circa 1934).

2.1.2 Early Injection Molding Processes

While compression molding and transfer molding processes remain popular today for molding thermoset plastic materials, they are not widely used for molding thermoplastic materials. Again, this is largely due to the time and energy required to heat and cool the compression mold for each and every molding cycle when compression molding thermoplastic materials as shown in Figure 2.1. The injection molding process was developed specifically to mold thermoplastic materials, many of which were invented during the same timeframe (early 1900s).

"While compression molding has its definite place in the sun for many jobs, it lacks the versatility afforded by injection molding and hence has limited scope of creative ability (for plastic part designers)." From The Hydraulic Press, HPM Corporation newsletter (1937).

The fundamental concepts of the injection molding process date back to Hyatt's patented stuffing machine of the late 1880s. While injection molding machines can be used to mold both thermoplastic materials and some thermosetting plastic materials, the primary focus of this text is on the more commonly utilized thermoplastic injection molding process.

Early injection molding machines all used heated cylinders to soften or melt the thermoplastic materials by conductive heat transfer. After melting, the thermoplastic melt was then injected into a closed, relatively cold mold cavity through flow channels known as runners. Very early injection molding machines were manual, using mechanical leverage to generate the high forces needed for melt injection and mold clamping. Hydraulically powered injection molding machines were introduced in the 1930s. Hydraulic power is still very common for today's injection molding machines, but all-electric injection molding machines are state-of-the-art today. Several other sources provide a more detailed history of injection molding process development [1–3].

Early injection molding machines were relatively crude by today's standards. The most significant issue with early injection molding machines was their limited ability to thoroughly melt and mix (i. e., plasticate) the thermoplastic material being molded. These plunger molding machines had a heated cylinder or barrel having very limited melting capability and little or no mixing capability. Mixing is generally beneficial when preparing a thermoplastic melt, although in some cases, too much mixing is not desirable. For example, at that time, it was desirable for molded plastic parts to have a marble-like appearance, to simulate the appearance of natural materials, such as tortoise shell or horn. These natural raw materials were popular for consumer items like decorative hair combs or eyeglass frames. However, the melt quality produced by plunger molding machines was not optimum for many other applications. In addition, the injection unit shot size that could be plasticated with a plunger injection unit was very limited, generally to less than a few hundred grams of material at the most.

The molding machine shown in Figure 2.4 is typical of an injection molding machine from the 1930s or 40s. Injection molding machines of that era were used to produce relatively small parts, such as the small automotive and appliance components shown in Figure 2.4 [3]. Also note the complete lack of molding machine safety guarding and worker personal protective equipment (such as eye protection). Today, safety is a priority for the injection molding industry. Injection molding machines (and auxiliary equipment) incorporate many required safety features in order to reduce the potential for worker injury.

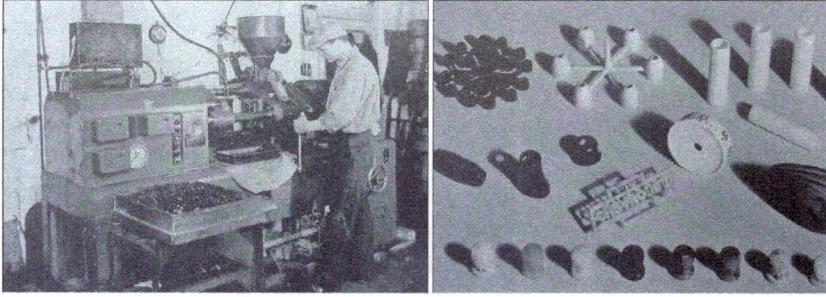

Figure 2.4 Early plunger injection molding machines were relatively small and had very limited plasticating capacity. The relatively small thermoplastic parts shown on the right are typical of early injection molded thermoplastic parts Note that there are no machine clamp safety guards or safety glasses for the operator. Safety is now a priority for the injection molding industry (HPM Corporation Newsletter, circa 1937).

The early injection molding machine shown in Figure 2.5 has one vertical clamping unit, and four injection units (i. e., four barrels and four plungers). Today's multi-barrel injection molding machines are used for multi-material over-molding or multi-material co-injection molding. However, in 1939, multi-injection unit molding machines were used to increase the plasticating capacity of the machine (i. e., total shot size). All four barrels would be fed with the same thermoplastic material grade. The use of four smaller barrels could produce a better-quality melt than one larger barrel, since the melting mechanism with plunger injection molding machines is primarily by conductive heat transfer. The use of multiple injection units allowed for the production of larger injection molded parts, such as the automobile dashboard panels shown in Figure 2.5.

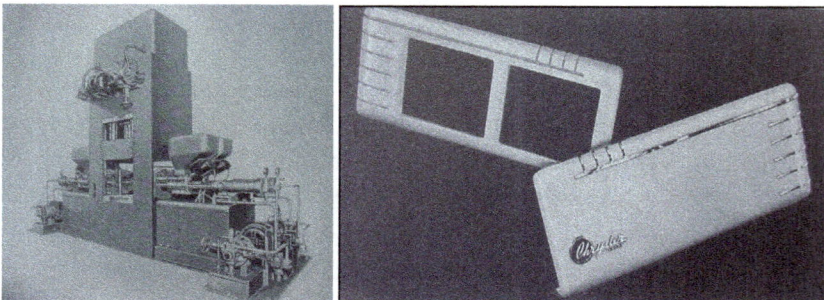

Figure 2.5 This 1939 HPM injection molding machine had four, nine-ounce injection units giving it a total shot capacity of thirty six ounces. It was used to mold the "very large" (for the time) automobile dashboard components shown to the right [3].

2.1.3 Improving Plasticating Capacity and Shot Size

While early injection molding machines were relatively crude by today's standards, they were able to produce large quantities of relatively tight tolerance thermoplastic parts economically. As a result, the injection molding industry was growing rapidly. The demand for injection molding machinery that could produce larger injection molded parts was increasing. Molders were also searching for injection molding machinery with improved process control and molding machines that could process a wider variety of thermoplastic materials. Many new thermoplastic materials were being developed at time, some of which were more difficult to process than the early cellulosic thermoplastics. For example, nylon 66 was invented in 1935. This new plastic material had both a far higher melting temperature and narrower melt temperature window than the cellulosic thermoplastics commonly used for injection molding at that time. A number of injection molding machinery developments were implemented to allow for the production of higher quality injection molded parts from a wider variety of thermoplastic materials.

2.1.3.1 Plunger Injection Molding Machines

The limited plasticating capacity of early injection molding machines was a plastics industry problem that needed to be solved. New product developers of the time were demanding larger plastic parts which would require larger injection molding machine shot sizes. Multi-barrel machines did help in that respect, but the resulting melt quality was still poor due to the limited melting and mixing capabilities of plunger injection units. In addition, a wide variety of new thermoplastic materials were being developed in this same timeframe. Over time, a number of injection unit plasticating capacity and mixing improvements were implemented as discussed in the following section.

The plasticating capacity of a standard plunger injection molding machine could be improved significantly by adding what is known as a torpedo towards the front of the heated injection molding barrel. The torpedo concept is shown in Figure 2.6. The addition of the torpedo to the plunger molding machine barrel improved the machine's plasticating capacity and the melt quality for several reasons. First, the torpedo increased the surface area in contact with the thermoplastic, which improved conductive heat transfer. It also created shear forces acting on the material during the injection phase of the process, resulting in viscous heating as the softened plastic melt flowed through the narrow annular gap between the O.D. of the torpedo and the I.D. of the barrel. This also resulted in some, although very limited, mixing. The addition of the torpedo did increase the pressure required to fill the injection mold, but the streamlined shape of the torpedo was designed specifically to minimize the added injection pressure requirement. While the addition of the torpedo did improve the melt quality compared to a standard plunger injection molding machine barrel (without a torpedo), the degree of improvement was inadequate for many applications, especially those applications that require improved mixing, large shot sizes, or materials that have a narrow melting temperature window.

Plunger (Only) Plasticating Unit (Early 1900s)
- Limited melting capacity and shot size
- No capability for mixing
- Acceptable for thermoplastics of the time

Plunger and Torpedo Plasticating Unit (1930s)
- Limited melting capacity and shot size
- Larger shots required multiple barrel machines
- Minimal mixing capability
- Not suitable for newly developed thermoplastics

"*Reciprocating Screw*" Plasticating Unit (1950s)
- Significant increase in plasticating capacity (+)
- Significantly greater mixing capability (+)
- Leakage flow results in shot variation (-)
- Limited ability to generate packing pressures (-)

Reciprocating Screw with a "*Non-Return Valve*" (1960s)
- Good plasticating capacity (+)
- Moderate mixng capability for many applications (+)
- Non-return valve (e.g. check ring) limits leakage (+)
- Greatly improved ability to provide packing pressure (+)

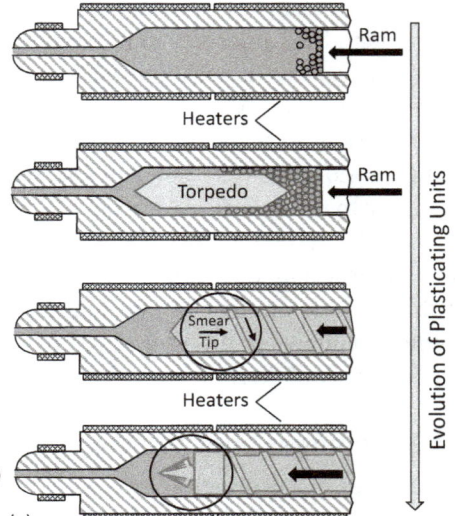

Figure 2.6 The plasticating unit of an injection molding machine must perform number of functions. These functions include; melting, mixing, injection (i. e., mold filling) and packing. Molding machine plastication units have evolved over the years in an effort to improve both the quantity and the quality of the plastic melt that can be generated during the plastication phase of the process.

Another major injection unit machinery development that resulted in greatly improved plasticating capacity for plunger injection molding machines was the development of the two-stage plasticating unit. The first patent for a plunger injection molding machine equipped with an extrusion-like plasticating screw was filed in 1938 [2]. While many different two-stage machine designs are possible, a piggy-back design is most popular where an axially fixed plasticating screw is mounted above the injection molding machine's conventional barrel and plunger. The axially fixed screw, also known as a pre-plastication screw, communicates with the shot chamber via a two-way nozzle valve. The nozzle valve allows the pre-plasticated melt to fill the shot chamber with the appropriate volume of melt during the plastication phase of the process. The pre-plasticating screw then stops rotating, the valve allows melt to enter the mold when the injection ram is energized. The use of the pre-plastication screw completely separates the plastication phase of the injection molding process from the injection and pressurization phases of the molding process. The advantages associated with the pre-plastication machinery developments are significant and numerous. The two-stage plasticating unit could have a significantly greater plasticating capacity compared to conventional plunger machines of the day. The use of a two-stage machine also improved the melt quality and degree of mixing. The pre-plasticating screw would allow for thermoplastic melting by both conductive heating and viscous dissipation (i. e., mechanical shearing during screw rotation). The two-stage injection units also resulted in improved physical and temperature mixing, particularly distributive mixing. In addition, two-stage injection units were capable of melt-processing all of the thermoplastic materials that were being developed. The only real downside associated with two-stage machines is the added complexity of the molding machinery which results in added capital cost and machine maintenance. Two-stage injection molding machines are still used today, but their use has declined, due primarily to the development of the reciprocating screw injection molding machine plasticating unit. Ironically, some state-of-the-art micro injection molding machine technologies utilize two-stage injection units.

2.1.3.2 Reciprocating Screw Injection Molding Machines

The invention of the in-line reciprocating screw plasticating unit was more of a revolutionary development than an evolutionary development. Most of the injection molding machines in use today continue to use a reciprocating screw plastication technology, even though this developments dates back to about 1950. The reciprocating screw, is essentially an extrusion type plasticating screw that can move back and forth axially, in order to allow for both (i) shot plastication and (ii) injection of the thermoplastic melt into the injection mold. The reciprocating screw sits within an electrically heated cylinder or barrel as shown in Figure 2.6. During the plastication phase of the molding process, the screw rotates and pumps melted plastic into the shot chamber (i. e., into the front section of the barrel). Pressure builds as the melt is pumped from the rotating screw, because the molding machine nozzle outlet is blocked by the previous shot. Melt pressure from the rotating screw cause the screw to push or pump itself backwards. This backwards movement of the screw as it rotates is described as reciprocation. The screw will continue to reciprocate and pump melt until it reaches a pre-set axial distance that is set based on the shot volume required to fill the injection mold. The shot volume that is generated by the reciprocating screw is slightly greater than that required to fill the mold. The extra plastic melt serves as a cushion that allows the screw to pressurize the melt during the packing and holding phases of the molding process.

Molded Part and Runner Volume + The Cushion Volume = Shot Volume

Shortly after plastication is complete, the mold open, and the solidified part from the previous molding cycle is ejected. The mold then closes, and the reciprocating screw is used to inject the melt into the injection mold by causing the screw to move forward (without rotating). The screw pushes the melt into the closed mold and then continues to apply packing and holding pressure against the cushion to help compensate for mold shrinkage. The non-rotating screw is essentially acting like a ram or plunger. Once the packing or holding pressure phases time out, the reciprocating screw will begin rotating and reciprocating as it builds the shot volume for the next molding cycle. The reciprocating screw plastication concept is simple and offers tremendous versatility in terms of shot capacity, melt quality, and degree of mixing.

The main issue with this early and basic reciprocating screw concept is backflow of melted thermoplastic material along the screw channel during the melt injection and pressurization phases of the process. Ideally, the reciprocating screw would act like a positive displacement ram during the injection and pressurization phases of the molding process. However, in reality, the melt pressure ahead of the screw can cause the melt to flow back along the screw channel. This can lead to shot size inconsistency and the screw will eventually bottom out. Once the cushion dissipates and the screw bottoms out, the melt can no longer be pressurized to compensate for mold shrinkage. Various screw tip designs, like high surface area smear tips, can be used to minimize backflow, but other more positive anti-backflow devices would need to be developed in order to improve the pressure generating capabilities of the reciprocating screw.

2.1.3.3 Reciprocating Screw Equipped with a Non-Return Valve

The material backflow that occurs with the reciprocating screw during the injection, packing and holding phases of the molding process, has a negative impact on molded part quality. The leakage limits the ability to pack out the part and compensate for mold shrinkage. The amount of material backflow can also be variable, which reduces shot to shot inconsistency. The solution to the leakage or backflow would be to add a non-return valve to the discharge end of the reciprocating screw. This development of anti-backflow valves for the reciprocating screw evolved during the 1960s.

As stated earlier, the shot size that is generated by the reciprocating screw should be somewhat greater than the volume of plastic melt that is required to fill the mold. During the injection phase of the molding process, the mold runner system and cavity are filled with compressible melt. Plastic melts are more compressible than may other fluids at the pressures associated with injection mold filling. The melt is compressed at the end of fill to ensure that the plastic melt fully replicates the mold cavity geometry. It is important to have a cushion of material in front of the reciprocation screw (which is now acting like a plunger) so that the screw can transmit packing pressure from the barrel into the mold cavity.

After the mold is filled, the process engineer can adjust the pressure on the screw during the packing and holding phases of the molding process to help control and compensate for molding shrinkage (and therefore molded part dimensions). The screw pushes on the cushion, pressurizing the melt in the sprue, runners, and mold cavity. This ensures good part surface replication and dimensional control. If leakage or backflow occurs, and the cushion dissipates (i. e., the screw bottoms out), the screw can no longer pressurize the material in the mold cavity.

It is helpful to introduce an analogy here to help explain why cavity pressure history control is so important for the injection molding process and molded part quality. The injection mold filling process is in many ways analogous to the operation of an automotive brake system. The mold cavity is analogous to the brake caliper (or wheel cylinder). The mold runner system and brake lines are the conduits for pressure transfer. The brake system's master cylinder functions much like the molding machine's injection screw. The brake fluid in the brake system's master cylinder is analogous to the molding machine melt cushion. If there is no brake fluid in the master cylinder, the brakes will not function. Likewise, if there is no cushion (due to melt backflow along the screw channel) the compressible melt in the mold cavity cannot be pressurized during the packing and holding phases of the injection molding process.

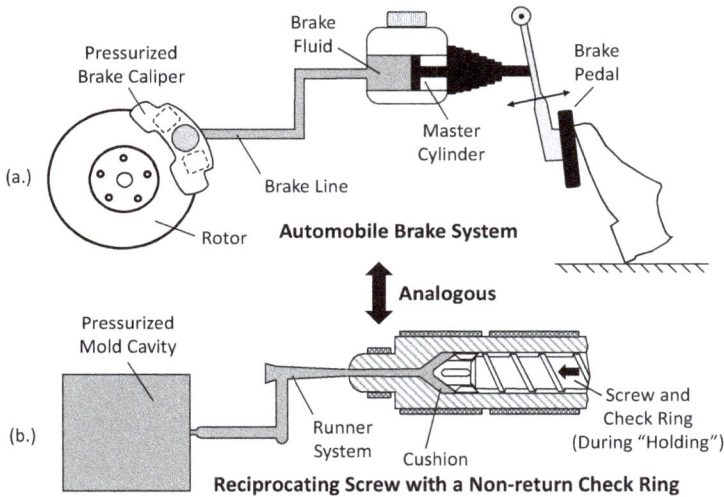

Figure 2.7 Pressurization of the mold cavity during the packing and holding phases of the molding process is required for good dimensional control of the molding. The reciprocating screw and check assembly (i. e., ram) need a cushion of melted polymer (i. e., extra melt not injected into the mold) to push on in order to transfer pressure from the barrel to the injection mold cavity. Losing the cushion due to melt backflow is like having no brake fluid in the master cylinder. No brake fluid, no brakes. The purpose of the non-return valve located at the discharge end of the reciprocating screw is to prevent backflow along the screw channel when the melt is pressurized.

Adding a non-return valve (or backflow preventer) to the discharge end of the reciprocating screw is the most common solution to minimizing or eliminating the backflow issues that occur when injecting and pressurizing the melt with a reciprocating screw. The one-way backflow preventer allows melt to flow through the valve during plastication so that the shot chamber can be filled with molten plastic. However, during injection, the backflow preventer creates a seal that prevents plastic melt from flowing past the check valve, and back along the screw channel. The addition of the backflow preventer allows for significantly greater shot-to-shot consistency and pressure transmission capabilities. These process improvements both lead to improved dimensional control for the molded plastic parts.

There are many different types of non-return valves that can be used with a reciprocating injection molding screw, but sliding check ring valves, such as those shown in Figure 2.6 or Figure 2.7 are most common. A more detailed description of check valves is given in Section 2.2.2.5 The check valves used with reciprocating screws are not perfect, in that some leakage can still occur, especially at the time when the check valve closes at the start of injection. This closure is usually initiated by the forward movement of the screw at the start of injection. Leakage over or under the check valve can also occur, especially if the valve or barrel are worn or damaged. However, as fluids go, many thermoplastic melts have relatively high viscosity, which helps minimize leakage issues. Leakage issues do become more significant with very high melt flow rate (low-viscosity) thermoplastics.

Ideally, the check valves used with a reciprocating screw should:

- Be streamlined and free flowing during plastication.
- Have a relatively low pressure drop during plastication.
- Seal tightly to prevent leakage flow during the injection, packing and holding phases.
- Have good wear resistance and prevent molding machine barrel wear.

The vast majority of injection molding machines in use today, large and small, utilize a reciprocating screw that is equipped with a non-return valve. The reciprocating screw and non-return valve designs will be discussed later in this chapter.

2.2 Major Components of a Reciprocating Screw Injection Molding Machine

Injection molding machines are available in a wide variety of sizes from a large number of molding machinery manufacturers. Machine designs vary from manufacturer to manufacturer, and they vary with machine size as well. That said, most injection molding machines do have certain common features and machine components. Injection molding machines are to a large extent self-contained in that they incorporate most of the machine elements necessary to mold parts once a mold has been set into the machine. While mostly self-contained, additional auxiliary equipment is also required depending on the application. The injection molding machine, along with the auxiliary or support equipment, together constitute an injection molding cell. The major components of an injection molding cell are shown in Figure 2.8.

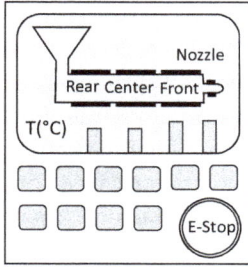

Injection Molding Machine Controls (Human-Machine Interface)

- Temperature Set Points and Control (Barrel...)
- Controls for Plastication Functions (RPM...)
- Controls for Injection Functions (Pressures...)
- Controls for Clamping Functions (Forces...)
- Controls for Part Ejection Functions
- Controls for Machine Power and E-Stop
- Digital Storage for Mold Specific Processes
- Real Time Process Monitoring / SQC Functions

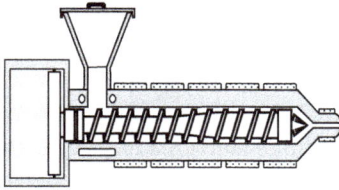

Injection / Plasticating Unit

- Short Term Pellet Storage and Feeding
- Plastication (Melts, Mixes and Generates Shot)
- Injects Molten Shot Into Mold Runners & Cavities
- Builds-Up Packing and Holding Pressures
- Pull-In Cylinders → Injection Unit Movement
- Pull-In Cylinders → Nozzle/Sprue Contact Force

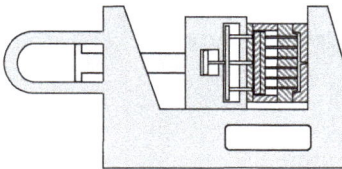

Mold Clamping and Ejection Unit

- Provide for Mold Mounting to Machine Platens
- Provides for Mold Opening and Mold Closing
- Clamp Has "Low Pressure" Mold Protection
- Clamp Force Generation After Mold Closes
- Includes Machine Ejector Plate & Motions
- Core Motion Connections (Hydraulic Machines)

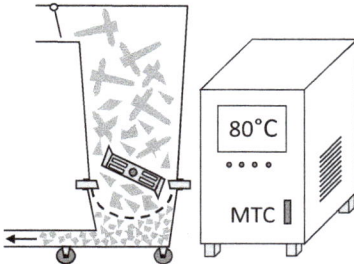

Auxiliary and Support Equipment

- Mold Temperature Controllers
- High Temp Mold Temperature Controllers
- Parts Conveyers and Runner Augers
- Runner Scrap Granulators
- Dryers and Vacuum Loading Systems
- Volumetric or Gravimetric Feeders
- Material Handling Ferrous Metal Magnets
- Parts/Runner Handling Automation (Robotics)

Figure 2.8 Reciprocating screw injection molding machine designs vary depending on the machinery manufacturer. However, injection molding machines do have common features that include the molding machine's injection unit, the mold clamping unit, machine controls, and various types of auxiliary or support equipment.

2.2.1 Injection Molding Machine Controller

Historically, early injection molding machines had discrete control systems. Each injection molding machine variable or function was controlled by a separate timer, a temperature controller, or a position limit switch working in conjunction with ladder logic electronics. Today, machine controllers are integrated computer-based controllers that regulate each and every function of the injection molding machine. They also have the option to monitor machine variables for quality control purposes. The computer-based controller facilitates operation of the complex injection molding machine and molding process. The controllers typically have software and a graphical user interface that helps the process engineer or machine operator visualize the process on the controller's screen. For example, injection molding machine controllers typically have several modes of operation, including manual, semi-automatic, and fully automatic. Manual mode is normally utilized during machine startup and shutdown. Semi-automatic mode is sometimes used in applications where operator assistance is required (e. g., manual insert loading) while fully automatic operation is the normal mode of operation. The user interface keys or touch screen points are typically presented as a series of intuitive pictograph symbols as shown in Figure 2.9.

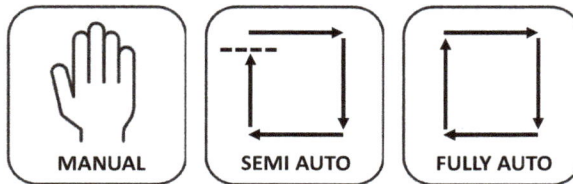

Figure 2.9
Intuitive pictograph symbols are typically used for the injection molding machine's user interface keys.

Computer-based machine controllers also offer the ability to store and automatically call up molding conditions from previous production runs. Access to the machine controls can also be fully or partially restricted based on management practices. The machine controllers are typically capable of controlling the on-board injection molding machine variables (e. g., barrel temperatures), as well as some auxiliary equipment variables, when standardized protocols allow for machinery communications (e. g., mold temperature controllers). The injection molding industry has also embraced manufacturing advances set out in the Industry 4.0 movement. Such protocols allow for the complete control of a networked manufacturing cell from a single human interface.

The user interface and computer-based machine control strategies vary from machine supplier to supplier [5]. This can make process set up and employee training difficult for production plants that have molding equipment from different machine suppliers or even different generations of machinery. There seems to be some standardization of the symbols or pictographs associated with each molding machine function, although this is one of the reasons that third-party monitoring and control systems have become so popular. For example, the *eDART®* system (short for electric Dispatcher Applications and Reporting Tool) is a very widely used add-on computer-based injection molding monitoring and control system [4].

2.2.1.1 Basic Injection Molding Process Steps

Injection molding is a cyclic process involving a number of sequential process steps that are controlled by the machine control system. The auxiliary equipment controls may be integrated with those of the molding machine, or each piece of auxiliary equipment may be controlled separately. A basic injection molding process sequence overview or process flow chart is shown in Figure 2.10. In reality, the injection molding process is more complicated than that shown in Figure 2.10. A more detailed injection molding Gantt type process chart for the injection molding process is given in Chapter 3.

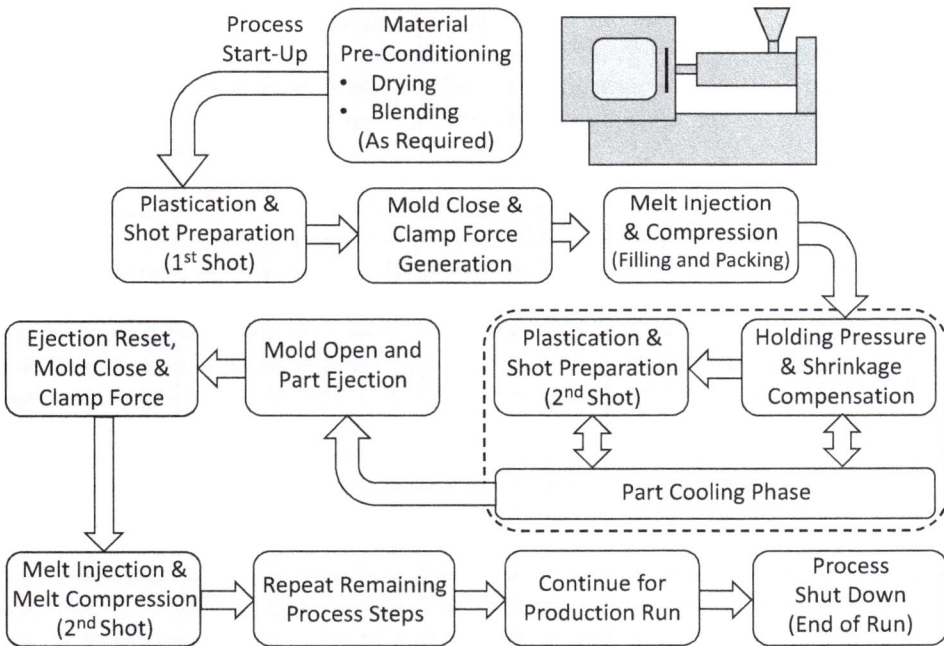

Figure 2.10 Flow diagram showing the main process steps associated with the reciprocating screw injection molding process.

Before describing the injection molding process sequence, it is assumed that a properly sized and equipped injection molding cell is being used for production. It is further assumed that the injection mold has been set and a process engineer has preheated the molding machine and mold, and has established an appropriate molding process. If the material is hygroscopic, it would need to be pre-dried prior to molding using drying conditions recommended by the material supplier. Additives, such as a color concentrate, could be pre-blended with the plastic material, or introduced using a volumetric or gravimetric feeder. feeder. Regrind (i. e., reground runner scrap) could also be added to the hopper at the appropriate ratio in applications where regrind use is permitted.

- In the first phase of the molding process, the reciprocating screw rotates, causing plastic material from the machine hopper to travel along the screw, melt by viscous heating and conduction, and fill the shot chamber. The amount of melted polymer in the shot chamber should be equal to the volume of material that is required to fill the mold, plus an additional amount of material to account for melt compressibility and to provide a cushion that allows for the transfer of pressure into the mold cavity. One rule of thumb suggests that the shot size should be somewhere in the range between 20–80% based on the molding machine's shot capacity. Once the required shot volume is generated, the screw stops rotating.

- The mold is then closed and clamp force (tonnage) is generated. Clamp force is required to counteract the forces associated with melt injection, packing and holding pressures. If clamp force is insufficient, parting line flash can occur. Excessive clamp force, on the other hand, could cause injection mold damage and parting line venting issues.

- The injection unit (carriage) is then brought forward so that the nozzle seats against the injection mold. The reciprocating screw injection cylinder (or motor) is then energized causing the screw and check valve assembly to move forward. The forward movement causes the non-return valve to close, and the melted shot is pressurized to fill the mold runners and cavities. The injection or mold filling phase is typically velocity controlled until the mold is about 95% full. Mold fill time is typically about a second (+/–) depending on the size of the part. Once the mold filling phase is complete, the melt compression phase (i. e., packing phase) is initiated immediately after. The mold filling phase is regulated using velocity control, while the compression (or packing) phase is regulated using pressure control. The final fill and compression phases typically have a relatively short duration, typically about 1/10 of the fill time. The packing pressure that is utilized should be high enough to complete mold filling (final few percent fill) and to ensure that the cavity is pressurized enough for the part to replicate all mold features, including the surface texture.

- Once the mold is filled and packed, an additional follow-on pressure, or holding pressure (or second stage pressure) is then applied by the screw and non-return valve assembly (acting on the cushion). If holding pressure were not used, the compressed (packed) melt could flow back from the pressurized cavity and runners, back into the machine shot chamber. If holding pressure were not used, the part would be full since the skins have solidified, but material would flow through the still molten core regions. The loss of the core material would cause the skins to collapse and sink marks would be observed. In normal practice, holding pressure is applied in order to compensate for mold shrinkage and ensure that the parts meet the dimensional requirements. Holding pressure and holding time are typically adjusted based on the part's dimensional requirements and surface appearance (e. g., sink marks). Holding time is typically limited by the part's gate solidification time, as the gate is typically thinner than both the runner diameter and the part's nominal wall thickness. Once the gate has solidified, material can no longer flow into or out of the mold cavity, even if holding pressure continues to be applied. Deeper gates therefore allow for greater dimensional control by extending the holding pressure process window.

- Once the holding pressure time has expired, screw rotation or plastication of the shot for the next molding cycle can begin. This is also described as dosing. The reciprocating screw will rotate until the shot volume set point has been reached. The mold will stay closed while the screw rotates to generate the shot. While the gate of the molding may have solidified by this time, it takes additional time for the thicker wall molding to cool and solidify. Once the next shot has been generated, *and* the molded part has solidified, the mold can be opened. It is common practice to set plastication conditions such that the screw reaches the shot size set point just before the part has cooled to the point where the mold can be opened and the part is ejected. The sum of the holding time and the plastication time should be less than the part cooling time as shown in Figure 2.10.

- The molded part and runner system (in the case of a cold runner mold) are then ejected or demolded. The ejector system then resets, the mold is closed, and the molding cycle repeats for the duration of the production run.

The injection molding process sequence described above is an oversimplification of the process, but this basic process description does help facilitate the discussion of molding machine components that follows.

2.2.2 Injection and Plasticating Unit

The injection and plasticating unit (injection unit for short) of a reciprocating screw injection molding machine is designed to perform a number of different functions. Reciprocating screw injection units have many components, most of which are common to all reciprocating screw injection molding machines. Injection unit designs do vary from machine supplier to machine supplier, but the same basic components are used in most cases. The injection units can be energized using either servo-hydraulic systems or they may be all electric or even hybrid. The injection units may be small or large, but the basic concepts associated with their operation are common. The discussion here is limited to typical reciprocating screw injection units. Other specialty injection units, such as multi-material injection units or micro molding injection units are discussed in Chapter 4.

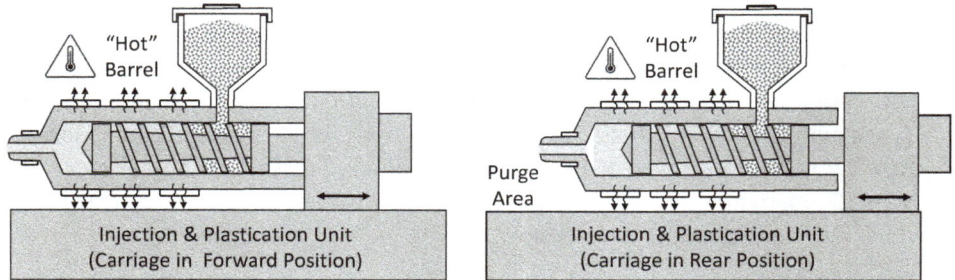

Figure 2.11 The injection and plasticating unit of an injection molding machine is designed to perform many functions ranging from shot generation to melt injection. The injection unit is comprised of a fixed or stationary base and a movable carriage or sled. The sled is moved to the rear position for purging. Pull-in cylinders (or ball screws) are used to move the sled to the forward position, bringing the nozzle tip (i. e., the barrel discharge orifice) in contact with the injection mold's sprue bushing (mold inlet orifice) for part production.

The basic functions of the injection unit are as follows:

- Include provisions for plastic material feeding to the reciprocating screw.
- Plasticate (melt and mix) the thermoplastic material along with any additives.
- Generate the shot of molten thermoplastic to be injected into the mold (including the cushion).
- The carriage (sled) moves back and forth as required, using a pull-in mechanism.
- Generate contact force between the injection unit nozzle and the mold sprue bushing.
- Inject the molten material (shot) into the mold using velocity control (i. e., mold filling).
- Apply packing pressure for the final fill and melt compression phase using pressure control.
- Apply follow-on or holding pressure for a portion of the cycle as the material cools and shrinks.

Injection units are generally categorized based upon their plasticating capacity and maximum shot size. Other variables of interest are their pressure capability, screw design, and maximum injection velocity. Shot size is the maximum amount of plastic material that can be injected into the mold in one molding cycle. The maximum shot size or volume is dictated by the reciprocating screw diameter and maximum reciprocating stroke. An injection unit's shot size can be described based on maximum possible shot volume (cm^3) or based on maximum possible shot weight (typically ounces), using neat general-purpose polystyrene as the reference material. Neat polystyrene has a specific gravity of about 1.05 as a reference that can be used to determine the injection unit's maximum shot size for other plastic materials. Very small injection units may have a maximum shot size of less than 1.0 ounce, while very large injection units can have a maximum shot size of several 1000 ounces.

The molding machine's maximum shot capacity is one of several key machine variables when selecting a molding machine for a new application (i. e., a new mold and plastic material). Under normal circumstances, the shot capacity of an injection unit must be greater than the volume or weight of plastic material required to fill the injection mold. As a first cut, one rule of thumb is that the mold's shot volume should fall somewhere between 20–80% of the injection unit's shot capacity. Using less than 20% of the injection unit capacity can lead to excessive material residence times (i. e., possible material degradation), and more shot-to-shot part dimensional variation.

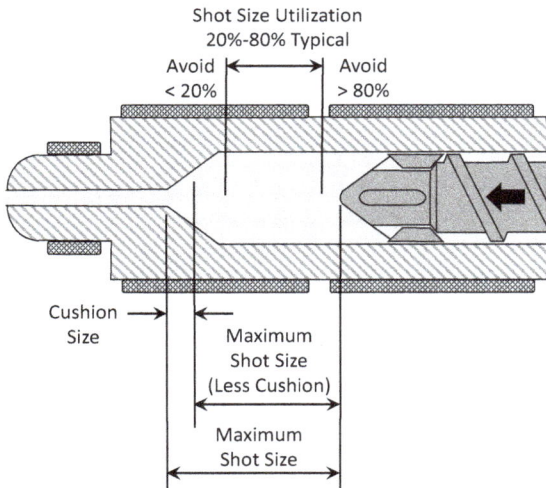

Figure 2.12
A common rule of thumb is that shot utilization should fall somewhere between 20–80% of an injection unit's maximum shot capacity.

There are many other features that differentiate one injection unit from another. All injection units have a maximum injection velocity that determines how fast a mold can be filled. High injection velocities are required in some injection molding applications, particularly for very thin-walled parts (e. g., thin-walled packaging). High injection velocities minimize skin solidification during mold filling. In such applications, an injection unit having an accumulator to increase the range of possible injection speeds would be a major advantage.

Once an injection mold is filled (or nearly filled), packing pressure is required to complete the final fill and squeeze more compressible melt into the mold cavity. Historically, injection units had a melt pressure limit of 140 MPa. Over the years, injection unit melt pressure capabilities have increase to as much as 280 MPa or more. Machine pressure capabilities have increased over the years due in large to trends in plastic part design. Larger plastic parts with long flow lengths and thinner wall parts are generally harder to fill. Higher melt pressure capabilities of an injection unit are typically achieved by using smaller diameter screws and barrels, thereby increasing the pressure magnification ratio for the hydraulic injection cylinder (or all-electric equivalent). An injection unit's maximum pack pressure limit is also its maximum hold pressure limit.

Another important injection unit variable is the type of reciprocating screw used for plastication and melt injection. While the screw is a component of the injection unit, it is a component that can be changed out as required. Ideally, the reciprocating screw should have a design that has been optimized for the thermoplastic material being molded. In practice, this is rarely the case. Many injection molding machines are supplied with general-purpose plasticating screws that allow for the processing of a wide variety of thermoplastics, but not in an optimized manner.

Using plasticating screws and barrels having the appropriate metallurgy is also important. This is particularly important in applications where the plastic material formulation is chemically corrosive or mechanically abrasive. There is no alternative to using corrosion- or abrasion-resistant screws and barrels in these applications.

As stated above, many (perhaps most) injection molding machines are equipped with a general-purpose reciprocating screw. This term has no universally accepted definition, but a general-purpose reciprocating screw is one that can be used to process a wide variety of different thermoplastics. The general-purpose screw design is not optimized for any one material, but is thought to provide adequate melting and mixing for most thermoplastic materials and applications. Many molders, especially custom molders, use general-purpose screws so that they do not have purchase and inventory multiple screw designs and they do not have to change out the plasticating screws when changing materials. While this is not ideal from a melt quality point of view, injection molding variables like barrel temperature profile, screw recovery speed (RPM) and back pressure can be used to fine-tune melt quality when general purpose screws are utilized for plastication. Still, using a plasticating screw design that has been optimized for the material being molded is bound to improve the molded part quality. The use of optimized plasticating screw designs is most common for long production run captive injection molding operations. The various components (and auxiliary equipment) associated with an injection molding machine's injection unit are discussed below.

2.2.2.1 Feed Material Handling Equipment

Material storage: Plastic materials are typically stored in a warehouse area or material storage silos adjacent to the manufacturing floor. Most thermoplastics are considered to be shelf-stable for relatively long periods of time so the time associated with material storage is probably not a critical variable. However, it is best if the material can be stored in an area where the environmental conditions are controlled and consistent over time. For example, there can be issues such as condensation when plastic thermoplastic materials are stored in a very cool environment before they are transferred to a warmer production area. Material storage areas may also include provisions for material drying, additive blending, regrinding, and regrind blending. Batch material handling (i. e., central material handling) is most common for applications where production volumes are relatively low, and the cost of more automated material handling equipment cannot be justified. Material handling is typically more automated for higher production applications [6,7].

Material transfer: The plastic material to be molded must also be transferred from the material storage area to the molding area. This can be accomplished manually or in a more automated way using vacuum or pneumatic conveyance piping and tubing. The relatively low specific gravity and physical size of most plastic pellets make them relatively easy to convey using vacuum or pneumatic loading systems. The plastic material conveyance equipment options are too numerous to discuss in detail here but some of the primary considerations associated with pre-production material handling are given in the following sections.

Consider the material handling examples given in Figure 2.13. In the simplest case, a virgin neat or pre-colored non-polar hydrophobic thermoplastic is to be injection molded. In such a case, there is no need for material drying since the material is not hygroscopic, and there is no need for additive blending. The feed material is simply manually loaded (or vacuum loaded) into the hopper of the molding machine. As with any molding process, the hopper should be equipped with a free flow magnet that is designed to remove any ferrous metal contamination from the feed stream before it enters the molding machine barrel for plastication.

The material handling requirements for the molding application shown to the right in Figure 2.13 are far more complex. In this application, a hygroscopic thermoplastic is to be molded, which means the material will need to be dried before it is molded. The application also involves the addition of a masterbatch additive and the use of runner scrap (i. e., regrind). The mixing of these feed material components can be conducted in a batch manner (i. e., off-line) to produce what is known as a salt and pepper blend. Alternatively, the mixing can be done in-line using automated feeding and blending equipment. It should be noted here that any additives that are to be blended with the base plastic material are usually supplied as masterbatches in pellet form to simplify material handling. The use of dry powder additives is uncommon for injection molding applications, largely due to the material handling issues associated with fine powders. Liquid additives, including liquid colorants, are gaining popularity in some injection molding applications, especially high production volume applications.

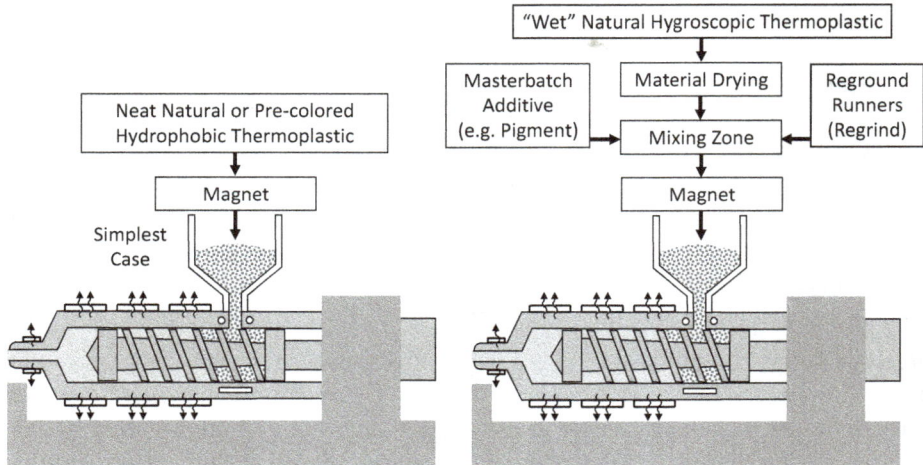

Figure 2.13 The feed material handling equipment required for an injection molding process can vary significantly from application to application. In the simplest case, only a feed material hopper and a magnet are required (left). At the other extreme, applications involving hygroscopic plastics, masterbatch additives (e. g., color concentrates) and the addition of regrind require far more complex material handling equipment (right).

2.2.2.1.1 Flood Feeding and Starve Feeding

One of the first material handling considerations is to determine if the injection molding process will utilize flood feeding or starve feeding as depicted in Figure 2.14. Most injection molding processes utilize flood feeding where the machines feed throat is fully charged with the feed material. In such a case, the screw's feed zone channels will be filled with plastic pellets and the screw's output rate will be determined by the screw's geometry, its rotational speed and the amount of back pressure utilized. Maintaining a consistent hopper level is important with flood feeding since the screw's feed rate can also be influenced by the forces associated with the material head. Flood feeding is very simple, and is appropriate for most injection molding applications.

However, there are injection molding applications where more control over the material feed rate is required [1]. With starve feeding, the material feed rate is set to be less than the feed rate associated with flood feeding as shown in Figure 2.14. A volumetric or gravimetric feeder is used to provide control over the starve feeding rate. With starve feeding, the output rate for the molding screw is dictated by the feeding rate, not by the screw's rotational speed. Starve feeding adds a level of melt quality control not possible with flood feeding.

As an example, starve feeding allows for better control over mixing since the screw rotational speed is independent of output rate. Starve feeding can also be used to control material residence time since the degree of fill for the screw channels can be managed. This can be helpful in applications where flood feeding would result is a residence time that is excessively long (e. g., for a molding machine with a barrel that is oversized for the application). Starve feeding can also be used in applications where a reduction in screw torque is required. Lastly, starve feeding is almost always used in vented-barrel injection molding applications as a means of minimizing vent bleed. In vented-barrel molding, the starve feed rate is set to be less than the feed rate for the second stage portion of the vented screw (see Figure 2.31).

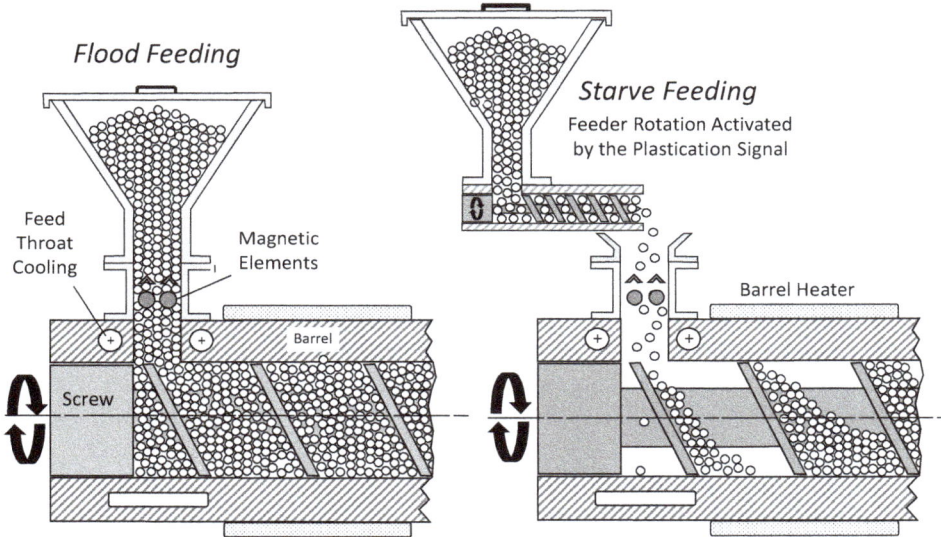

Figure 2.14 Most injection molding applications involve flood feeding where the feed throat is fully charged with feed material. However, there are injection molding applications where starve feeding is utilized. Starve feeding allows for improved control over material residence time, degree of mixing, and screw torque.

2.2.2.1.2 Injection Molding Machine Hopper

Injection molding machines generally utilize original equipment or specialty aftermarket feed material hoppers that sit directly above the barrel feed throat. The hoppers should facilitate a uniform and uninterrupted feed material mass flow rate as the screw rotates during plastication, as shown in Figure 2.15. Most hoppers have a funnel-like geometry and are available in a variety of sizes and construction materials. Stainless steel hoppers are very common due to their corrosion resistance and ability to operate at elevated temperature when material pre-drying (or pre-heating) is required. There is an industry trend towards the use of smaller size hoppers that are fed with a vacuum loader.

Regardless of the hopper size, the hooper design and shape should provide for a free flowing and uniform mass flow rate. This is typically described as first in – first out flow behavior. Improper hopper design can lead to feeding issues that range from stagnation to bridging [8]. Stagnation is a particular problem with hygroscopic plastics as stagnation results in non-uniform drying residence time. Bridging or caking tends to be more of an issue for low bulk density feed materials (e. g., thin wall regrind). Hoppers also have a cover to help eliminate the potential for contamination. Hopper covers are typically gasketed or sealed in applications where material drying is required in order to limit moisture reabsorption. It is also common to design hoppers with a slide shut off and discharge chute to facilitate a material change.

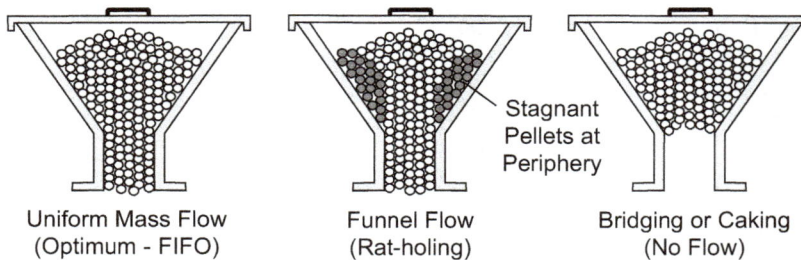

| Uniform Mass Flow | Funnel Flow | Bridging or Caking |
| (Optimum - FIFO) | (Rat-holing) | (No Flow) |

Figure 2.15 Most injection molding processes utilize original equipment or aftermarket feed material hoppers. These funnel shaped hoppers are available in a variety of shapes, sizes and materials of construction, but they all should be designed such that they allow for a uniform and uninterrupted material feed rate as the screw rotates during plastication.

2.2.2.1.3 Vacuum Loading Equipment

Vacuum loaders are used to facilitate the automated transfer of feed material from the material storage container (e. g., gaylord or silo) or material dryer to the molding machine hopper. The vacuum loading system uses a pick-up wand and conveyance tubing to transport the feed material to the molding machine hopper. A hopper level sensor signal is typically used to initiate the feed material loading process. Pneumatic or vacuum material conveyance systems can also include de-dusters or fines separators as shown in Figure 2.16. Fines can be particulate contamination, such as dust or package residue, or they can be small plastic particles or angel hair. While small plastic particles are not contamination per se, they can cause various processing issues. For example, the shear forces experienced by fines during plastication are different than those experienced by plastic pellets. Un-melted fines can cause issues such as small gate or pin gate clogging. Fines separators use ionized air to break the electrostatic attraction between the fines and the plastic pellets. Fines tend to be more commonly found with brittle thermoplastics and in applications involving the use of regrind. Removing dust and fines prior to plastication is generally considered to be a good manufacturing practice for the injection molding process.

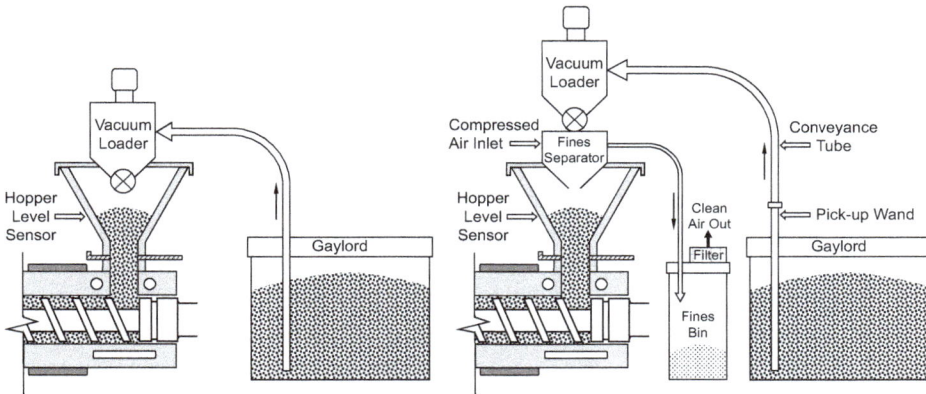

Figure 2.16 Vacuum loading systems are commonly used to transfer material from the material storage container to the molding machine hopper. The vacuum loading system can also incorporate an optional de-duster or fines separator to remove small particulates.

2.2.2.1.4 Plastic Material Pre-Drying Equipment

Many plastic materials require pre-drying prior to injection molding. Plastics that have a polar chemical structure tend to absorb moisture from the ambient environment. These thermoplastic materials are described as being hygroscopic. Material dryers (or pre-dryers) are used to remove the absorbed moisture from the plastic pellets before injection molding in order to avoid molding issues such as splay or hydrolytic degradation as discussed in Section 1.4.2.2.

Material suppliers typically provide values for both (i) maximum allowable moisture content, and (ii) recommended drying conditions (e. g., drying temperature, drying time, air dew point). The drying conditions provided by material suppliers are recommended guidelines since the drying equipment type and dryer design are also variables that impact the dryer variable setpoints. Having the ability to measure the moisture content after a material is dried is important in order to confirm that moisture level is below the acceptable moisture limit for the material. The dryer capacity should also be matched as closely as possible to the molding machine output so that the material residence time withing the dryer is within acceptable limits. Both underdrying due to insufficient dryer residence time and overdrying due to excessive residence time are generally unacceptable. However, coordinating the dryer output with that of the molding process can be challenging depending on equipment availability.

There are many different types of material dryers, but most dryers function by circulating hot, dry air through the mass of thermoplastic pellets contained within the dryer hopper. The hot dry air should pass uniformly through the gaps or void space between the pellets in order to extract the moisture contained within the plastic pellets. In some cases, it is only the base plastic resin that is dried, while in other cases, the plastic pellets and pre-blended additives (i. e., salt and pepper mixture) are dried after blending. Drying can be accomplished using air circulating ovens, vacuum ovens, compressed air dryers or most commonly using a desiccant dryer as shown conceptually in Figure 2.17.

Desiccant dryers circulate dry heated air through a hopper containing the hygroscopic feed material. The air is typically filtered to remove dust or fines. The dryers have a desiccant bed that removes the absorbed moisture from the hot and moist return air. An aftercooler is typically used in higher temperature drying applications (e. g., PET drying) to improve the efficiency of the desiccant. Desiccant dryers actually have at least two desiccant beds so that one bed is regenerating (i. e., removing the absorbed moisture from the desiccant) while the active bed is removing moisture from the return air. The desiccant beds are typically designed and maintained such that the delivery air dew point is about –40 °C. Small desiccant dryer hoppers can be mounted directly above the molding machine feed throat, but it is more common to use portable floor level dryers that deliver material to the injection molding machine hopper using a vacuum loader.

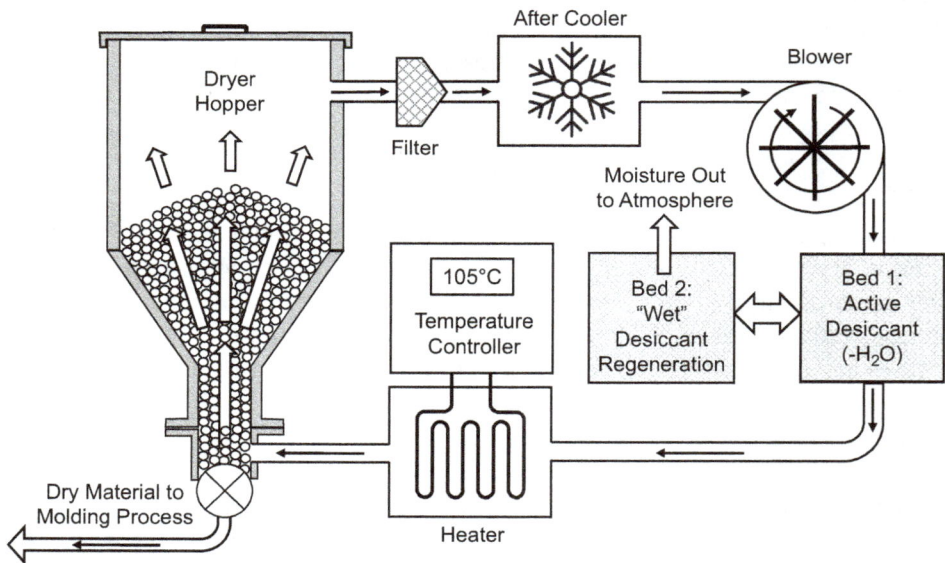

Figure 2.17 Desiccant dryers circulate hot, dry air through the mass of plastic pellets contained within the dryer's hopper.

2.2.2.1.5 Gravimetric and Volumetric Feeders

The use of at-the-throat auxiliary material feeders provides an alternative to batch blending in applications where the injection molding material feed stream involves multiple materials. Auxiliary feeders allow each feed component material to be metered at the appropriate concentration in a consistent and automated manner. The feeders typically have hoppers that can be loaded manually or using automated vacuum loading systems. The possible feeding conveyance mechanisms include belt feeding, vibration feeding, disc feeding, or screw feeding (single screw or twin screw). Variable rate screw feeders are most widely used for injection molding applications. Feeder screw geometries include solid core screws and spiral screws. The interchangeable feeder screw geometries are typically selected based on the required output rate and the pellet or granule geometry that is associated with the application. Feeder control strategies vary with feeder type, but feeder activation is typically triggered by the molding machine's screw rotate control signal with deactivation at the end of screw rotation (i. e., end of plastication).

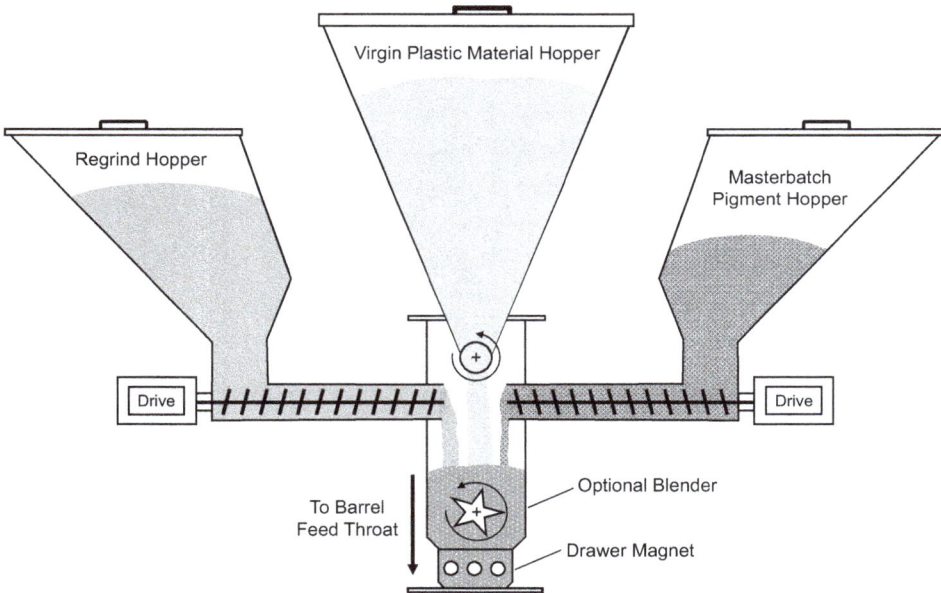

Figure 2.18 Above-the-throat material feeders are commonly used as an alternative to batch blending for multi-material feed streams. In this example, the feeders would be calibrated to accurately meter the plastic material, masterbatch pigment additive and regrind, all in the desired ratio. A mixing blender can be used to improve distributive mixing for the metered component materials just before they enter the molding machine's feed throat.

Feeders are further characterized as either volumetric feeders or gravimetric feeders (or loss-in-weight feeders). Volumetric feeders have a variable speed drive that causes the feed screw to rotate at a constant rotational speed during melt plastication. While volumetric feeders have a very consistent rotational speed, there is no feedback loop to ensure that the weight of material dispensed each cycle is consistent. The output consistency of a volumetric feeder is dependent upon factors such as the consistency of the feed material granule geometry, material temperature and bulk density. Volumetric feeders are simple, but they do require a calibration study in order to establish the relationship between feeder rotational speed and output rate. Gravimetric feeders are more sophisticated than volumetric feeders as they include provisions for measuring the weight of material that is metered each cycle. This feedback loop greatly improves the feeder dosing accuracy.

While screw feeders are the most widely used type of volumetric feeders, disc feeders provide an alternative to screw feeders in applications where the additive concentrations are relatively low. Disc feeders have a cylindrical hopper and a thin, slotted cylindrical disc that sits at the base of the disc feeder hopper as shown in Figure 2.19. The slotted disc is driven by a variable-speed stepper motor. Pellets or granules drop into the slots and eventually drop through the feeder discharge port as the slotted disc rotates. Disc feeders can meter material at feed rates as low as one pellet per plastication cycle [9].

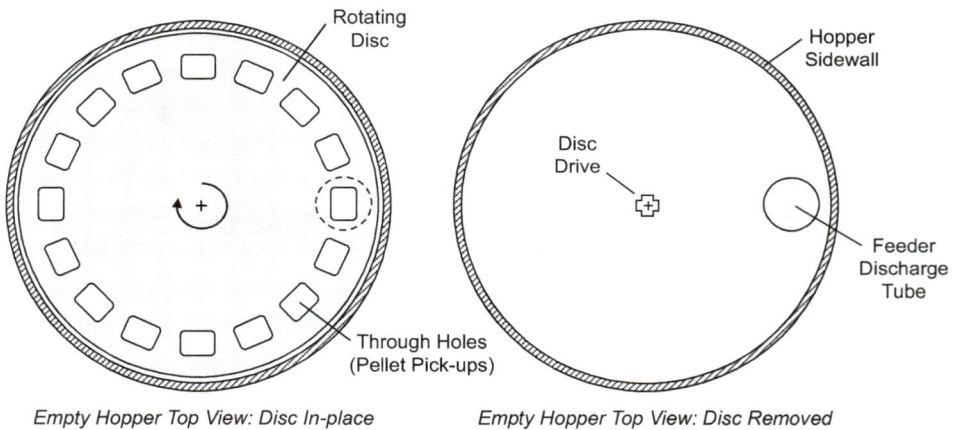

Empty Hopper Top View: Disc In-place   Empty Hopper Top View: Disc Removed

Figure 2.19 Disc feeders are low-rate feeders that consist of slotted horizontal disc sitting at the base of a cylindrical hopper. The slotted disc is driven by a variable speed stepper motor. Additive pellets or granules drop into the discs through slots. The pellets filling each slot eventually fall into the molding machine's feed throat when the filled slot becomes positioned over the feeder's discharge port.

2.2.2.1.6 Liquid Additive Feeders

While most masterbatch additives, such as masterbatch pigments, are supplied in pellet form, the use of liquid additives is gaining popularity. The increased use of liquid additives, especially liquid colorants, has been attributed to advances in both liquid carrier technology and liquid additive dispensing equipment. Liquid-based additives have been shown to be cost effective for injection molding applications, especially for long production runs. They have also been shown to be effective when tinting or coloring transparent plastics [10].

Liquid additive concentrates can be added directly into the molding machine feed throat as shown in Figure 2.20. The liquid injection nozzle can be inserted through a machined adapter plate that is situated between the molding machine's hopper and barrel. The liquid additive is pumped through the nozzle using a positive displacement pump, such as the variable speed peristaltic pump shown in Figure 2.20. The molding machine's screw rotation signals are used to initiate and terminate pump rotation.

The use of liquid additives is non-traditional in comparison with pelleted masterbatch additives and can require a bit of a cultural change. Color changes are somewhat more difficult than they are for traditional masterbatch pigments.

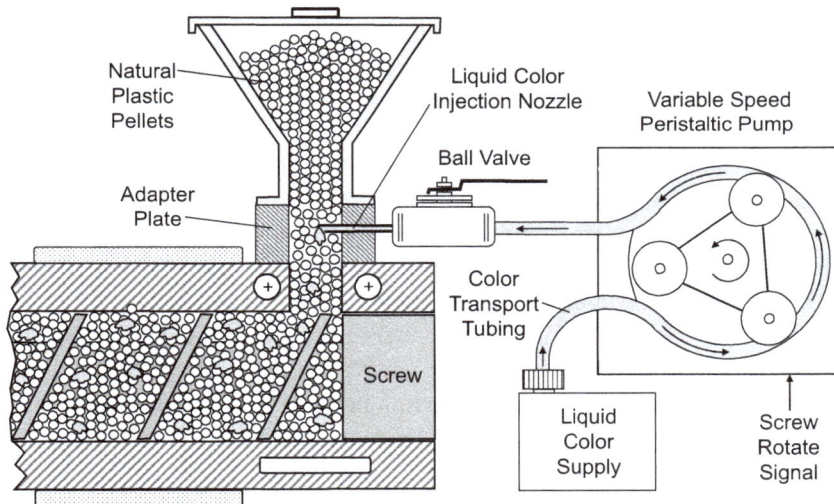

Figure 2.20 Additives such as pigments and dyes can also be dispersed as concentrates in a liquid carrier. Liquid additives are typically pumped through a liquid injection nozzle located just above the molding machine's feed throat. The liquid additive is pumped through the nozzle using a positive displacement pump, such as a variable speed peristaltic or roller pump. The molding machine's screw rotation control signals are used to initiate and terminate pump rotation.

2.2.2.2 Injection Molding Machine Barrel and Barrel Heaters

All conventional thermoplastic injection molding machines have a heated steel barrel that contains the reciprocating screw and non-return valve assembly. The barrel works together with the reciprocating screw and drive system to support the plastication and injection of the thermoplastic melt. Barrels are available in a very wide variety of sizes. Their geometry is typically characterized by both the barrel's bore diameter, and the length to bore diameter ratio of the barrel. Barrel size is normally an option when purchasing an injection molding machine. Having a smaller bore diameter will reduce the barrel's shot capacity, but it will usually increase the molding machine's injection pressure capability. Barrels having a smaller internal diameter will have a higher melt pressure magnification or intensification factor, which accounts for their higher injection pressure capability. In most cases, the barrel is a permanent injection molding machine component, although barrels can be replaced when worn or when another barrel size is required. Barrel changes can be time-consuming so, in recent years, quick-change injection molding barrel and screw combinations have been developed to shorten the barrel changeover time.

The materials used for the construction of an injection molding barrel are also important. Injection unit barrels must be; (i) very wear resistant, (ii) able to withstand high internal melt pressure (up to 300 MPa), (iii) able to withstand high temperatures, and (iv) corrosion resistant for some applications (i. e., when molding a corrosive thermoplastic formulation). The barrel construction can actually be quite complex due to all of these functional requirements. Barrels are produced using tool steels that range from 4140H to D-2. Some barrel bores are nitrided, some have a bi-metallic liner, and others a centrifugally cast and honed liner. Nitriding does little to improve corrosion resistance, while some wear-resistant liners can provide both abrasion and corrosion resistance. Abrasion resistance is important for injection molding barrels to ensure that the reciprocating screw, and especially the non-return valve (see Section 2.2.2.5), have consistent and proper clearance for consistent production. The barrel bore will experience abrasive forces when there is relative movement between the reciprocating screw and the barrel. These abrasive forces will occur during both the plastication and injection phases of the injection molding process. The wear tends to be most significant when molding abrasive thermoplastic formulations, such as a short or long glass fiber-reinforced plastic

A warm barrel will cause a reduction in reciprocating screw's plastication rate due to an increase in leakage flow, and allow more non-return valve leakage during the melt injection and packing phases of the molding process. One philosophy is to ensure that the barrel is more wear resistant than the reciprocating screw and non-return valve assembly since these components are less expensive and easier to replace than the barrel. Barrels also have an end cap at their discharge end. The end cap's internal passage is a funnel-like tapered transition between the barrel discharge diameter and the molding machine nozzle's inlet diameter.

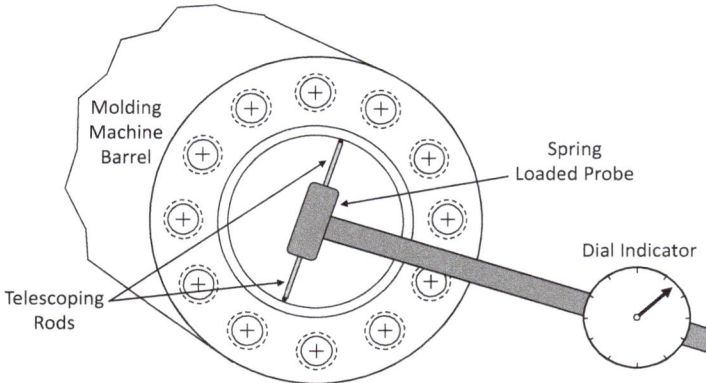

Figure 2.21 An inside diameter bore gauge can be used to quantify the amount of injection molding barrel wear. The bore gauge is used to make inside diameter measurements over the entire length of the barrel.

Injection molding barrels are normally heated using electrical resistance band heaters. Both mica band heaters and ceramic band heaters are used for molding machine barrel heating. Ceramic barrel heaters can have a longer life and higher temperature capability. Ceramic bands are recommended when processing high performance thermoplastics due to their ability to operate at high processing temperatures (up to 350°C). The heaters are wired or grouped such that the barrel normally has at least three barrel temperature control zones, and at least one nozzle temperature control zone. A machine having three barrel temperature control zones would normally describe these zones as rear, center and front, which roughly correspond the reciprocating screw's feed, compression and metering zones. Having multiple temperature control zones improves process flexibility by allowing a process engineer to input a variety of different barrel temperature profiles [11].

Each temperature control zone also includes a temperature sensor, such as a type J or K thermocouple, that will generate the temperature signal required for the zone's PID temperature control feedback loop. It is worth noting here that, while almost all extruder barrels have both heating and cooling capabilities, injection molding barrels rarely incorporate provisions for barrel cooling, except for the water cooling at the feed throat. This is likely due to the fact that viscous heating is more of an issue with extrusion processing since screw rotation is continuous for extrusion processing, while screw rotation for injection molding is intermittent.

Insulating blankets can also be used in conjunction with injection molding barrel heaters. Insulating blankets save energy by cutting down on barrel heat loss to the surrounding environment. While their use improves energy efficiency, they may also cause overheating issues, particularly for applications where too much viscous heating causes the barrel temperature to exceed limits.

The injection molding barrel is also protected with a sheet metal safety guard designed to protect employees from the barrel heat and electrical hazards. Even though the barrel is guarded, it is best for all heaters to have closed electrical junctions. Protective ceramic caps should be used if the barrel heaters have exposed electrical connections.

Historically, many injection molding machines used open-loop rheostat temperature control for the molding machine's nozzle temperature control. The lack of nozzle temperature feedback can minimize temperature swings, but this is, in some ways, analogous to driving blind. Today, most injection molding nozzles do have thermocouple feedback and PID nozzle temperature control. Both the nozzle thermocouple placement and uniformity of nozzle heating are critical for reliable nozzle temperature control.

Injection molding barrels also have a feed throat (opening) that is located directly above the first and second turn of the reciprocating screw's feed section. This is typically a circular or rectangular opening that allows for the free flow of thermoplastic granules into the screw channels below the opening. The axial length of a rectangular feed opening is typically about 1.5 times the screw diameter. The feed throat can be machined directly from the barrel (integral) or it can be contained within a separate feed casting that is mounted to the rear of the barrel. In either case, the feed throat region of the barrel or feed casting is water cooled to prevent the plastic granules from melting or softening prematurely. If the feed throat becomes too hot, granules can stick to the feed throat walls, preventing free granule flow due to bridging. Historically, feed throat temperature control has been open-loop, and sometimes blind (i. e., with no temperature sensor). With an open-loop system, feed throat temperature is controlled by manually adjusting the feed throat cooling water flow rate. Today, most molding barrels have closed-loop feed throat temperature control based on the use of a feed throat thermocouple and an automated cooling water flow control solenoid.

2.2.2.3 Reciprocating Screws

2.2.2.3.1 The General-Purpose Reciprocating Screw

The reciprocating screw has been described as the heart of the injection molding machine. The reciprocating screw must perform many critical injection molding process functions, all of which have an effect on the molded part's quality. While there are many different reciprocating screw designs that can be used for thermoplastic injection molding, the general-purpose reciprocating screw is most common. The general-purpose screw does not have a single specific definition or geometry. Rather there are many different general-purpose screw designs, however, they all have certain common features. While it is well known that general purpose screws are *not* the best possible screw design for thermoplastic injection molding, their use is popular because these screws can be used to plasticate a wide variety of different thermoplastics. This statement could be amended to say that general purpose screws can be used to plasticate a wide variety of materials, but probably not in an optimum way for any of these plastics [12]. The limited performance of the general-purpose screw is offset by the fact that a plastic material change may not necessitate a reciprocating screw change. This is particularly important for custom molding applications, particularly short production run custom molding. Changing out the reciprocating screw (or the screw and barrel combination) is time consuming and capital intensive.

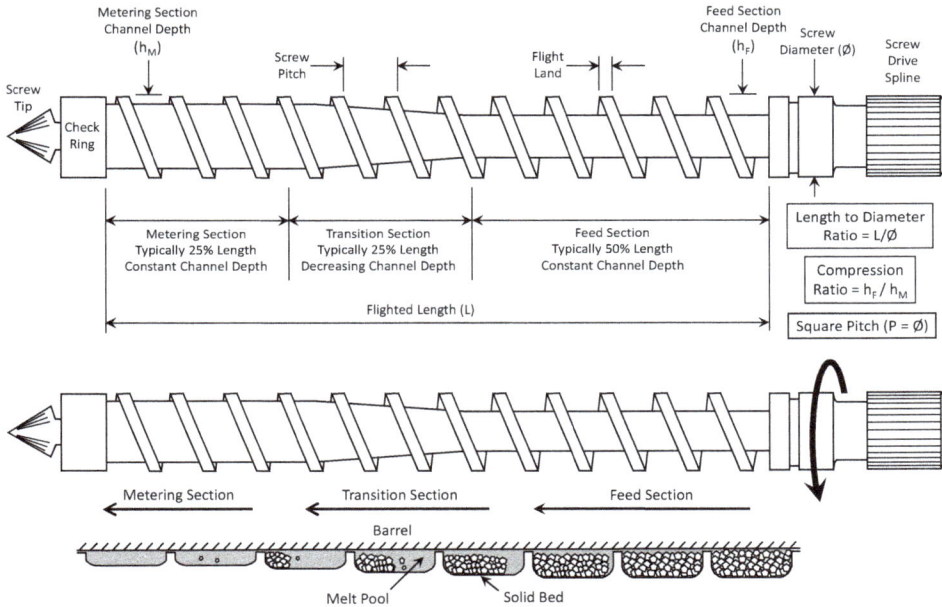

Figure 2.22 A typical general-purpose square pitch reciprocating screw geometry used for injection molding. General purpose screws are typically characterized by their outer diameter, length to diameter ration, and compression ratio.

A variety of different general-purpose reciprocating screw designs are available from screw manufacturers, but all of these screws have certain common features. Most general-purpose molding screws are single-flight, square pitch metering screws meaning that the flight-to-flight distance (i. e., the screw's pitch) is equivalent to the screw's outer diameter (Ø). This results in the screw flights having a 17.7° helix angle. The screw diameter and the screw's length (L) are both dictated by the molding machine's barrel dimensions. The reciprocating screw's L/Ø ratio is the ratio of the flighted length of the screw to its outside diameter. Most injection molding screws have a significantly lower length-to-diameter ratio (L/D) compared to extrusion screws. Single screw extruders have L/Ø ratios as high as 40:1, while the reciprocating screws used for injection molding have a shorter L/Ø ratio, typically in the range from about 20:1 to 26:1. One issue that limits the L/Ø ratio possible with injection molding screws is the end loading that occurs during melt injection.

General-purpose reciprocating screws also have three distinct zones: (i) the feed zone, (ii) the compression or transition zone, and (iii) the metering or pumping zone, as shown in Figure 2.22. The feed section has a constant channel depth and usually represents about 50% of the general-purpose screw's flighted length. The relatively long feed section that is characteristic of injection screws is needed because the screw reciprocates as it rotates. The effective length of the screw's feed section decreases as the screw reciprocates during plastication. The main functions of the screw's feed section are to convey, compress and pre-heat the solid thermoplastic granules or pellets. The optimum feed section depth is dictated by factors such as the bulk density of the material to be plasticated.

The second zone of a general-purpose screw is the compression or transition section. As these terms suggest, material is compressed and sheared withing this region of the screw as the screw channel's depth transitions from deep to shallow. Most general-purpose reciprocating screws have a compression zone with a gradual channel depth transition that represents about 25% of the screw's flighted length, as shown in Figure 2.22. By definition, all solid pellets or granules should have reached a melt state before they exit the transition zone of the screw. Gradual transitions are best for plastics that have broad melting or softening temperatures. Reciprocating screws having a shorter compression zone with a more abrupt transition (and a corresponding longer feed zone) can be advantageous when processing highly crystalline semi-crystalline thermoplastics that have a relatively narrow melting temperature range.

The third section of the general-purpose screw is the metering or pumping zone, which also represents about 25% of the screw's flighted length. The screw channel depth is shallow and constant in the metering section. The metering zone drag flow provides the pumping action required to "push" or pump the reciprocating screw rearwards during screw rotation (i. e., during plastication). The metering section also causes additional shear heating and provides the bulk of the screw's mixing action, which is limited at best with general-purpose screws. General-purpose reciprocating screws have very limited additive and temperature mixing capabilities. Other reciprocating screw designs that incorporate mixing elements are more appropriate for applications where enhanced melt mixing is required.

The compression ratio for a plasticating screw is the ratio of channel volume (or depth) in the first turn of the feed zone to the volume (or depth) in the last turn of the metering zone. General purpose reciprocating screws typically have a compression ratio in a range of about 2:1 or 3:1. Both the compression ratio and absolute value of the screw channel depths are important. Screws having a shallower metering zone channel depth are most suitable for lower viscosity melts. When high compression ratio screws are used to plasticate more viscous melts, they can generate too much viscous heating. Excessive viscous heating can cause barrel temperature zones to overheat (particularly in the compression and metering zones) and can cause material degradation. Unlike extruders, most injection molding machines do not have provisions for barrel cooling that would help bring the barrel temperatures back down to the setpoint. Thermoplastic material suppliers can normally provide guidance or recommendations regarding the compression ratio that is best for the material grades they supply.

The relatively short L/Ø ratio common for general purpose injection molding screws is not optimum in terms of producing a good melt quality. Plasticating screws are required to melt, mix and convey the plastic material being molded. Plasticating screws having a high L/Ø ratio are better at all three of these functions than shorter L/Ø ratio screws. The short L/Ø ratio for injection molding screws is even more problematic than it first appears, since the screw reciprocates (rearward) during screw rotation. The actual or effective L/Ø ratio that the plastic material sees is highest at the start of plastication and lowest at the end of plastication. While there is no question that plasticating screws having a higher L/Ø ratio produce a better melt quality, injection molding screw L/Ø ratios are limited by factors such as column bending during injection, which is not an issue for extruders since extruder screws have no axial movement.

While injection molding screws do have shorter L/Ø ratios than extrusion screws, their use can still produce a high-quality melt since the overall melting and mixing mechanisms for single screw extrusion are actually very different than those for the injection molding process. Single screw extrusion is a continuous process where the screw rotates continuously. The plasticating screw used for the injection molding process rotates for only for a portion of the molding cycle. Both processes use a heated barrel to help control melt temperature, although the vast majority of the energy input during plastication is from viscous dissipation or frictional heating caused by relative movement between the plastic material in the screw channel and the barrel. With injection molding, there is also a heat soak period during the time in which the reciprocating screw is not rotating. This heat soak time helps to equalize the barrel and plastic melt temperatures. In addition, injection molding machines all have the ability to control melt pressure during plastication. Adding this back pressure or resistance increases the amount of time it takes for the reciprocating screw to generate the shot, which increases overall energy input to the material. This back pressure variable is a tool that allows processors to fine tune melt quality. Lastly, the injection molding screw is a reciprocating screw (unlike an extrusion screw), which results in additional relative movement between the material in the screw channels, and the barrel during the injection phase of the molding process. This forward movement causes an additional shearing action for the plastic material in the screw channels.

The construction materials for reciprocating screws can be very variable depending upon the supplier and application requirements. Screws can also have a multi-material construction, especially for the flights which can have their sides, root or top surface welded with hard face cobalt or nickel alloy for improved wear resistance. Screws are produced from tool steels or alloy steels that can be ion nitrided or have thick chrome plating. Most plastic materials can be processed with screws made from a variety of different metals. However, reciprocating screw metallurgy becomes particularly important when processing abrasive plastic material formulations (e. g., glass fiber reinforced plastics) or corrosive plastic material formulations (e. g., chlorinated or fluorinated polymers).

2.2.2.3.2 Melt Pressure During Plastication (Back Pressure)

Back pressure is a process variable that allows molders to fine tune the plastic material energy input that occurs during the plastication phase of the injection molding process. More energy input can cause both an increase in the temperature of the melt and the degree of melt mixing. The mechanism for creating the back pressure varies for hydraulic and all-electric injection molding machines, but the result is the same in both cases (from the plastic material's perspective). The concepts associated with back pressure are most easily explained for hydraulic plasticating units as shown in Figure 2.23.

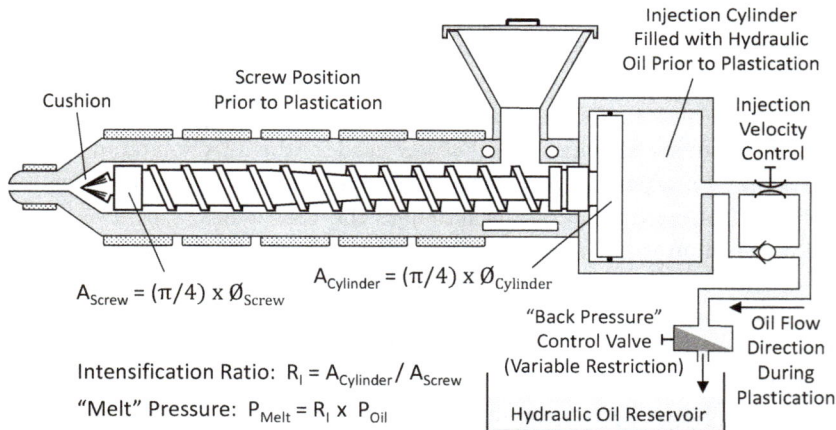

Figure 2.23 Back pressure is the melt pressure generated by the pumping section of the reciprocation screw during plastication. Increasing back pressure adds resistance during the plastication phase of the process. A higher back pressure means that the screw will need to rotate for a longer period of time in order to generate the required shot volume (for the next molding cycle). Adding back pressure results in more energy input and better mixing.

Consider the molding machine schematic in Figure 2.23. During the injection phase of the molding process, hydraulic oil is pumped into the injection unit's injection cylinder(s). During plastication, the reciprocating screw rotates and pumps material into the front portion of the barrel (i. e., the shot chamber). The sprue from the previous shot blocks the nozzle discharge, so the screw must reciprocate as it rotates in order to make room for the plastic melt exiting the last screw channel. The molding machine hydraulic system includes a variable-restriction hydraulic valve between the hydraulic injection cylinder and the molding machine's oil reservoir. If this hydraulic valve is fully open, the melt pressure during plastication (i. e., back pressure) is very low. Adding hydraulic valve resistance (or back pressure) makes it harder for the screw to reciprocate, causing the melt pressure during plastication to increase. Reciprocating screws are *not* positive displacement pumps, so adding back pressure means that the screw will need to rotate for a longer time (at a given RPM) in order to generate the same shot size.

The theoretical output of an injection molding screw can be determined using Equation 2.1. The (positive) drag flow output of a general-purpose reciprocating screw is determined by the screw's metering section geometry and rotational speed. The maximum possible screw output occurs when there is no back pressure, and there is only drag flow (the positive term in Equation 2.1). Adding back pressure, causes both screw channel pressure flow and flight leakage flow (both negative terms in Equation 2.1). The negative pressure flow and leakage flow terms both increase as back pressure increases. Adding back pressure reduces the screw's volumetric output per revolution.

The net effect of adding back pressure (for a given screw rotational speed) is that the screw rotates for more revolutions in order to reciprocate the same distance. The added number of screw revolutions will increase mixing and will cause some increase in melt temperature, especially for more viscous polymer melts. There is some debate as to the extent to which back pressure influences melt temperature. This debate is due in part to difficulties in obtaining reliable melt temperature measurements. Back pressure will also cause a variation in melt temperature distribution within the shot volume. The back pressure-induced melt temperature increase is also masked by the heat soak period when additional conductive heat transfer occurs for the melt. Adding back pressure also compresses the melt which helps squeeze entrapped air from the shot. The use of at least a relatively low back pressure is recommended in most molding applications as it helps to improve repeatability of the shot weight from cycle to cycle. Higher back pressures are typically used for applications where more shear history is desirable, such as applications that involve the use of masterbatch pigments or other additives. High back pressures should be avoided for applications such as glass fiber reinforced polymers in order to promote a gentler shear history and limit the amount of fiber damage during plastication.

$$Q = [AN] - [B\Delta P/\mu] - [C\Delta P/\mu] \tag{2.1}$$

Where:

Q = Screw Volume Output (mm^3/rev)

A, B, C = Screw Geometric Constants

ΔP = Metering Zone Pressure Differential (MPa)

μ = Melt Shear Viscosity (Pa-sec)

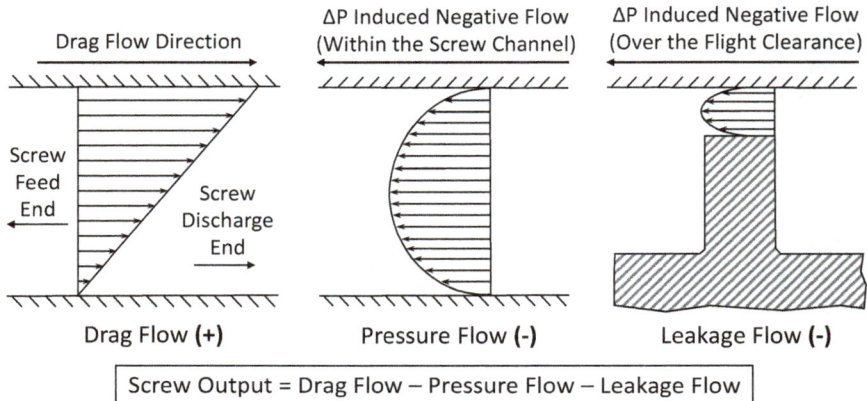

Drag Flow Direction

ΔP Induced Negative Flow (Within the Screw Channel)

ΔP Induced Negative Flow (Over the Flight Clearance)

Screw Feed End

Screw Discharge End

Drag Flow **(+)**   Pressure Flow **(-)**   Leakage Flow **(-)**

Screw Output = Drag Flow − Pressure Flow − Leakage Flow

Figure 2.24 The maximum output rate for a given general purpose reciprocating screw is achieved when there is only drag flow and no pressure flow. Adding back pressure reduces the screw's output (per revolution) by introducing pressure flow and flight leakage flow. While adding back pressure reduces output per revolution (and increase screw recovery time), its use does allow a molder to fine tune melt temperature and improve the degree of melt mixing.

When back pressure is used during plastication, any residual back pressure (after plastication is complete) will tend to cause melt drool from the nozzle (or hot runner gates) when the mold opens or when sprue break (carriage retract) is used. This residual back pressure can be relieved using rear decompression or "suck back" after plastication is complete. Rear decompression increases the volume of the shot chamber, which relieves the back pressure. The recommended decompression distance is the minimum distance that will result in zero residual back pressure.

2.2.2.3.3 Barrier Reciprocating Screws

Reciprocating screws having a barrier screw design are an improvement over the general-purpose reciprocating screw described above. The feed section and the metering section of a barrier screw are essentially equivalent to those of the general-purpose screw. The main difference in these two screw types relates to the design of the compression zone. Barrier screws introduce a secondary or barrier flight in the first turn of the compression zone. The purpose of the barrier flight is to separate the melt from the solid pellets or granules. Adding a barrier flight within the compression zone channel essentially creates two channels, separated by the barrier flight. The solid channel is open to the feed section, while the melt channel is open to the metering section. The volume of the melt channel increases over the length of the compression zone while the volume of the solids channel decreases as shown in Figure 2.25. The solids channel volume eventually disappears at the end of the compression zone that transitions into the first turn of the screw's metering section.

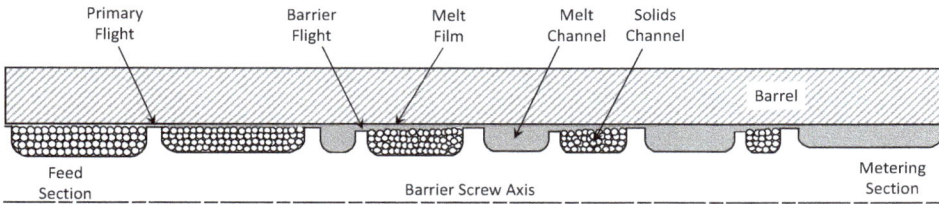

Figure 2.25 Barrier screws have both primary flights and barrier flights within the compression zone of the reciprocating screw. The barrier flight separates the solid pellets from the melt. The volume of the melt channel increase over the length of the compression zone while the volume of the solids channel decreases, eventually disappearing.

The volume changes for the melt and solid channels can be achieved by varying either the channel's width or depth, depending on the barrier screw design. The barrier flight clearance is large enough to allow the melt material to spill over the barrier flight clearance, while solid pellets cannot pass over the narrow gap. This separation of the solid pellets from the melt results in far more efficient melting of the solid pellets. This is not the case when a general-purpose reciprocating screw is used, as some pellets end up dispersed within a melt pool. Barrier screws allow for greater plasticating rates, improved melt temperature control and melt temperature uniformity. A properly designed barrier screw can be used with a variety of different plastic materials, in much the same way that a general-purpose screw is typically used with a variety of different plastic materials. Barrier screws are available with a wide variety of L/Ø ratios and compression ratios [13,14].

2.2.2.3.4 Reciprocating Screw Mixing Elements
The reciprocating screws used for the injection molding process must perform a number of functions that include feeding, melting, mixing, and pumping the thermoplastic material being molded. The relatively short length to diameter ratio of injection molding screws limits the screw's ability to perform each of these functions optimally, and tradeoffs are often required. For example, reciprocating screws that incorporate elements to improve mixing, may have a reduced plasticating capacity or plasticating rate. The importance of mixing for an injection molding process is largely dependent

upon the material formulation to be molded. For example, molding a pre-colored thermoplastic material formulation that has been melt-compounded by the material supplier prior to molding may not require further mixing. However, a material formulation containing a masterbatch additive added while molding would require an additional mixing action to ensure that the pigment and carrier polymer are both uniformly distributed and dispersed throughout the matrix material. Even neat plastic materials can benefit from improved mixing, since temperature mixing will result in a more uniform melt temperature for the plasticated shot.

Having a basic understanding of mixing theory is helpful when evaluating the mixing capabilities of an injection molding screw. Mixing theory is helpful since the degree of mixedness associated with a particular material formulation and process can be difficult to quantify in practice. In some cases, a molded part's appearance, such as color uniformity, may be an indication of the degree of mixedness on a macro-scale. However, the degree of mixedness can be very difficult to determine on a micro-scale where a molded part's quality may be influenced by the factors such as the degree of agglomerated particle dispersion for a filler or pigment.

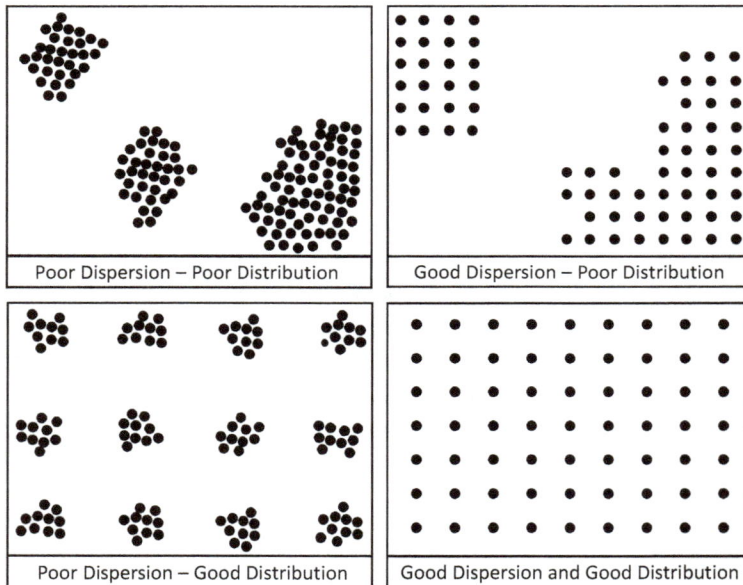

Figure 2.26 Some injection molding screw mixing elements function as dispersive mixers, while others function as distributive mixers. Mixers that provide both good distributive mixing and dispersive mixing are rare. The desired degree of mixing can sometimes be achieved using both a mixing screw element and a static mixing nozzle.

The concepts associated with both distributive and dispersive mixing are shown in Figure 2.26. Dispersive mixers reduce the physical size of the dissolved solutes or domains that are dispersed within the thermoplastic matrix material. These domains are typically solid or liquid additives that are agglomerated and require an applied shear stress to break down their physical size into smaller particles. Distributive mixers are designed to redistribute the different materials in an effort to obtain a more uniform additive concentration throughout the matrix. Some injection molding screw mixing elements result in improved distributive mixing, while others have dispersive mixing capabilities. Mixing elements that provide both distributive and dispersive mixing capabilities are rare.

Most general-purpose or barrier reciprocating screws have some distributive mixing capability but very limited dispersive mixing capability. The degree of mixing that can be obtained with a given reciprocating screw can be improved by adjusting both screw rotational speed and by adding back pressure. Adding back pressure increases the time required for screw recovery (at a given RPM) and therefore the amount of mechanical work that is applied to the melt. The general-purpose or barrier screw designs themselves can also be modified in order to improve their dispersive or distributive mixing characteristics. This modification typically involves adding a mixing element or mixing section somewhere within the metering zone of the reciprocating screw, where the material should be fully plasticated (i. e., completely in the molten state). When mixing sections are incorporated into the metering zone, their length-to-diameter ratio is usually limited ($\leq$ 2–3 L/Ø) in order to limit any negative effect on the screw's pumping capability. It should also be noted here that adding certain types of mixing sections to a reciprocating screw to improve the screw's mixing capability, may limit the types of plastic that can be processed with the modified screw (i. e., it may no longer be considered general-purpose). For example, screws having very high shear mixing sections may not be appropriate for processing glass fiber reinforced plastic materials due to the glass fiber length degradation the mixing section would cause.

Two of the simplest and earliest reciprocating mixing elements are shown in Figure 2.27. The blister ring is a raised section within a channel of the metering zone. Melt must pass over the narrow gap between the molding machine barrel's inside diameter and the blister ring's outside diameter. The melt must accelerate as it flows over the thin gap causing high shear forces that can help reduce the size of agglomerated particles. The added shear forces caused by the blister ring will improve dispersive mixing. Blister rings do add flow resistance during plastication, which can reduce plasticating rate and increase melt temperature. Another relatively common and simple mixing modification is the addition of one or more rows of mixing pins. Mixing pins help to break up the laminar flow pattern as the screw rotates and reciprocates. Mixing pins have been popular due to their simplicity and ability to improve distributive mixing, although other more efficient distributive mixers have been developed over the years.

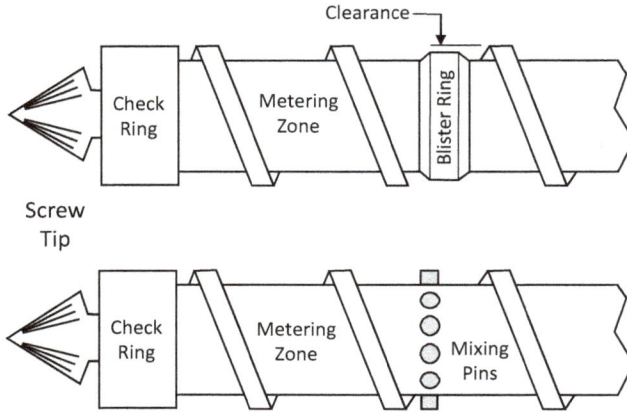

Figure 2.27

Screws that incorporate a blister ring element have a narrow gap that leads to improved dispersive mixing, while screws incorporating mixing pins result in improved distributive mixing.

Another common injection molding screw mixing section is the "pineapple mixer" shown in Figure 2.28. The pineapple mixing section functions in much the same way as a screw that incorporates mixing pins. The pineapple mixer has offset rows of mixing teeth that result in the repeated splitting and recombination of the laminar flow field during plastication. The pineapple mixer achieves a more aggressive mixer action than that achievable with mixing pins, making it a very good option when improved distributive mixing is required. Pineapple mixers do add pressure drop during plastication, which can reduce the screws plasticating capacity and increase melt temperature.

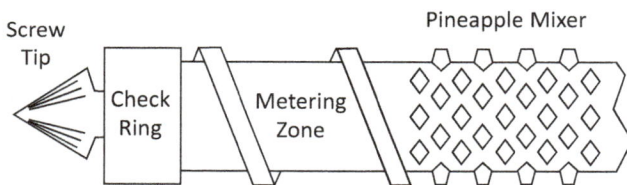

Figure 2.28

Reciprocating screws that incorporate a pineapple mixing section exhibit an improved distributive mixing capability.

Injection molding screws that incorporate fluted mixing sections are an option when a high degree of dispersive mixing is required. A popular fluted mixing section is the Union Carbide mixer or Maddock mixer, which was invented and patented by Gene LeRoy. While LeRoy was the inventor of the fluted mixer, its use and design was optimized by Maddock for commercial applications [15]. The straight Maddock mixer shown in Figure 2.29 has pairs of inlet and outlet channels, separated by a barrier land. The barrier land has a narrow gap which prevents un-melted thermoplastic plastic pellets, gels, or large agglomerates from reaching the outlet channels. The barrier holds these solids back until they become fluid enough to pass over the barrier land. In addition, the material flowing over the barrier land experiences high shear forces as it travels over the thin barrier gap, providing the dispersive mixing action.

While the fluted mixer is a good dispersive mixer, the mixing section does not provide much in the way of distributive mixing action. Straight Maddock mixers provide no pumping action, causing a decrease in plastication rate along with a corresponding increase in melt temperature. Many fluted mixers have a helical design (rather than the straight design shown in Figure 2.29) that provides dispersive mixing action along with some pumping capability.

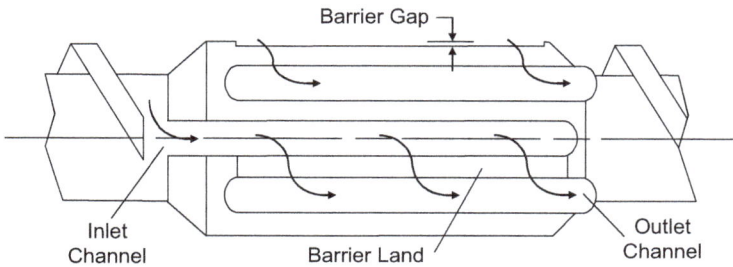

Figure 2.29 The fluted Union Carbide or Maddock mixer has pairs of inlet and outlet channels separated by a barrier land. The melt experiences high shear forces as it passes over the narrow barrier land gap resulting in a dispersive mixing action.

There are also a number of mixing section designs that are based upon variable-depth metering zone screw channels. These mixers are essentially multi-channel mixers where the shallow and deep portions of the screw channel are out of phase with one another. The deep channels transition to shallow channels, which introduces side flows or cross-channel flows that superimpose on the normal drag flow during plastication. Most variable depth mixers are distributive mixers that can be used for most thermoplastic material formulations with little or no negative impact. They improve color uniformity and help create a more uniform melt temperature. Variable depth mixers that include a barrier land between the deep and shallow sections may also improve dispersive mixing.

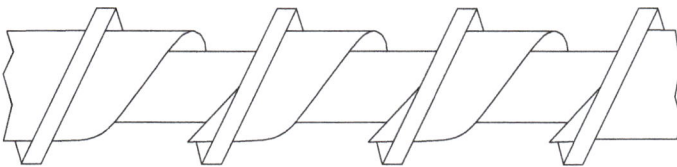

Figure 2.30 Variable channel depth mixing sections are essentially multi-channel mixers having out-of-phase (adjacent) deep and shallow sections within the screw channel. This helps cause cross-channel flows, which improve distributive mixing.

2.2.2.3.5 Two-Stage Devolatilizing Screws

Injection molding machines that are equipped with a vented-barrel injection unit can be used to remove volatiles from a thermoplastic plastic melt. The most common application for vented-barrel injection molding machines is the removal of water vapor from hygroscopic plastics, as an alternative to material pre-drying. Vented-barrel injection molding machines use a two-stage screw for both plastication and devolatilization. The two-stage screw is in essence two integrated general-purpose screws arranged in series, as shown in Figure 2.31. The first stage of the screw functions as a normal general-purpose screw, feeding, melting, mixing, and pumping the thermoplastic material. The material that exits the last turn of the first-stage metering section during screw rotation should be fully mixed and melted by that point in the process. The melted material then enters the deeper first turn of the feed section of the second-stage screw segment. This results in a sudden decompression of the melt, which allows volatiles (e. g., water vapor) to escape through the vent opening. The volatiles either (i) vent to the atmosphere, or (ii) they are caught in a trap if a vacuum pump assist is used. Proper ventilation above the decompression area is required if the volatiles are vented to the atmosphere. The second stage portion of the two-stage screw has a very short compression section (since melting has already occurred) and a metering section that is used to build the now devolatilized shot of thermoplastic. The L/D ratio for vented-barrel molding machines is generally greater than that of conventional non-vented barrel molding machines (which do not incorporate a devolatilization zone). The L/D ratio of a typical vented barrel injection unit is typically 26–32:1 [2].

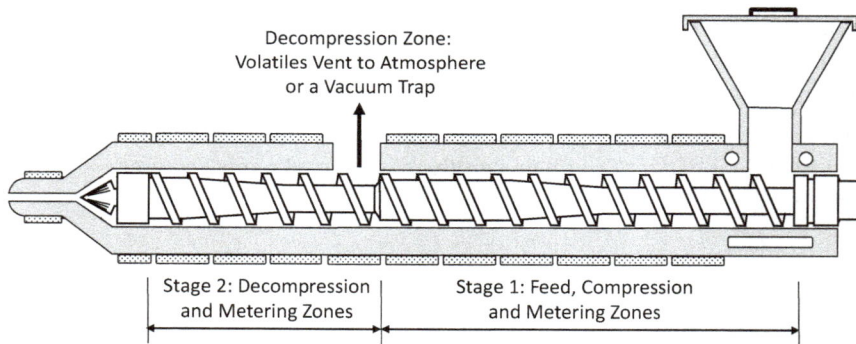

Figure 2.31 Schematic of a two-stage injection molding screw used in combination with a vented-barrel injection unit. This screw and barrel combination can be used to perform both melting and devolatilization functions. The vent opening is shown in a vertical orientation, but vented injection molding barrels typically have a horizontal orientation.

While vented-barrel injection units would seem to offer a very significant advantage by eliminating the need for hygroscopic thermoplastic pre-drying, their popularity has actually declined in recent years. This is likely due in large part to advancements in drying equipment technology, but it is also due to the fact that vented-barrel processing is somewhat more complicated than non-vented processing. Vented-barrel processing is also not appropriate for all hygroscopic plastics. Vented-barrel processing is most appropriate for hygroscopic addition polymers such as ABS or SAN or acrylics. These hygroscopic plastics do absorb water from the surrounding atmosphere, but they do not tend to react with the water when they are heated and melted.

Condensation polymers, such as PET or a nylon, will react with absorbed water if the water is present when these materials are heated and melted. This reaction can occur when a condensation polymer containing residual moisture is melted in the first stage of the two-stage screw prior to devolatilization. This reaction can cause a reduction in the average molecular weight of the condensation polymer and therefore a reduction in its mechanical performance. As a result, vented-barrel processing is most common for additional polymers like ABS.

Vent bleed is another complication that can occur with vented barrel injection units. Ideally, only volatiles enter the barrel vent opening. However, if the second screw stage is pumping at a rate that is lower than the first stage of the screw, melted material will enter the vent opening, thereby reducing venting efficiency (and eventually clogging the vent). As a result, starve feeders are typically used in conjunction with vented-barrel injection units so that the screw's output rate (at a given screw rotational speed) can be independently controlled. Having the ability to reduce the output of the screw's first stage metering section helps prevent vent bleed.

With vented-barrel injection units, there can also be a tendency for degraded or oxidized material to deposit on the screw in the pressureless decompression zone where the melt is exposed to oxygen. These degraded deposits can eventually break free causing cosmetic or performance issues for the injection-molded plastic parts. It should also be noted that the vent opening can also be plugged for applications where a conventional (non-vented) molding process is more appropriate.

2.2.2.4 Non-Return (Check) Valves

Injection molding reciprocating screws are designed to perform a number of functions that include shot generation (dosing) as well as melt injection and post fill pressurization. Most reciprocating screws are equipped with a non-return valve assembly located at the discharge end of the reciprocating screw to prevent backflow of melt during the injection, packing and holding pressure phases of the molding process [16]. While reciprocating screw injection units can function without a non-return valve, the leakage that occurs during injection would be significant, variable, and would limit the screw's ability to apply or transmit packing and holding pressures. While molding without a non-return valve is not appropriate for most injection molding applications, it is used for some very shear-sensitive materials, especially rigid polyvinyl chloride which is often molded using a high surface area smear tip.

Non-return valve assemblies are also described as check valves. When a check valve is used, the reciprocating screw can function as both a plasticating screw, and as a plunger or ram. There are many types of non-return valve that can be used with a reciprocating screw, but ring type check assemblies and ball type check assemblies are the most common. Most check valves can be used with a variety of different thermoplastic materials and the same check valve can normally be used for different production runs. However, there are times where a check assembly may need to be changed out based on the material to be molded.

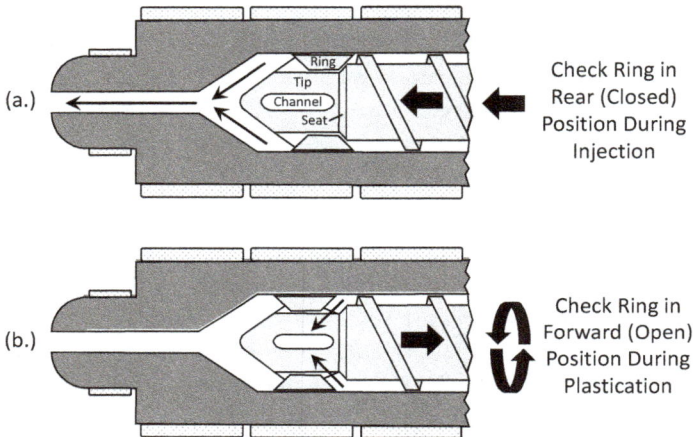

Figure 2.32 Sliding ring check valve assemblies are a very common type of non-return valve. Melt flows under the ring during plastication to allow for shot preparation. During injection, the ring seats against a rear seal, effectively turning the reciprocating screw into a ram or plunger.

Sliding ring type check assemblies, such as that shown in Figure 2.32, are used to prevent backflow and leakage flow during the melt injection and pressurization phases of the molding process. The sliding ring check assembly is made up of at least three components, a screw tip, a ring and a rear seat. The assembly threads into the discharge end of the reciprocating screw. During plastication (screw rotation), the melt that is pumped from the end of the screw forces the ring to its forward position against the screw tip. Melt flows under the ring during plastication until screw rotation stops, once the appropriate shot volume has been dosed. During the melt injection or mold filling phase of the process, the reciprocating screw and check valve assembly are pushed forward. The check ring, which is a sliding fit within the barrel, tends to remain stationary during the initial portion of the injection stroke (i. e., as the screw begins its forward movement) until the ring seats against the rear seal. During this short period of time before the ring seals, some melt can leak back under the ring and along the screw. After seating, the now sealed check valve assembly helps the reciprocating screw to act as a ram. After sealing, there can still be some material leakage over the top of the ring (i. e., over the gap between the ring and barrel bore) and between the ring and its rear seal. These leakage flows tend to become more significant if the check valve assembly or the barrel are worn. Leakage flows also tend to be more significant for low viscosity (or higher melt flow rate) thermoplastics.

Check ring assemblies of various designs are available commercially, but those described as free flowing are popular, especially when molding more shear sensitive thermoplastic formulations or fiber reinforced thermoplastics. With some check ring assembly designs, the ring is not keyed to the screw tip and therefore the ring does not rotate along with the screw during plastication. Other check ring assemblies are keyed. Self-locking check ring assemblies have also been developed by molding machine manufacturers. A self-locking check ring seals tightly against the rear seat at the end of plastication (i. e., before melt injection) by having the screw counter-rotate at the end of screw rotation. Self-locking check valves require a molding machine control system that is capable of this control strategy. The counter rotation pre-seals the check assembly to minimize the backflow of material at the start of the injection stroke. Convention check ring assemblies have some forward screw movement, which results in some material backflow, before the ring seals [16,17].

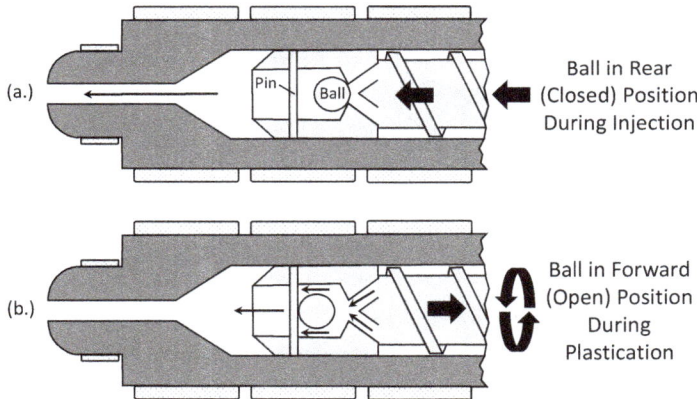

Figure 2.33 Ball check non-return valve assemblies perform the same basic function as ring check assemblies. The ball check valve allows melt flow into the shot chamber during plastication. During the melt injection phase of the process, the melt pressure causes the ball to seat, effectively allowing the reciprocating screw to function as a plunger.

Ball check valve assemblies of various designs are also used to perform the non-return valve function for reciprocating screws. Injection molding ball check valves function much like the ball check assemblies used for many other one-way fluid flow applications. The ball check assembly has a cylindrical body that is a sliding fit within the injection unit's barrel bore. During plastication, melt flow forces the ball forward to allow for shot generation. During injection, melt pressure forces the ball in the opposite direction, closing the valve. Ball check non-return valve assemblies are commonly used for lower-viscosity (high melt flow rate) thermoplastic plastic materials.

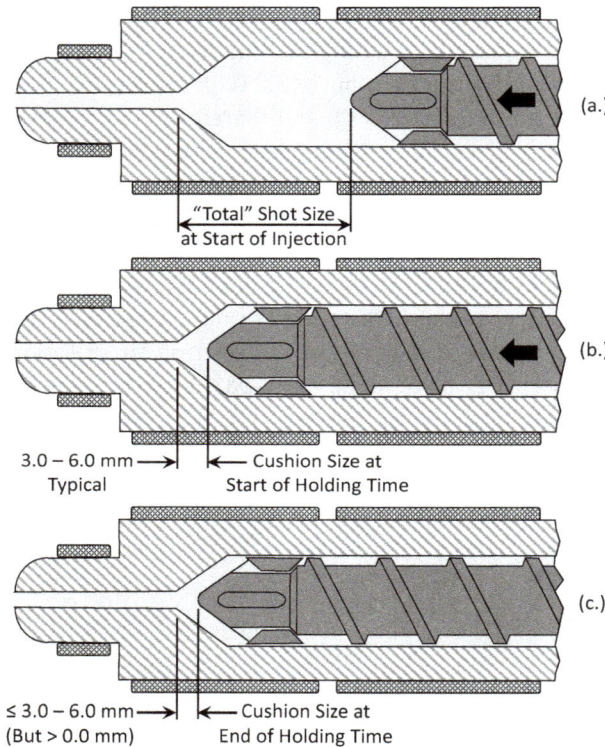

"Total" Shot Size at Start of Injection

(a.)

3.0 – 6.0 mm Typical → | ← Cushion Size at Start of Holding Time

(b.)

≤ 3.0 – 6.0 mm (But > 0.0 mm) → | ← Cushion Size at End of Holding Time

(c.)

Figure 2.34
Check valve leakage can be due to a variety of factors including excessive wear, or poor sealing due to particulate or char contamination. While a small amount backflow or leakage is inevitable, the amount of leakage must be less than the cushion volume. Under no circumstances should the molding screw and check assembly reach the zero position (i. e., the screw should not bottom out) before the holding pressure time has expired. A cushion is required so that the screw and check valve assembly can transmit pressure into the mold cavity.

Check valve assemblies should be monitored for wear and leakage flow. A worn check valve, or a check valve that is clogged with particulates (contamination, char), will allow leakage flow to occur during the injection and pressurization phases of the injection molding process. Monitoring the rate at which the screw creeps forward during the holding pressure phase of the molding process is indicative of check valve leakage. For conventional injection molding, the molding machine's reciprocating screw should *not* bottom out before the holding timer has expired. The ability to apply packing and holding pressure to the melt within the mold cavity is lost when the screw bottoms out. Using a larger than normal cushion size can be a temporary solution if the amount of leakage flow during the holding pressure phase is significant. Either purging or manual cleaning are common solutions when a check valve does not seal properly. Check valve assemblies are also a subject to repeated frictional and abrasive forces during production. These abrasive forces will eventually lead to wear

123

and leakage flow such that a replacement valve assembly may be required to elimi-nate excessive backflow leakage.

2.2.2.5 Injecting Molding Nozzles

The discharge end of an injection molding machine's barrel has both an end cap and a nozzle. The barrel end cap is a cylindrical steel cap bolted to the discharge end of the barrel. It serves as the flow transition component. The end cap has a circular rear opening diameter that is equal to the inside diameter of the molding machine barrel. The end cap typically has a lip that pilots into the barrel for alignment and to make a leak proof seal. The inside diameter of the end cap tapers gradually over its length, such that the discharge diameter of the end cap is significantly smaller than its inlet diameter. A barrel end cap discharge diameter is typically about 20 mm for small to medium size injection units. The front portion of the end cap is also threaded to re-ceive a nozzle. Like the barrel, the end cap is heated using PID temperature-controlled electrical resistance band heaters. The barrel end cap is normally a permanent ma-chine component that is changed out only if it is worn or damaged, not when a new material or mold is to be run.

The molding machine's nozzle threads into the discharge end of the barrel end cap. Molding machine nozzles have an internal flow channel that transitions or tapers down to the nozzle discharge diameter. A variety of different nozzle designs are used for injection molding processes. Some of the more common molding machine nozzle types are described in the following sections.

2.2.2.5.1 One-Piece Injection Molding Nozzle

The simplest nozzle designs are one-piece; as shown in Figure 2.35. The key features of the one-piece nozzle are highlighted in the figure. The nozzle threads into the bar-rel end cap and has a smooth face seal designed to prevent molten plastic leaking when the melt is pressurized. The nozzle discharge end has a standard radius (usually either ½-inch or ¾-inch radius with a minus tolerance) that will pilot into the injection mold's sprue bushing. The mold's sprue bushing will have a matching radius (typi-cally with a plus tolerance). These matching radii are designed to provide a leakproof seal between the nozzle and injection mold sprue bushing when the injection unit carriage is moved forward. The one-piece nozzle also has a discharge diameter that should be equal to (or slightly less) than the injection mold's sprue bushing inlet diam-eter.

The optimum nozzle discharge dimensions are then dictated by the injection mold sprue bushing dimensions. It is therefore best if the nozzle is changed when the injec-tion mold is changed (unless the new mold happens to have the same sprue dimen-sions as the previous mold). The advantage of the one-piece nozzle is both its simplic-ity and completely streamlined nature of the nozzle's internal flow channel. It is a single piece so there are none of the seams or gaps that can occur with multi-piece nozzle constructions. However, having to change the entire one-piece nozzle when a new nozzle discharge diameter is required is inconvenient.

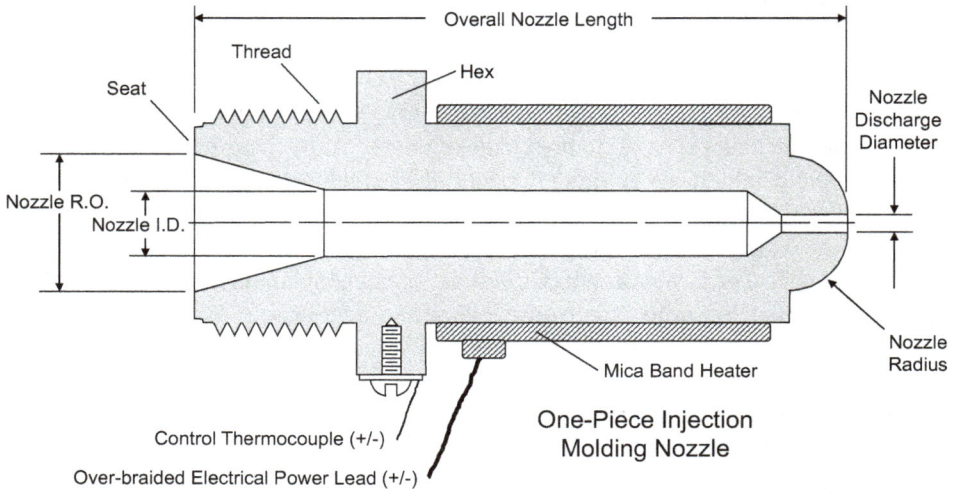

Figure 2.35 One-piece injection molding nozzles are simple and streamlined. The main limitation of the one-piece nozzle construction is that the entire nozzle must be replaced if a different nozzle discharge diameter or nozzle radius is required.

2.2.2.5.2 Modular or Two-Piece Injection Molding Nozzles

Modular injection molding nozzles are common because they offer geometric flexibility. The two-piece nozzle is the most widely used modular injection molding nozzle design. Two-piece nozzles have an electrically heated nozzle body that threads into the barrel end cap in exactly the same way as a one-piece nozzle. However, the discharge end of a two-piece nozzle is also threaded to accept a threaded nozzle tip. The rear opening diameter of the nozzle tip is the same as that of the nozzle body bore. Nozzle tips are small, inexpensive components that take only minutes to change. Molders will typically have a wide variety of tip designs in stock so that a nozzle tip having an appropriate radius and discharge diameter can be installed when a different injection mold is installed. A recommendation for the relationship between the nozzle tip discharge diameter and the mold's sprue bushing inlet diameter is given in Figure 2.36.

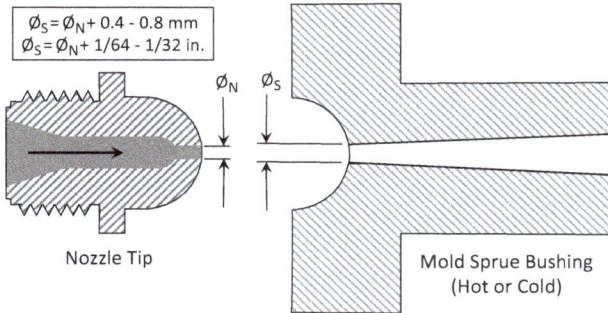

$\emptyset_S = \emptyset_N + 0.4 - 0.8$ mm
$\emptyset_S = \emptyset_N + 1/64 - 1/32$ in.

$\emptyset_N$ $\emptyset_S$

Nozzle Tip

Mold Sprue Bushing
(Hot or Cold)

Figure 2.36 A nozzle tip's radius and discharge diameter should be sized based upon the injection mold's sprue bushing dimensions. One rule of thumb is to use a nozzle tip that has a discharge diameter that is slightly smaller than the sprue bushing's inlet diameter.

Two-piece nozzle tips are available with different internal geometries so that the tip can be optimized based on the material being molded. Common two-piece nozzle tip styles include (i) a general-purpose nozzle tip, (ii) a full-taper nozzle tip and (iii) a reverse-taper nozzle tip, as shown in Figure 2.37. Full internal taper nozzle tips are commonly recommended due to their streamlined internal flow passage geometry.

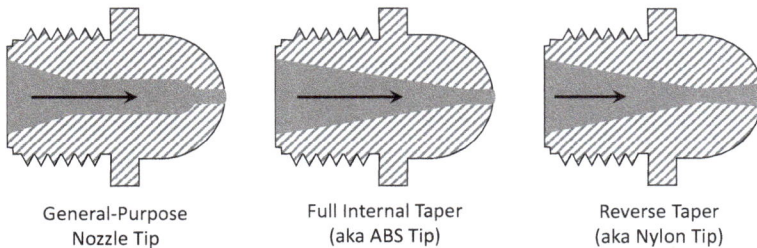

General-Purpose
Nozzle Tip

Full Internal Taper
(aka ABS Tip)

Reverse Taper
(aka Nylon Tip)

Figure 2.37 A variety of nozzle tip geometries are commercially available for use with two-piece injection molding nozzles. The type of plastic material being molded is the primary consideration when selecting the general type of nozzle tip. The nozzle tip's radius and discharge diameter are dictated by the mold sprue bushing radius and sprue inlet diameter.

Reverse-taper nozzle tips are recommended for cold runner mold applications where nozzle tip freeze-off is encountered. Nozzle tip freeze-off can happen when molding thermoplastic materials that have a sharp or narrow melting temperature range, such as a nylon 66. The reverse taper helps the frozen polymer to release from the nozzle tip when the sprue is pulled from the sprue bushing during mold opening.

Nozzle temperature control issues are common and deserve some discussion. Injection molding nozzles, such as the nozzle shown in Figure 2.35, are electrically heated using one or more electrical resistance band heaters or resistance heater cartridges. Mica band heaters are typically used for nozzles because nozzles have a relatively small diameter and mica band heaters have a low profile. This is especially important when using an extended nozzle for molds that have a recessed cold sprue bushing. Recessed sprue bushings are typically used to shorten the length of the cold sprue and reduce runner scrap. The nozzle body itself is electrically heated, but the nozzle tip itself is generally not electrically heated, unless it is an extended nozzle tip. Most nozzle tips rely on heat conduction from the nozzle body into which they are threaded.

Older generation injection molding machines typically used open loop, rheostat type temperature control for the nozzle. Open loop temperature control was likely common for its simplicity. The nozzle body is actually heated by both conduction from the barrel end cap and by the nozzle band heater. As injection molding machines have become more sophisticated, most nozzles now have PID feedback temperature control, which necessitates the inclusion of a control thermocouple. Nozzles have a limited wall thickness compared to a molding machine barrel. As a result, bayonet type thermocouples commonly used for barrel temperature control are not always appropriate for nozzles. Many injection molding nozzles use a washer thermocouple and machine screw located on the hex region of the nozzle body as shown in Figure 2.35. Other nozzles use a spade type thermocouple that is sandwiched between the nozzle band heater and the nozzle body. That said, it is best not to drill into or remove material from a nozzle wall in order to maintain the nozzle's structural integrity. Nozzles are exposed to extremely high internal melt pressures, especially during the packing phase of the molding process. It is also important to make sure that all nozzle heater wiring has both primary and secondary electrical insulation since these electrical wires may be unguarded. The electrical and thermocouple wires should be neatly arranged and properly supported. These wires must travel freely with the injection unit carriage as it travels. Nozzle wiring should also be positioned such that the wiring cannot be encapsulated by molten plastic when the molding machine barrel is purged.

In most cases, the nozzle tip itself is heated *only* by heat conduction from the nozzle body and heat from melt flow during mold filling. In cold runner applications, heat loss to the relatively cold mold sprue bushing can be an issue. In a cold runner application, the nozzle is also in contact with the relatively cold mold sprue for either the entire molding cycle, or for at least a portion of the molding cycle, if sprue break (carriage retract) is utilized. This continuous flow of heat from the nozzle tip to the mold sprue can cause the nozzle tip to cool to the extent that some plastic material within the nozzle tip solidifies. Note that this is true only for cold runner mold applications.

Figure 2.38 shows a cold sprue that has been ejected from a cold runner mold after cooling. The sprue has solidified along with some solidified material that has been pulled from the nozzle tip when the mold opened. The actual transition from solid material to molten material occurs some distance into the nozzle tip. If this distance becomes too great, the solidified material within the nozzle tip may cause the sprue to stick in the sprue bushing. The reverse taper nozzle tip shown in Figure 2.37 is designed to solve this problem. The tip's reverse taper reduces the force required to pull solidified material from the nozzle tip. Nozzle freezing is most likely to happen with semi-crystalline thermoplastic materials having a very narrow melt temperature range. Increasing nozzle temperature helps to reduce the possibility of nozzle freezing, but can lead to material degradation if too much heat is applied. Using sprue break is perhaps the most effective means of preventing nozzle tip solidification by allowing the nozzle tip temperature to recover during the time that the carriage and nozzle are retracted from the mold. However, sprue break can lead to other issues, such as nozzle drool, which will be a separate discussion. Nozzle tip insulators can also be used to reduce heat transfer between the nozzle tip and the sprue bushing [18]. Shut-off nozzles can also be used to eliminate nozzle drool.

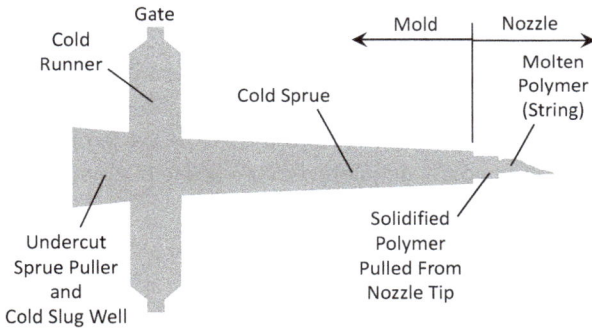

Figure 2.38 Most nozzle tips are heated by conductive heat transfer from the nozzle body. In a cold runner molding application, the section of the nozzle tip that is in contact with a relatively cold mold sprue bushing can cool to the extent that the plastic material within the tip solidifies. The solidified plastic in the nozzle tip can cause the sprue to stick if the nozzle tip does not have a reverse taper.

Most nozzle tips have a radius that is designed to seat against a radiused sprue bushing as shown in Figure 2.39. Nominal nozzle tip radius values are typically ½ or ¾ inch with a minus tolerance, while sprue bushings have a matching nominal radius with a plus tolerance. The matching radii and axial nozzle pull-in carriage forces combine to create a good seal between the nozzle tip and sprue bushing so that there is no melt leakage when the melt is pressurized. An advantage to using a radiused seal area (as opposed to a flat seal area) is that the nozzle will self-align with the sprue bushing. Even so, the injection unit carriage should also be adjusted such that the nozzle tip aligns precisely with the mold's sprue bushing. A misaligned carriage can cause damage to the nozzle tip and/or sprue bushing. Damage to the seat surfaces can result in material leakage when the melt is pressurized, or sprue sticking when the mold opens. Machinist bluing and visual inspection can help determine if there is any nozzle tip damage or misalignment. The nozzle tip and sprue bushing radii will normally allow for a slightly misaligned carriage to align itself when the nozzle tip engages with the sprue bushing as the carriage moves forward.

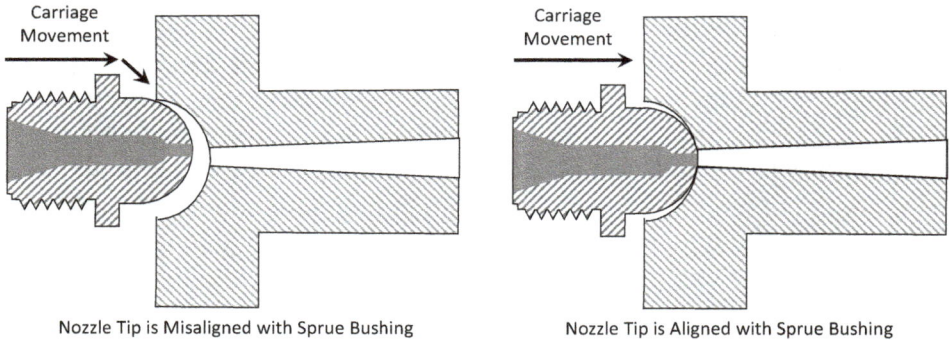

Figure 2.39 The injection unit's carriage alignment should be adjusted so that the nozzle tip is precisely aligned with the mold's sprue bushing as shown above (right). Most injection molding nozzle tips have a radius, which helps the nozzle tip compensate for any carriage misalignment as the nozzle tip pilots into the sprue bushing's radius.

2.2.2.5.3 Mixing and Melt Filtration Nozzles

Some injection molding nozzles have an internal element that has been incorporated into the nozzle body to perform an additional function (above and beyond that of a straight flow through nozzle). Static mixers and melt filters are both examples of internal nozzle elements added for a specific purpose. Static mixers, such as that shown in Figure 2.40, are particularly useful in laminar flow plastic applications like injection molding since there is no natural mixing or turbulence with laminar flow. The injection molding machine's reciprocating screw is the primary mixing device for the process. However, the relatively short length-to-diameter ratio associated with most injection molding screws limits their mixing capability. Static mixers provide an additional degree of distributive mixing for both temperature mixing and additive mixing (i.e., overall melt homogeneity). Static mixers are particularly useful as a means to improve the appearance of thermoplastic plastic parts that are pigmented using a masterbatch color additive. Static mixers can be used with most thermoplastic materials unless they are sensitive to shear degradation, or if they contain reinforcing fibers.

Static mixers are cylindrical mixing devices that fit tightly within the internal bore of a nozzle body. There are many possible static mixer designs, but all static mixers have no moving parts (i.e., they are static not dynamic). The mixing function is accomplished by repeatedly dividing and recombining flow streams as the melt passes through the static mixing elements during melt injection. They function much like the static mixer that is used to mix a two-component double syringe epoxy adhesive. While static mixers do improve distributive mixing, they create additional mold filling pressure drop. Static mixing nozzles are typically purchased as a complete unit or, in some cases, the mixing element can be retrofitted into the bore of an existing nozzle. Most static mixing nozzles have a modular construction to simplify removal of the mixing element when cleaning is required.

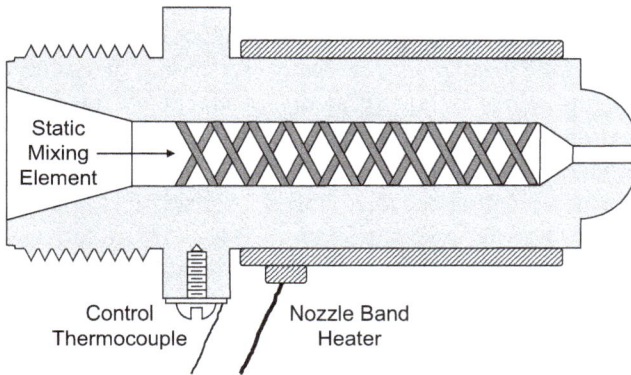

Figure 2.40 Some injection molding nozzles incorporate a static mixer that improves melt homogeneity and distributive mixing. They have no moving parts and function by repeatedly dividing and recombining the laminar flow stream layers.

Nozzles containing a melt filter are also used in some injection molding applications. While almost all extrusion processes utilize melt filtration (usually screen filters), very few injection molding processes involve melt filtration. Injection molding melt filters are primarily used to filter out non-thermoplastic contamination. They can also hold back plastic granules that have not melted by the time they reach the molding machine's nozzle. The filter causes an increase in residence time for the un-melted thermoplastic granule that should eventually pick-up enough heat to soften and pass through the filter. Melt filters are used in a variety of applications, especially in applications where (i) the injection mold has very small gates (that are easily clogged by contamination) and (ii) regrind is used. The potential for plastic or metallic contamination is significantly greater in applications where regrind is utilized. A reminder that hopper magnets, not nozzle filters, should always be the first line of defense against ferrous metal contamination. Nozzle filters provide an added measure of safety to keep contamination from entering the injection mold. This is especially important for mold having very small diameter gates.

As with static mixers, there are many different nozzle melt filter designs [19]. Some are based on supported screen filtration, while others are described as gap filters. The key to selecting the best filter is to have an understanding of both the physical size and the concentration of the particulate contaminates that are associated with the process. Screen filters are best for filtering smaller particles, while gap filters are suitable for larger particles. A gap filter is typically a metal cylinder that fits tightly within the internal bore of a nozzle. The cylinder has axial inlet flow channels, and discharge flow channels, separated by a dam having narrow slots. Contamination that is too large to fit through the narrow slot will then collect within the inlet channel. Injection molding nozzle filters only work well if the concentration of contamination is very low. Over the course of a production run, the quantity of captured contamination increases continuously and it begins to clog or restrict the amount of surface area that the melt can flow through. This added flow restriction adds a pressure drop which increases over the course of the production run. Many extrusion screen filters allow for backflushing or changing screens on the fly so that production is never interrupted due to a clogged screen. Most injection molding nozzle filters are *not* self-cleaning or automated in any way. The filter element located within the nozzle must be removed and cleaned when the filter becomes clogged. A steady increase the molding machine's velocity-to-pack transfer pressure (over time) is an indication that the nozzle filter is clogging. At some point, the nozzle filter will need to be cleaned. It is helpful to have multiple melt filter nozzles to minimize the downtime associated with a clogged nozzle filter that needs cleaning.

2.2.2.5.4 Shut-Off Nozzles

All of the nozzles discussed in the sections above are described as open-flow nozzles as they have no mechanical mechanism to stop the flow of melt. Shut-off nozzles differ from open-flow nozzles in that they include a mechanical means of preventing or shutting-off the flow of melt through the nozzle tip orifice during portions of the molding cycle when melt flow is unwanted [20]. Shut-off nozzles do have a more complex construction than flow-through nozzles but they offer numerous advantages. They can be used routinely or only for applications where a shut-off nozzle is absolutely required. The use of a shut-off nozzle is advantageous in a number of injection molding applications situations. Shut-off nozzles:

- Prevent melt drool while the sprue is being ejected from a cold runner mold (even if decompression or suck-back is not used).

- Prevent the stringing that can occur as the sprue is pulled from the sprue bushing.

- Prevent melt drool if sprue break is used (sprue break helps prevent nozzle tip freezing).

- Prevent melt drool if shot plastication or shot preparation occurs after sprue break.

- Prevent melt drool for applications where the plastic melt contains a foaming agent.

Melt drool can be caused by a number of factors that include thermal expansion of the melt, residual moisture (i. e., gas pressure), and the use of back pressure. Melt drool can be eliminated or minimized using rear decompression (i. e., screw suck-back) after plastication. However, even when decompression is used, melt drool can still occur. In fact, in cold runner applications where sprue break is not used, melt drool from a conventional flow-through nozzle can occur undetected, as shown in Figure 2.41. Melt drool that extrudes into the sprue bushing (after the mold opens) is not visible from the operator station. If this does occur, the melt drool will at least partially solidify before the injection phase of the molding process. The partially so-lidified drool can cause gate clogging, filling imbalances for multi-cavity molds, or cosmetic defects for the injection molded part surface. Cold runner systems typically incorporate a cold slug well to capture solid or semi-solid nozzle drool that may occur before entering the mold cavity.

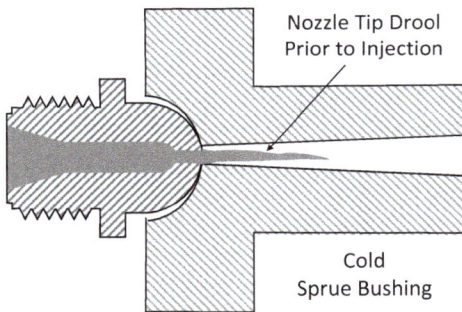

Figure 2.41 Nozzle tip drool can be difficult to detect for cold runner molds when sprue break is not used. Any nozzle drool that occurs prior to injection begins to solid-ify and can cause mold filling imbalances or surface defects on the molding parts. Cold runner systems typically incorporate a cold slug well to capture solid or semi-solid nozzle drool. Using a shut-off nozzle is also one way to eliminate this issue.

The use of a shut-off nozzle is a failsafe means of preventing melt drool. There are many types and designs for shut-off nozzles, but they all include some means to me-chanically shut-off the melt flow.

Figure 2.42 shows two different versions of a needle type shut-off nozzle. In both cases, a bullet-shaped pin or needle is used to temporarily block the nozzle tip orifice, in much the same way that a hot runner valve gate operates. The needle valve is made to open only during the mold filling, packing and holding pressure phases of the molding process. The valve closes for the other phases of the process (e. g., during plastication, cooling and ejection). The shut-off nozzles like the one shown on the left in Figure 2.42 are activated using an externally mounted hydraulic or pneumatic ac-tuating cylinder. Other shut off valves, such as the one shown on the right in Fig-ure 2.42 are fully self-contained. They use an internally or externally mounted spring to keep the needle valve closed to prevent drool during and after plastication. Nozzle melt pressures higher than a specific threshold value (dictated by the spring constant) will cause the valve to open. The spring-loaded shut-off nozzle will open during the high-pressure injection, packing and holding phases of the molding process but will be closed during plastication, even if a low back pressure is used. Spring-loaded shut-off nozzles are reliable and simple, but shut-off nozzles having on demand external hydraulic or pneumatic cylinder actuation provide for more processing flexibility since the needle actuation is independent of other process variables. While needle type shut-off nozzles are most common, sliding bolt or gate type shut-off nozzles are also available. A nozzle tip with a spring-loaded shut-off mechanism (i. e., a mini-shut off nozzle) is also commercially available and has the benefit of preventing nozzle drool without the use of decompression or suck-back [21].

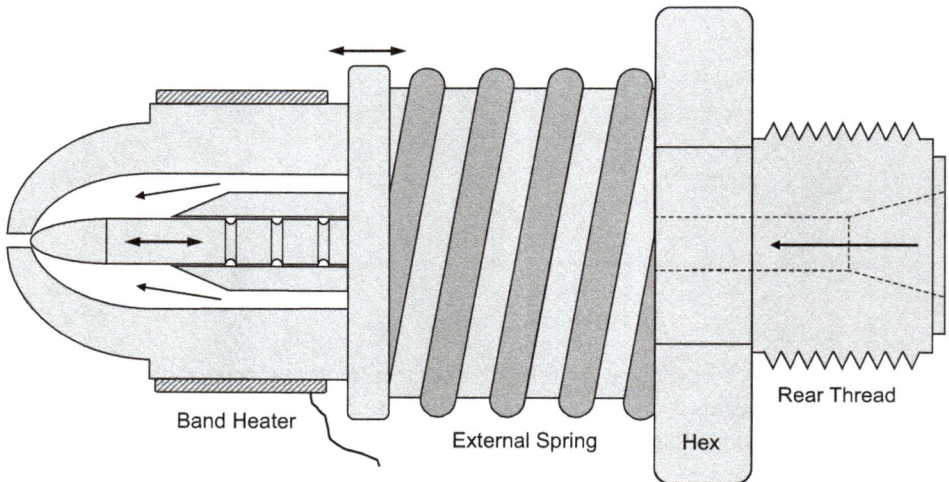

Figure 2.42 Shut-off nozzles have the ability to temporarily block the flow of melt through the nozzle tip orifice during portions of the molding cycle when melt flow is unwanted (e. g. to prevent nozzle tip drool). The needle valve of a shut-off nozzle can be externally activated using a hydraulic or pneumatic cylinder, while other shut-off nozzles are spring activated (shown in the closed position). The spring force is sufficient to prevent melt drool, but it is overcome and the valve opens when mold filling and holding pressures are applied.

Shut-off nozzles can also be used in applications where the plastication time required to build a shot is greater than the additional cooling time required to cool the part. The time required for shot plastication is normally less than the additional cooling time. Ideally, the cycle time for an injection molding process should be determined by part cooling requirements, not by the time required for plastication. In such cases, the time available for plastication may be increased by allowing the screw to rotate during the time the mold is opening and the part is being ejected (if the machine controller allows for this). In such a case, the time available for plastication is extended, and the shut-off nozzle prevents nozzle drool during plastication after the sprue has been pulled from the sprue bushing.

2.2.3 Injection Molding Machine Clamping Unit

The discussion above focuses on the functions and hardware that are associated with the injection molding machine's injection unit. Summarizing, the injection unit is responsible for shot preparation and for generating the mold filling, packing and holding pressures required for molding thermoplastic materials. The second major hardware component of the molding machine is the mold clamping unit. Like the injection unit, the mold clamping unit must perform a number of different functions that can be accomplished using a variety of different machine designs. Unlike most compression molding machines, most injection molding machines utilize clamping systems that are based on a horizontal clamp design. However, molding machines with vertical clamp designs are also available and are most popular for insert molding applications where gravity helps with insert positioning (see Chapter 4). The discussion here focuses on horizontal injection molding machine clamping systems. An injection molding machine clamp system is designed to perform at least the following functions:

- The mold clamping unit must locate and support the injection mold halves.

- The clamp creates the B-side mold movement during the mold closing and mold opening phases of the molding process.

- The clamp must include provisions for low-pressure mold protection prior to initiating the application of the full clamp force.

- The clamp system must generate the force required to keep the mold closed to counteract the mold's internal melt pressure.

- The clamp must include a means of activating and re-setting the mold's ejector system.

- The clamp must include a redundant pinch point and other safety features to protect the workers.

Selecting an appropriate injection molding machine for a given application involves many considerations related to both the molding machine's injection unit and its clamp unit. In terms of the clamp system, the are many dimensional issues that relate to the physical size of the mold (i. e., mold size envelope). Clamp platen size, tie bar spacing, daylight range, and ejector stroke are all dimensional considerations that are associated with the injection mold to be run. Even clamps that are physically large enough are suitable only if they can generate the required clamp tonnage associated with the application. It can be challenging to locate an available injection molding machine which has *both* optimum injection unit capacity and optimum clamp capacity and size. Fortunately, both the injection unit and the clamping unit have process parameters that allow for some flexibility in terms of capacity utilization.

2.2.3.1 Directly Acting Hydraulic Clamp Unit

There are many different molding machine clamping unit designs available commercially, but directly acting straight hydraulic clamps have historical significance and continue to be widely used today. Figure 2.43 shows is a simplified schematic of a three-platen, straight hydraulic clamp shown in the closed position.

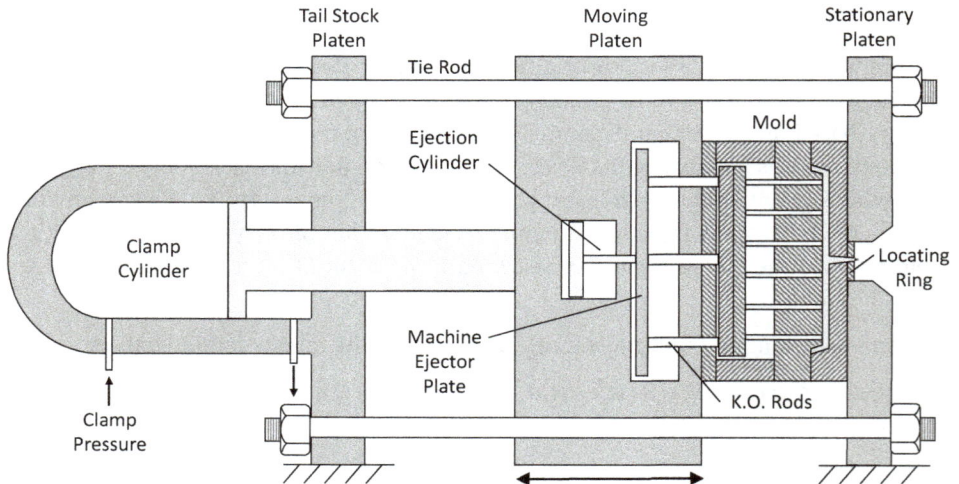

Figure 2.43 Simplified schematic of a directly acting, single cylinder, horizontal hydraulic injection molding machine clamping system. The clamp force for a hydraulic clamp system will be directly proportional to the applied hydraulic oil pressure. The injection molding machine clamp system safety gates are not shown.

The simplified hydraulic clamping unit schematic shown in Figure 2.43 has a total of three parallel platens, two of which are stationary and one of which is movable. The parallelism is critical to ensure that the injection mold faces have a uniform parting line contact pressure when the mold is closed. Parallelism is also required to limit abrasive wear for mold half alignment features (e. g., leader pins and bushings). The distance between the front stationary platen and the rear stationary platen (i. e., the tailstock) is determined by the length of the four tie bars (or tie rods) and tie bar nuts that connect the platens. The front stationary platen also has a central circular hole that is used to help support and centrally locate the injection mold by mating with the injection mold's protruding locating ring as shown in Figure 2.43. The stationary and moving molding machine platens also incorporate a plethora of threaded holes for mold clamp bolts.

The B-side of the injection mold is bolted or clamped to the unit's moving platen. The back-and-forth movement of the platen is guided by the clamp's tie bars and replaceable lubricated shoes riding in rails that help support and guide the heavy moving platen and B-side mold half. The molding clamp shown in Figure 2.43 has a single hydraulic cylinder, which is integral with the tailstock. The directly acting clamp cylinder is used to both open and close the mold and to generate the required clamp force. One advantage of a hydraulic clamp is that the clamp force is infinitely adjustable from zero to maximum tonnage simply by adjusting the magnitude of the applied hydraulic pressure.

Hydraulic clamping systems, like all injection mold clamp systems, incorporate a function for low-pressure mold protection that is designed to eliminate or minimize mold damage. Injection mold damage can occur when the mold closes under high pressure if debris, hot runner strings or un-ejected parts interfere with mold closure. Low-pressure mold protection is set to be high enough to overcome the leader pin friction and ejector plate return force when the mold closes, but low enough such that any additional resistance (associated with the debris or an un-ejected part) will prevent complete mold closure and clamp force generation. If the injection mold is unable to fully close in this low-pressure mode of operation, the clamp then reverses (opens) triggering an alarm indicating that corrective action is required to clear the issue.

Other more complex hydro-mechanical clamp designs (not shown) are common on larger clamp capacity injection molding machines. Large directly acting hydraulic cylinders have a very large hydraulic oil capacity. Reducing the volume of hydraulic oil required for clamp movement allows for faster clamping speeds and reduces energy consumption. Hydro-mechanical clamps have separate long stroke hydraulic traverse cylinders that can move the platen quickly and efficiently, and a short stroke, large diameter clamp cylinder for generating the final clamp tonnage.

The moving platen of the injection molding machine shown in Figure 2.43 also incorporates a molding machine ejector plate that is driven forward and rearward by its own hydraulic or electric power source. The molding machine's ejector plate is typically connected to the mold's ejector plate using a series of knock-out rods that pass through clearance holes within the moving platen. The ejector plate movement facilitates molded part ejection and resetting of the mold's ejector system. Both the moving platen and the front stationary platen have numerous threaded holes or T-slots so that mold clamps can be used to support and locate the A and B sides of the injection mold. Both the knock-out rod hole pattern and the clamp bolt hole patterns are typically standardized to allow injection molds to be more easily moved from press to press.

There are many dimensional and other variables that are used to describe a molding machine's clamp system. These include:

- Maximum clamp force (kN, U.S. tons or metric tons).

- Horizontal and vertical distance between the tie rods (mm)

- Minimum daylight – minimum distance between the moving and stationary platens (mm)

- Maximum daylight – maximum distance between the moving and stationary platens (mm)

- Maximum ejector stroke (mm)

- Maximum ejector force (kN, U.S. tons or metric tons).

The distance between the tie rods is one factor that helps determine maximum injection mold size in terms of maximum injection mold width and length as shown in Figure 2.44. Molds that are safely held closed with safety straps are typically loaded into the clamp from above using an overhead crane. The mold and any protruding water fittings or fixed water lines must fit within the horizontal tie bar spacing as shown in Figure 2.44. The overall length of the mold (i. e., vertical length) can be as long as the platen. It is best to avoid installing injection molds that overhang the platen area by even a small distance. Molds that have cooling lines that extend outside of the platen area are acceptable as long as the injection mold plates themselves do not overhang the platen area.

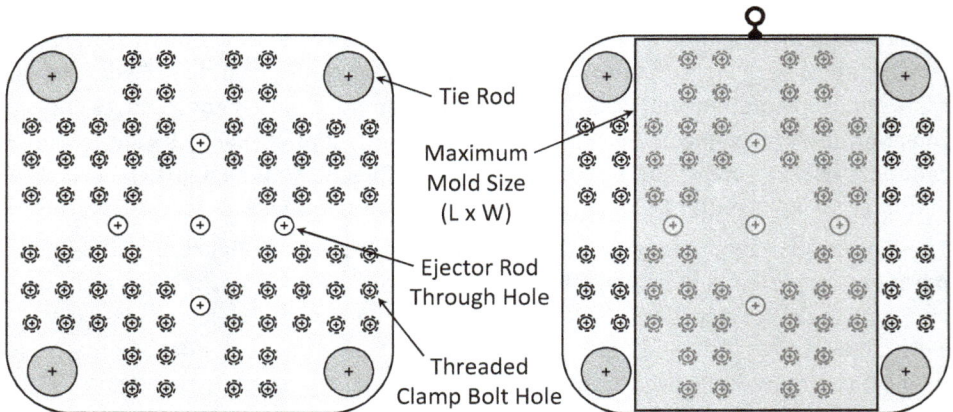

Figure 2.44 The maximum recommended mold size that can be used with a four tie bar clamp system would have a mold width (W) equal to the horizontal distance between tie bars, and a mold length (L) that is equal to the vertical height of the molding machine's platens.

Whenever possible, it is also good practice to recess quick-disconnect water fittings so that they are not exposed to potential damage during mold storage or during an injection mold change. Brass water fittings are relatively soft and can leak if they become deformed or nicked. It is also good practice to avoid placing water fittings, especially quick-disconnect water fittings, on the top side of the mold, to avoid the potential for water leaking between mold plates when the water lines are removed or if there is any water fitting leakage during production. Small water leaks can lead to corrosive damage to the injection mold and molding machine platen damage even when the mold cooling water is treated with bacterial and corrosion resistant additives.

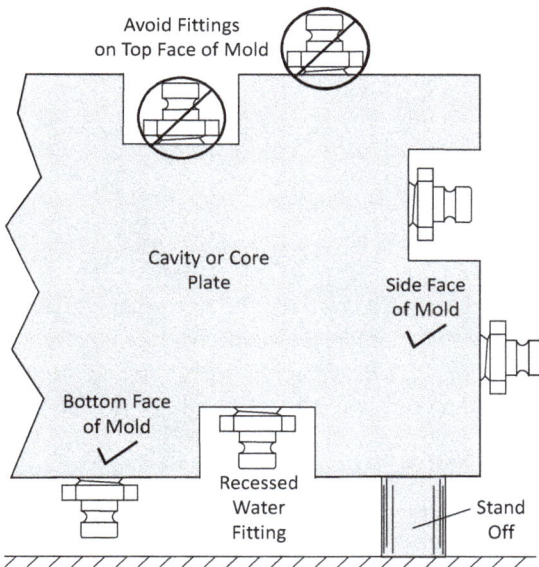

Figure 2.45
Brass quick-disconnect water line fittings that protrude from the mold surface can be easily damaged during mold changes, particularly when tie bar clearance is limited. Recessed mold fittings are less prone to damage and water leakage but may require the use of a larger mold base. It is also best to avoid quick-disconnect water fitting on the top surface of the injection mold to avoid mold damage in the event of water leakage occurring during mold changes or production.

A clamping unit's minimum daylight and maximum daylight values also help determine if the clamping system is compatible with an injection mold. On the low side, the injection mold must have a total stack height that is greater than the minimum clamping unit daylight value. Bolster plates can be added to a thin mold in situations where the injection mold does not have enough stack height. On the high side, the maximum clamp daylight value must be great enough to accommodate the total stack height of the mold, as well as the space required to eject the moldings from the open mold during the ejection phase of the molding process. This can be a problem for very deep-draw parts that require enough open space for the parts to clear the core during the part ejection phase of the molding process. It is relatively easy to set clamp position and clamp force variables for hydraulic clamps, such as that shown in Figure 2.43, as the clamp cylinder is in-line and acts directly.

All injection mold clamp units have a maximum clamp force value that also helps to determine if a clamping unit is suitable for a particular injection mold application.

Molding machine clamps with high clamp force tend to have larger platen sizes and daylight values, but there is no universal correlation between the clamping unit's maximum clamp force and the clamping unit's physical X-Y dimensions.

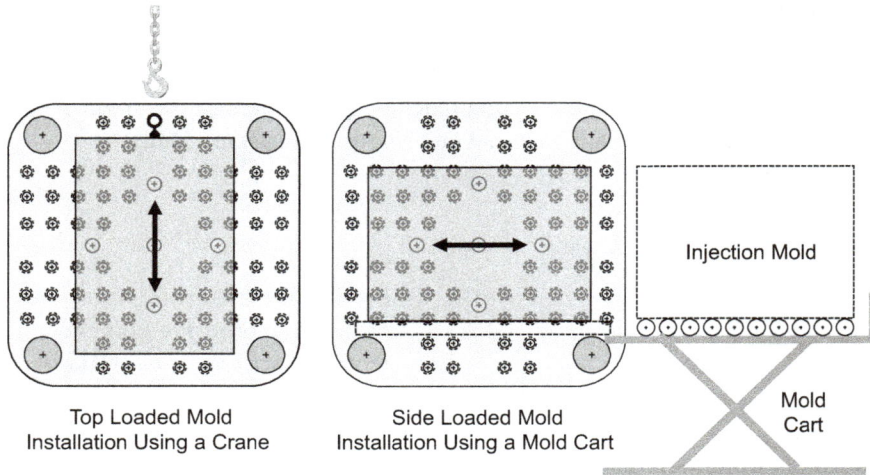

Top Loaded Mold
Installation Using a Crane

Side Loaded Mold
Installation Using a Mold Cart

Injection Mold

Mold
Cart

Figure 2.46 Injection molds can be installed vertically from the top of the clamp using a crane or from the side using an adjustable-height injection mold cart. Mold plate size standardization, especially clamp plate sizing, is very beneficial in reducing mold change time in both cases, but is generally required for horizontal mold installation.

It is obvious that the maximum injection mold dimensions that can be installed in a clamp unit are determined by a number of clamp dimensional variables (e. g., tie bar spacing, platen size). However, the guidelines associated with the minimum recommended mold size for a given clamping unit size are less clear. Injection molds that are physically smaller than the injection mold shown in Figure 2.44 can be installed in a mold clamp system. Smaller injection molds generally require less clamp force than larger injection molds. Fortunately, the clamp tonnage that can be generated with an injection molding machine's clamping unit is infinitely variable from a low (near zero) clamp force to the molding machine's maximum clamp force. It is important to use only enough clamp force to ensure that the mold does not flash during molding. The use of higher than optimum clamp force for a given injection mold can cause a number of issues including:

- Hobbing of the clamp platen face due to excessive compression force.
- Damage to the injection mold parting line due to excessive compression force.
- Clamp platen deflection or clamp wrap (especially the stationary platen).
- Decrease in mold cavity venting efficiency due to parting line compression.

Several sources suggest that the injection mold's X-Y footprint should cover at least 60–70% of the molding machines platen area [22,23]. This is a good general recommendation that is designed to limit platen wrap and the potential for compressive hobbing of the platen by the relatively small mold. However, smaller size injection molds can be used successfully if a correspondingly low clamp force is set properly (based on the smaller projected mold cavity area). Even so, platen wrap can still be an issue since the stationary platen transfers clamp force through the tie rods that are located only at the platen corners. Adding larger clamp plates to a smaller than optimum mold can also help with both the platen hobbing and platen wrap issues.

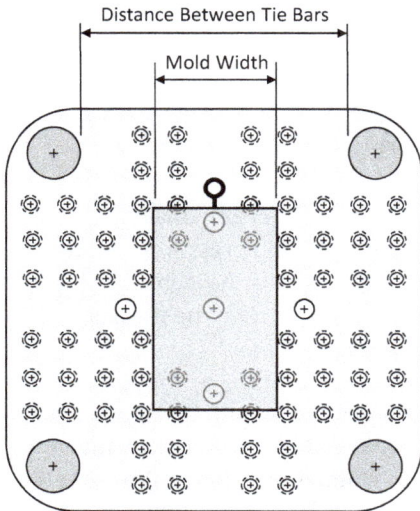

Figure 2.47
The maximum mold size that can be used with an injection mold clamp is dictated by the platen size and the tie bar spacing. The guidelines for the minimum recommended mold size are less obvious. When small molds are used, the molding machine's clamp force should be adjusted so that it is high enough to prevent flash, but not high enough to cause permanent damage to the molding machine platens or the injection mold parting line face.

2.2.3.2 Estimating Clamp Force Requirements

A primary function of the injection molding machine's clamp is to provide enough clamp force to prevent the parting line of the injection mold from opening during the pressurization phases of the molding process. The point in the process where flash is most likely to occur is during the relatively short packing phase of the molding process. During the packing phase of the molding process, the remaining compressible melt in the shot chamber (the cushion) and the melt withing the injection mold is pressurized to ensure the mold filling is complete and that the mold's surface features (e. g., texture) are fully replicated. Melt compression also helps compensate for mold shrinkage. High melt pressures are also applied during the holding phase of the molding process, however, typical holding pressures are normally less than the peak packing pressure. The molding machine clamp unit must provide enough force to counteract the injection force during these pressurization phases of the process. If the clamp's force is insufficient, the mold's parting line will open slightly and some or all of the pressurized melt cushion will flow into the thin planar gap along the now open parting line resulting in flash. It should be noted here that flash defects can be caused by other reasons, including mold damage or wear, as well as mold plate deformation. Full perimeter flash, such as that depicted in Figure 2.48, is the type of flash pattern that is normally associated with insufficient clamp tonnage.

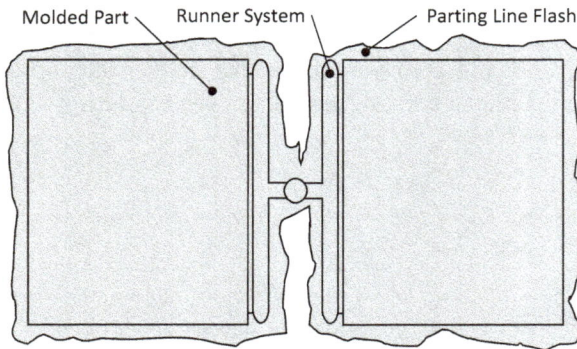

Figure 2.48 Injection molding machine clamping units must perform many functions that include counteracting the injection mold filling and packing pressures that can separate the mold halves. Moldings produced in clamp units that have insufficient clamp force will exhibit full perimeter flash caused by some or all of the pressurized melt cushion being squeezed into the thin parting line gap.

Commercial injection molding machines are available with clamp forces that range from several tons for micro molding machines to many 1000's of tons and anywhere in between. It is helpful to estimate the amount of clamp force that will be theoretically required for a new injection mold and material combination so that an injection molding machine with the appropriate clamp force can be utilized for production. Injection molds that have been run previously will hopefully have historical data that can be utilized for this purpose. However, when no historical data is available, the optimum clamp size and clamp capacity must be estimated. The clamp size requirement as it relates to the injection mold size is relatively straight forward as discussed above. Estimating clamp force or clamp tonnage requirements for a given injection mold and material combination can be less precise and somewhat more complicated. This is in large part due to the fact that the actual peak packing pressure that will be used in production can only be estimated prior to establishing the actual production molding process pressures.

The best estimates of the clamp tonnage requirements for a new injection mold and material combination can be obtained by conducting a mold filling simulation (or by reviewing previously conducted mold filling simulation results). Mold filling simulations can provide a theoretical estimate of the melt pressure required to fill a mold by considering the injection mold geometry, the thermoplastic material properties, and estimated molding process conditions. While this information is all useful, the required mold fill pressure is likely to be lower than the peak packing pressure that is used in the actual molding process. The actual peak packing pressure value used in production is set by the process engineer during process start-up. This peak pressure value is at their discretion and can be any pressure value up to the maximum injection pressure value for the injection unit being used. The peak packing pressure used in production is unrelated to the mold filling pressure output from a mold filling simulation. As a result, the fill pressure value provided by the mold filling simulation would help determine the minimum clamp force requirement, but additional clamp force will be required in most applications since the peak pack pressure used in molding is likely to be far higher than the molding simulation's fill pressure value.

In practice, rules of thumb are more commonly used to estimate clamp tonnage requirements for a new injection mold and material combination. The most common rules of thumb regarding the clamp tonnage required to keep an injection mold closed are based upon the concept of the injection mold's projected area. The mold's projected area is essentially a two dimensional "shadow" of the molded parts and runner system when viewed from the injection unit. The two moldings shown in Figure 2.49 have very different surface areas, but they both have the same projected area. The deeper-draw sections of the plastic part shown in Figure 2.49 (b.) involve additional forces acting on the injection mold, but these forces act on the mold's side walls, not on the injection mold clamp. While the deeper-draw parts may not have a greater projected area, they are likely to require a somewhat greater clamp tonnage since the pressure required to fill the parts will be greater due to its longer flow length (all other factors being equal).

Clamp force rules of thumb vary depending on the source. One common rule suggests that a clamp tonnage in the range of 2–5 tons of clamp force is required for every one square inch of projected area [24]. This corresponds to a peak cavity pressure of 4,000 to 10,000 psi (26–70 MPa). This is a very wide range that is usually further qualified to suggest that parts with thicker walls require 2–3 tons/in^2, while thinner wall parts require 3–5 tons/in^2. Thicker wall parts tend to be easier to fill and therefore require less force to keep the mold closed in order to prevent flash. While it is generally true that thin-wall parts do require faster fill times and higher injection pressures, injection mold filling dynamics are more complex than that.

One should also consider that most thin-wall moldings are produced with lower viscosity (i. e., higher melt flow rate) material grades. Lower-viscosity materials may require lower fill pressures on one hand, but they also have more of a tendency to flash, especially if the tool parting line surfaces or molding machine platens are not perfectly flat and parallel. In reality, thin-wall parts may need the higher clamp force to (i) prevent flash by preventing mold separation, and (ii) to ensure that the mold parting line shuts off with adequate and uniform compression force across the entire parting line shut-off area. A rule of thumb suggesting a 3–5 tons/in^2 clamp force requirement is probably more appropriate for thinner wall injection molded parts. This rule would suggest that a 9.0 in^2 thin-wall part, such as that shown in Figure 2.49 (a.), would require somewhere between 27.0 and 45.0 tons per cavity.

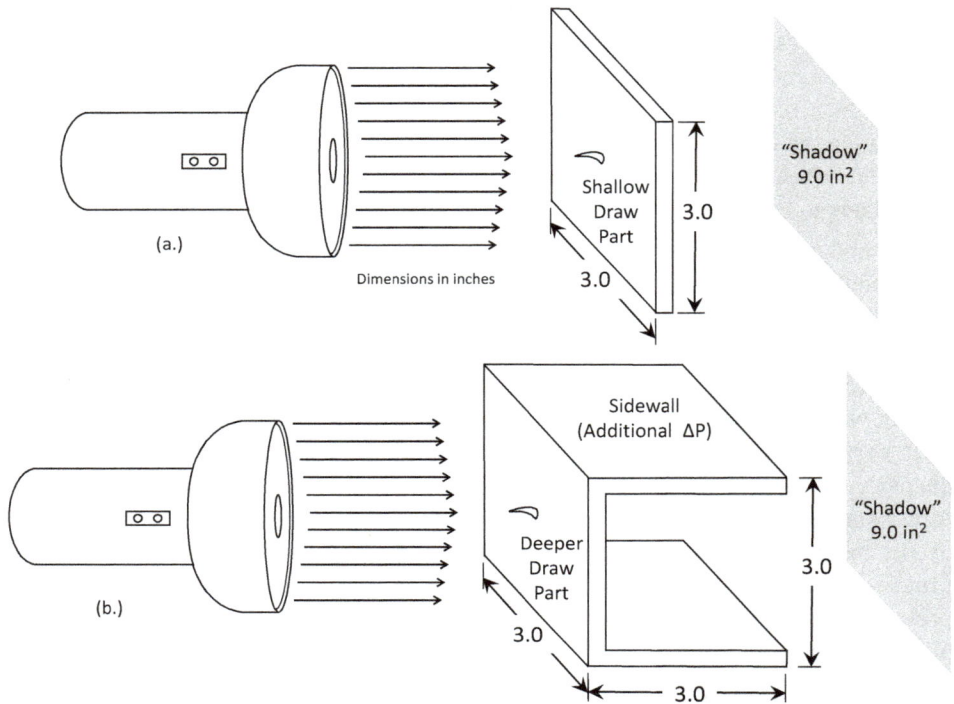

Figure 2.49 The two plastic parts shown have different geometries and surface areas. However, the clamp tonnage requirements for the two parts will be very similar since they both have the same shadow or projected area. The force required to keep a mold closed and prevent flash is determined by the peak cavity pressure multiplied by the molding's projected area.

An additional modification for the common clamp tonnage rule of thumb is the inclusion of an additional factor to account for deeper-draw molded parts [25]. This modified rule of thumb first calculates the clamp force requirement based on projected area as with the example given above, but then adds an additional 10% of this value for every 1.0 inch of draw. This modification is included to account for the higher injection pressure requirement for the deeper-draw part (not an increase in projected area). Using this method, a thinner-wall part, such as that shown in Figure 2.49 (b.), would require between 27 and 45 tons per cavity based on the 3–5 tons/in^2 rule of thumb for projected area, plus an additional 8.1 to 13.5 tons per cavity to account for the added fill pressure requirement for the deeper-draw sidewalls. The estimated total clamp force requirement for the part shown in Figure 2.49 (b.) is then somewhere between 35.1 and 58.5 tons.

The basic clamp force rule of thumb seems to have increased in both magnitude and range over the years. This also seems to be related to both a historical increase in injection unit pressure capabilities and injection rate capabilities, as well as a trend towards thin-walling that results in part designs having harder-to-fill thinner wall sections. Thick-wall parts of the past had very wide process windows with very limited mold fill pressure requirements. Very thick parts can be molded with clamp pressures even lower than 2.0 tons/in^2. Today's thinner-wall injection molded parts require higher injection rates and pressures and therefore far greater clamp tonnages. Two recently published clamp tonnage sources suggest using a rule of thumb of 2–8 tons/in^2 of projected area [26] and 2–10 tons/in^2 of projected area [27] respectively. There is clearly a trend towards higher clamp force estimates based on factors that increase injection pressure requirements, although rules of thumb with such a wide range are difficult to implement in practice. It should be noted here that a clamp tonnage estimate based on rules of thumb is exactly that, an estimate. Estimated clamp force values are used to assist with injection molding machine size selection. The actual clamp tonnage used in practice should be just slightly higher than that required to prevent flash. Higher clamp pressures can be used, but any excess clamp pressure has no real benefit and many disadvantages. Excess clamp force can damage the injection mold, reduce molding machine life, and result in higher energy consumption.

2.2.3.3 Toggle Mold Clamping Units

Injection molding machines having a mechanical or toggle clamping units are also very popular. Toggle clamps perform the same functions as a hydraulic clamp, but they accomplish these functions in a very different way. The toggle clamping units are energized using a servo-electric motor in the case of an all-electric molding machine clamp unit, or by a relatively small hydraulic cylinder for hydraulic molding machine clamps. These clamps offer fast opening and closing speeds, and relatively low energy consumption, compared to directly acting hydraulic molding machine clamp systems.

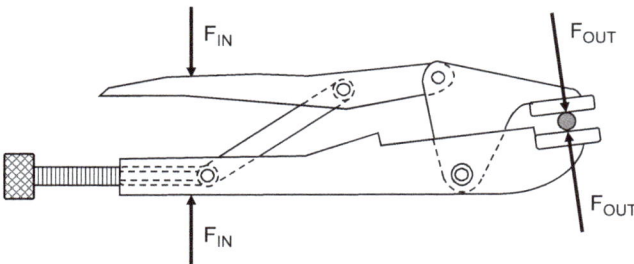

Figure 2.50 Toggle clamps function in much the same way as locking pliers. They have an over center cam action that provides the self-locking action. The clamp force is controlled using an adjusting screw that controls the amount of compression force applied to the jaws.

Injection Molding Machinery and Auxiliary Equipment

Toggle clamps operate in much the same way as a pair of locking pliers shown in Figure 2.50. The jaws of the pliers are analogous to the moving and stationary platens of the molding machine. The squeezing of the hand grips on the locking pliers is analogous to the hydraulic cylinder (or electric motor) that energizes the toggle linkage. The cam-adjusting screw controls the magnitude of the clamp force. Once clamped, both the pliers and a toggle injection molding clamp are self-locking. Toggle clamp units do not consume energy once they have been locked. The clamp closing action is reversed once the injection mold cooling time has expired.

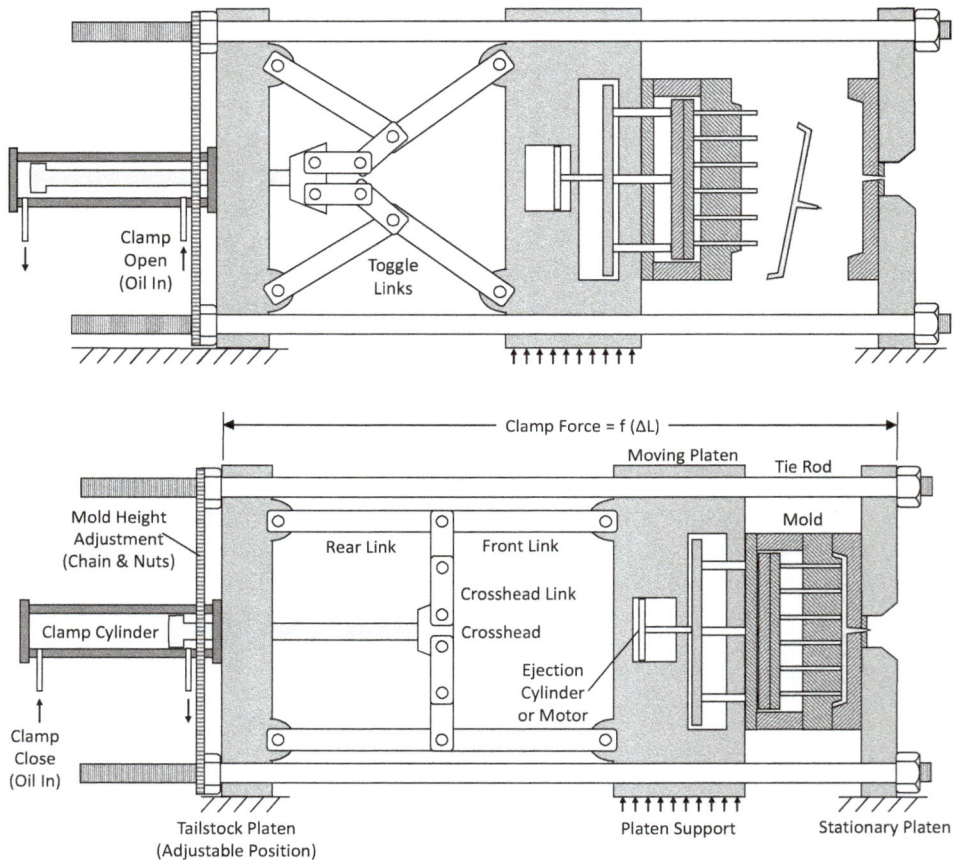

Figure 2.51 Toggle-molding machine clamp systems offer a number of advantages including fast traverse speeds and relatively low energy consumption. Toggle clamps have an over center cam action that makes them self-locking. The activating hydraulic cylinder or electric motor is energized only during the mold closing and mold opening phases of the molding process. The injection molding machine clamp safety gates are not shown.

Figure 2.51 shows a molding machine toggle clamp unit in both the open and closed positions. The clamp unit has three platens and four tie bars. As with a direct acting hydraulic clamp, the stationary platen is used to locate and support the A-half of the injection mold. The B-half of the mold is supported by the moving platen. The motion of the moving platen is activated by the toggle clamp system's linkage. The moving platen also incorporates a machine ejection system that connects to the injection mold's ejector system using knock out rods. The third platen is the clamp's adjustable tailstock. The tailstock's axial position can be moved forward or backward depending on the stack height (overall thickness) of the injection mold. Fine-tuning the tailstock axial position will determine the magnitude of the clamp force in the same way the adjusting screw on the locking pliers controls the plier jaw clamp force. This clamp force adjustment is typically accomplished using a motorized chain and gear system. The tailstock position is critical as its position ultimately determines the amount of tie bar stretch and the magnitude of the clamp force. If the tailstock is positioned such that the faces of the mold just touch, there is little or no clamp force. Moving the tailstock forward (while the mold is open) will increase clamp force when the mold closes by causing a degree of tie bar strain when the mold is closed. The distance between the stationary platen and the tailstock platen determines the amount of tie bar strain and therefore clamp tonnage. Toggle clamps are self-locking and maintain tonnage until the clamp activation is reversed.

Toggle clamps do tend to require more maintenance than hydraulic clamps due to their larger number of moving parts and stressed components. Directly acting hydraulic clamp units also tend to be easier to set up, and they do provide a direct indication of the clamp force they generate as their clamp force is directly proportional to the applied hydraulic pressure. Toggle clamp units have an indirect clamping mechanism and offer a significant mechanical advantage. Modern injection molding machines may provide a calibrated indication of clamp force, but tie rod strain sensors are sometimes used to gain a better indication of actual clamp force for toggle injection-molding machine clamps.

2.2.3.4 Tie-Bar-Less Mold Clamping Units

Another class of injection molding machine clamp units are tie-bar-less clamps (or C-frame clamps). These clamps simplify mold changes by completely eliminating the use of conventional clamp tie bars and the interference that they normally cause. The tie bar elimination allows for maximum flexibility with respect to injection mold size and mold complexity in terms of mold side actions, core pulls and cooling lines. The key to tie-bar-less molding machine clamp design is maintaining parallelism for the machine platens and injection mold parting line, so that there is not a tendency to develop flash along the injection mold's parting line. This is typically accomplished by having a very robust and stiff C-frame that resists the deformation caused by high mold cavity pressures. Some tie-bar-less molding machine clamp designs incorporate dynamic counter hydraulic cylinders that compensate for C-frame platen deflection [1]. Tie-bar-less molding machines are generally limited to lower capacity molding machine clamps but their popularity is expected to increase as their benefits are numerous.

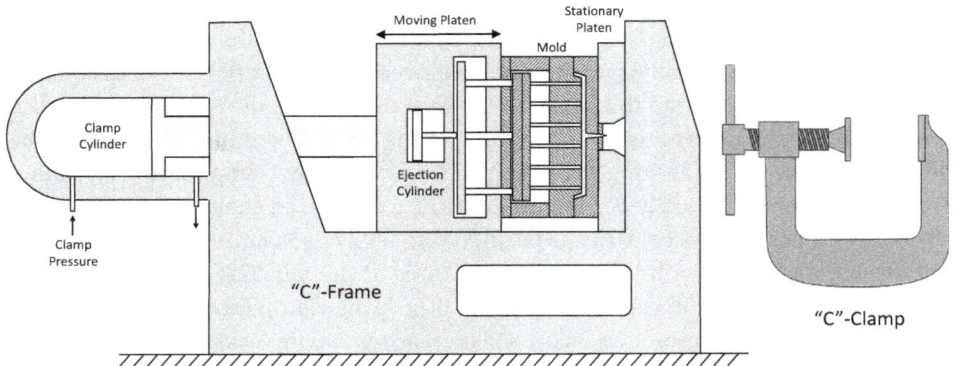

Figure 2.52 Tie-bar-less injection molding machine clamping systems have a very rigid C-frame that has no tie bar interference. This provides for maximum flexibility with respect to mold width and height, and more space for special injection mold side actions. The injection molding machine clamp safety gates are not shown.

2.2.3.5 Injection Mold Lifting Hardware

The clamp unit of an injection molding machine includes platens that must locate and support the two halves of the injection mold being used for production. Injection molds are subject to a very wide variety of forces over the course of the injection molding cycle. These forces include clamp force, internal melt pressure, thermal stresses, gravity, ejection forces, and leader pin friction during mold opening and mold closing. Ensuring that the injection mold is firmly clamped to both the stationary and the moving molding machine platens is critical in terms of proper mold alignment and worker safety. Injection mold changes represent one of the more critical and potentially dangerous activities associated with the injection molding process.

2.2.3.5.1 Mold Safety Straps

Before an injection mold is installed into the molding machine, it is important to inspect the tool to ensure that the injection mold halves are held securely closed so that the mold halves do not accidently separate when the mold is transported or lifted during mold installation. This is normally accomplished using one or more mold safety straps that span the mold's parting line(s) as shown in Figure 2.53. The safety straps should span all movable plates including stripper plates. All injection molds should have properly designed safety straps. It is also good practice for the safety straps to include a designation that is the same as the designation of the injection mold that it is used with (e.g., mold number), as shown in Figure 2.53. The safety straps should be robust and made from high quality steel and held in place with high quality bolts (e.g., grade 8 bolts). Once the mold has been clamped into the molding press, the safety strap *must* be removed and stored in the mold's accessory case or remain attached to the mold (but not spanning the parting line). It should be noted here that the use of nylon (or other thermoplastic) cable ties as an alternative to steel safety straps is *unacceptable*, even for small injection molds. Worse yet is the all-too-common practice of using thermoplastic cable ties and brass water fittings to keep an injection mold closed when it is being transported or changed. Neither the brass water fitting nor the thermoplastic cable tie shown in Figure 2.53 were designed for such an ill-advised purpose.

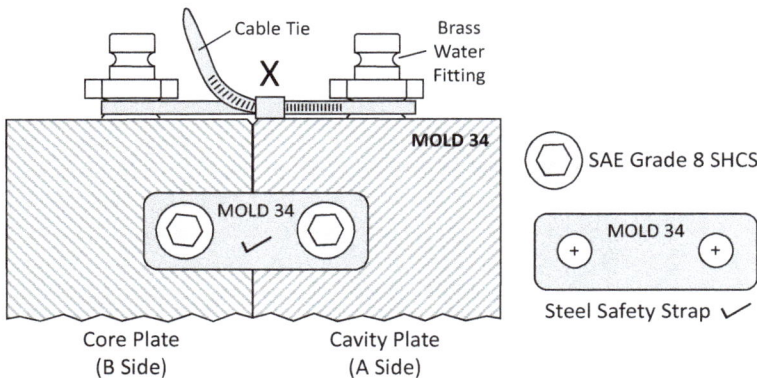

Figure 2.53 Injection molds *must* be held closed with steel safety straps during storage, transportation and mold installation to prevent accidental opening of the mold halves. The use of nylon cable ties secured around brass water fittings is *NOT* an acceptable alternative to the use of steel mold safety straps.

2.2.3.5.2 Mold Lifting Eye-bolts

Most injection molds are installed in the molding machine clamp from above using an overhead crane. It is important to ensure that the crane operator has been trained and certified, and that the crane itself has an up-to-date certification for its lifting capacity that must be greater than the weight of the injection mold to be installed or removed from the molding press. The injection mold will normally incorporate one (or more) machine thread eye-bolts (or lifting bolts) threaded into the top face of the mold. There are many important variables and safety issues associated with lifting bolts. Some of the more common issues associated with injection mold lifting bolts are discussed in the following section.

General purpose (or non-shouldered) eye-bolts should *not* be used as injection mold lifting bolts under any circumstances for a variety of reasons. First, the strength of a non-shouldered eye-bolt is a strong function of the lifting angle. The eye-bolt is strongest if the pull is straight upwards, but the eye-bolt strength is compromised when the eye-bolt is subjected to side forces as shown in Figure 2.54. Side forces can occur in situations where (i) the eye-bolt is off center, (ii) when multiple eye-bolts are used together with a chain, or (iii) when the injection mold is lifted from a horizontal position to a vertical position. These side forces and cause the eye-bolt to bend or even fail due to the bending moment as shown in Figure 2.54. Non-shouldered eye-bolts also provide no visual indication as to the number of threads (or number of turns) of engagement. An eye-bolt (of correct diameter and strength) should have a minimum engagement of 1.5 times the eye-bolt diameter. Depth of engagement is not readily obvious or easily determined when non-shouldered eye-bolts are utilized as injection mold lifting bolts.

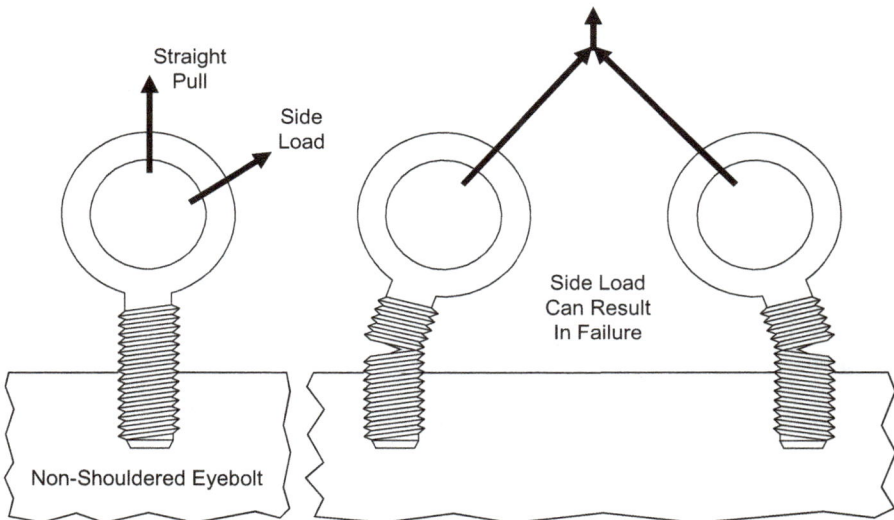

Figure 2.54 General purpose (or non-shouldered) eye-bolts should *not* be used as injection mold lifting bolts. The lifting capacity of these eye-bolts is greatly reduced if the lift involves side loading. In addition, there is no visible or obvious indication as to the depth of engagement (number of engaged threads or turns) for non-shouldered eyebolts.

Shouldered eye-bolts are more suitable for use as injection mold lifting bolts. There are many types and sizes of shouldered eyebolts available commercially. In the simplest case, the eyebolt is designed with a shoulder where the eye and the shank come together as shown in Figure 2.55 (left). The injection mold's weight must be known in order to determine the appropriate shouldered eyebolt diameter. It is good practice to consult with the eyebolt manufacturer when specifying eyebolt type and size since variables such as lift angle can still have an impact on the shouldered eyebolt's strength and therefore its safety factor for the application. It is also good practice to communicate with the tool designer during the initial injection mold qualification to determine what methodology the tool designer used to ensure the eyebolt size is appropriate for the injection mold's weight.

Shouldered eyebolts, such as that shown in are designed to have a specific and self-limiting engagement depth. When fully engaged, the eyebolt shoulder should seat tightly against the top face of the injection mold. The shouldered eyebolt should ideally be positioned above the center of gravity for the injection mold so that the mold remains level when lifted with the crane. The shoulder design minimizes the bending stresses that act on the eyebolt's shank and allows the eyebolt to be used even if side loads are present. While shouldered eyebolts are an improvement over non-shouldered eyebolts as injection mold lifting bolts, there are better alternatives. The multi-piece eyebolt, as shown in Figure 2.55 (right), has a shoulder, and an eye that is capable of both tilt and rotation. Shouldered swivel eyebolts with the appropriate lifting capacity are most suitable as injection mold lifting bolts.

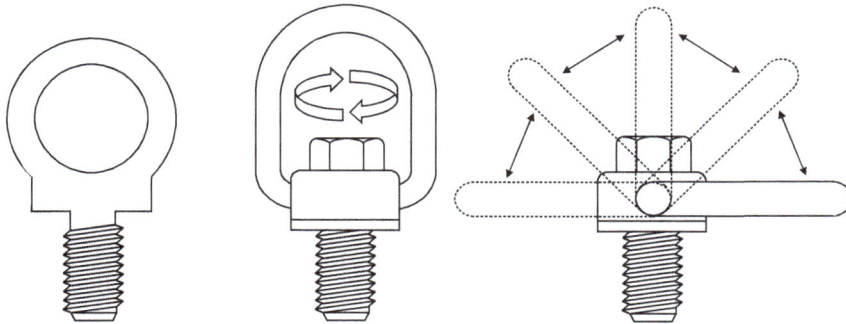

Shouldered Eyebolt   Rotating Eyebolt (360° Rotation, 180° Tilt)

Figure 2.55 Shouldered injection mold lifting eyebolts engage with the threaded receiving hole in the mold until the bolt's shoulder seats tightly against the mold's top surface. This minimizes the potential for damage from side loading and ensures full thread engagement. Shouldered lifting eyebolts having both rotation and tilt capability are most appropriate as mold lifting bolts.

While mold lifting bolts are typically threaded directly into the mold's cavity or core plate (or retaining plate), they can also be used in conjunction with an injection mold lifting bar as shown in Figure 2.56. The lifting bar is a solid, robust steel bar that is threaded to receive the mold lifting bolt. The lifting bar also has clearance holes that are aligned with threaded bolt holes in the mold so that bolts can be used to attach the lifting bar to the mold. The lifting bar allows the lifting bolt to be placed at any location along the bar without interference so that the lifting bolt can be placed directly above the mold's center of gravity thereby keeping the mold horizontal when it is lifted. This is not always possible when the eyebolt is threaded directly into the mold itself as there may be other mold components that interfere with the lifting bolt's placement. The lifting bar can also serve as the mold's safety strap as long as it spans the parting line of the injection mold.

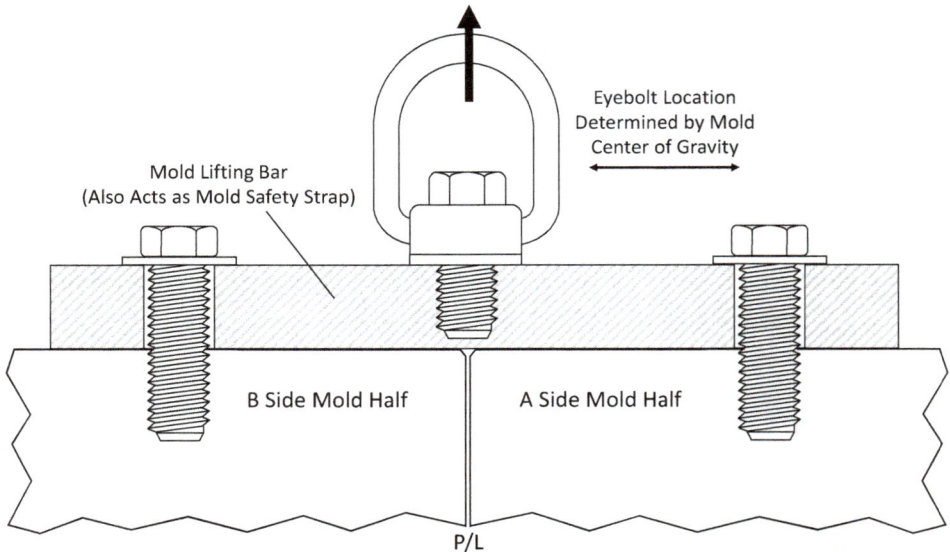

Figure 2.56 Mold lifting bolts can be used in conjunction with a mold lifting bar. The lifting bar allows the lifting bolt to be placed directly above the mold's center of gravity without interference. The mold lifting bar can also serve as the mold's safety strap as long as it spans the mold's parting line.

2.2.3.6 Injection Mold Mounting Clamps

The injection mold must be securely and safely mounted to the injection molding machine platens before molding can begin. Mold mounting is usually accomplished using mold clamps, although directly bolting the mold halves to the molding machine platens is also an option as shown in Figure 2.57. Direct bolting, using only bolts and washers, offers a number of advantages including added mold slip safety, but this method of mold mounting is only possible when the mold's clamping bolt pattern aligns with the molding machine's platen bolt pattern. Molds that are mounted directly may also require oversized clamp plates in order to provide space for the through holes or integral clamp slots as shown in Figure 2.57 (left). Molds that are bolted directly to the machine platens typically have at least four clamp bolt locations for each mold half.

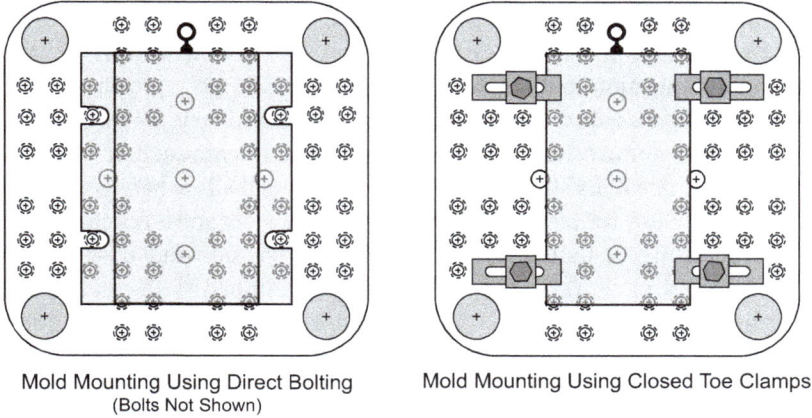

Mold Mounting Using Direct Bolting
(Bolts Not Shown)

Mold Mounting Using Closed Toe Clamps

Figure 2.57 Injection mold halves can be mounted to the molding machine platens by direct bolting or using mold clamps. Direct bolting offers a safety advantage in terms of mold slip, but this method of mold mounting is only possible when the mold's clamp bolt pattern aligns with the molding machine's platen bolt pattern. Slotted mold toe clamps allow for a greater degree of platen bolt pattern flexibility.

While direct mold mounting is an option, most injection molds are mounted using mold clamps. The use of mold clamps for mold mounting allows for more flexibility in terms of platen bolt pattern compatibility for a given mold geometry. Unlike direct bolting of an injection mold, the use of mold clamps allows for greater variability in the molding platen's bolt pattern. For example, the closed-toe manual mold clamps shown in Figure 2.57 have slots that provide for greater flexibility with respect to clamp bolt location. The use of mold clamps can also be advantageous in terms of maximum possible mold size (which is limited by tie bar spacing) since the mold can have recessed clamp slots instead of an oversize mold clamp plate typically used for direct bolting. Manual mold clamps are most common for mounting injection molds, although magnetic mold mounting systems and hydraulic mold clamps are gaining popularity as their use reduces mold changeover times. These more automated and advanced mold clamping systems can require clamp plate standardization and require additional capital expenditure.

The recommended procedures and best practices associated with injection mold changes have been documented in a recent publication [28]. Several of these recommended practices involve mold clamp installation procedures. The basic concepts associated with the use of manual injection mold clamps are discussed here. To begin, the injection mold platens and threaded bolt holes should be visually inspected prior to each and every mold change. Any raised areas on the platen faces should be stoned smooth and any debris removed by vacuum. Platens are typically sprayed or wiped with light oil or rust preventative prior to injection mold installation in order to minimize corrosion issues. Excess oil should be wiped off with a clean rag. Molds are normally installed using a crane as one unit that is held closed with safety straps, onto the stationary platen first. The mold's locating ring centers the mold within the stationary platen and the injection mold is then leveled. Mold clamps, typically closed-toe mold clamps, are then used to clamp the mold to the stationary platen. Course thread grade 8 bolts and no-turn washers are typically used in conjunction with the closed-toe mold clamps. The bolts should have a length that will allow between 1.5 to 2.0 times the thread diameter for proper thread engagement. A torque wrench should be used to tighten all of the clamp bolts to the appropriate torque value based upon bolt thread diameter and pitch. Once the mold has been clamped to the stationary platen, mold installation continues using the injection molding machine's mold set mode, where the moving platen is brought in contact with the B-side clamp plate for moving platen mold clamp installation, along with the ejector rod installations. Once the mold is safely clamped to both the stationary and moving molding machine platens, the crane and mold safety strap(s) can be removed.

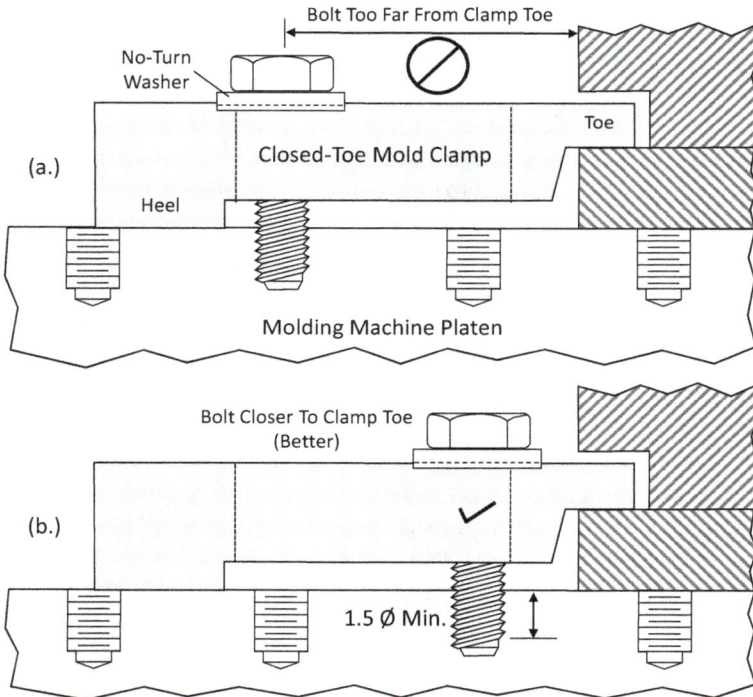

Figure 2.58 The mold clamp bolts should be located as close to the injection mold as possible to provide for maximum possible mold clamping force.

The clamp bolts should be installed in platen holes that are located as close as possible to the injection mold (and the toe end of the clamp) for maximum mold clamping force as shown in Figure 2.58. The mold clamps should also be installed such that the acting face of the closed-toe mold clamp is parallel with the molding machine platen and the recessed mold's clamp slot as shown in Figure 2.59. The angle of inclination θ should be zero degrees. Clamps that include a leveling bolt or height adjustment screw can be used to ensure that the mold clamp installation is parallel with the machine platen, regardless of clamp plate thickness. It is also important that the mold clamp bolts do not bottom out within the platen's threaded holes so that the bolt head is able to apply clamp force to the washer and clamp. It is helpful to tap the mold clamp with a dead blow hammer after torquing the clamp bolt to ensure that the clamp assembly is not loose (as a double check to ensure the clamp bolt has not bottomed out). *SmartBolts*® have also been developed and can be used with injection mold clamps to provide a visual indication that the bolt torques firmly against the washer and mold clamp, and that the bolt has not bottomed out in the threaded hole [29].

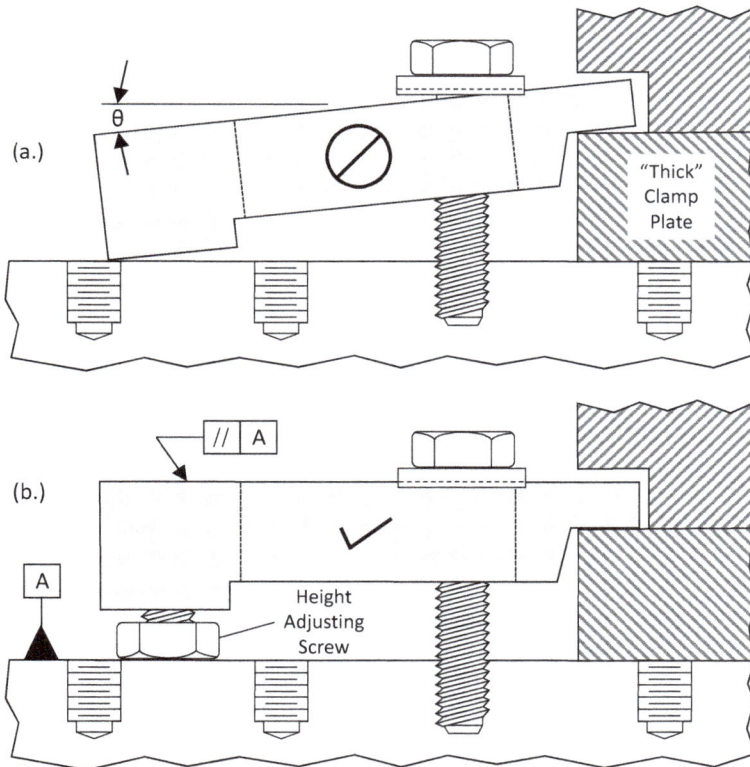

Figure 2.59 The active face of the mold clamp toe should be parallel with the machine platen and the mold's clamp plate slot. Mold clamps containing leveling screws or heels are used to compensate for different clamp plate thicknesses so that parallel clamp face installation is always possible.

In recent years there has been a trend towards the use of closed-toe arched injection mold clamps similar to those shown in Figure 2.60 [30]. Arched clamps are forged mold clamps that contain an integral, sliding washer that is interlocked with the mold clamp. The clamp can be used to clamp injection molds with mold clamping plates of different thicknesses *without* the need for shims, washers or an adjustable heel bolt. The arched clamps are essentially self-adjusting in that respect, which helps to simplify and reduce mold change time and improve injection mold clamping reliability.

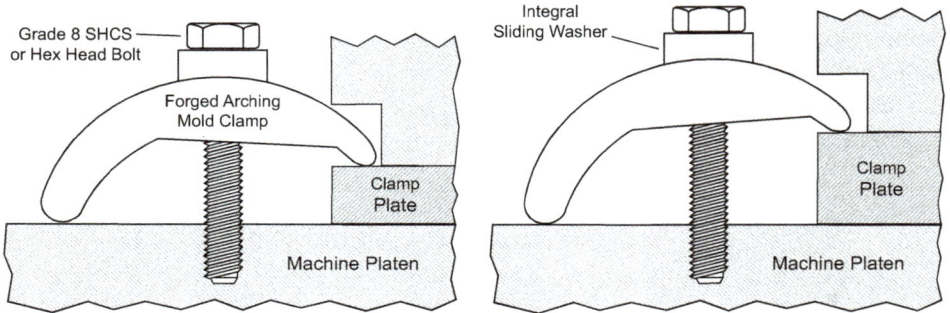

Figure 2.60 The use of closed-toe, arched mold clamps containing an integral sliding washer has increased in recent years. The clamps are essentially self-adjusting with respect to different mold clamp plate thicknesses without the need for shims or an adjustable heel bolt, thereby helping to reduce mold changeover times.

The number of mold clamps required for an injection mold depends on many factors, especially the injection mold's physical size and mold weight. For example, it is very common practice to use at least four mold clamps per mold half for smaller size molds. When four mold clamps are used, a bolt tightening order shown in Figure 2.61 has been recommended. As mold size increases, the recommended number of mold clamps increases to six per mold half and up to twelve clamps per mold half for even larger molds [31]. The clamp bolts should all be tightened using a click-type torque wrength with a torque setting that is based on bolt diameter and pitch [28].

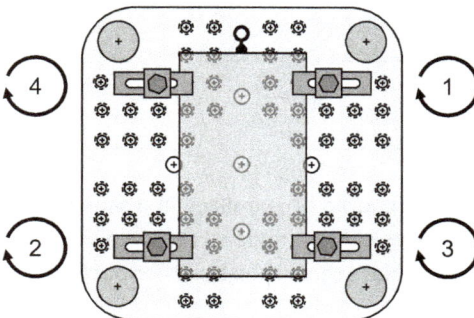

Figure 2.61
Consider the tightening order when torquing injection mold clamp bolts. One source suggests the tightening order shown above when using four mold clamps per mold half [31].

2.2.3.7 Mold Insulation Sheets

The mold temperatures used for the thermoplastic injection molding processes are cool relative to the injection unit's barrel temperature. As a result, conductive heat transfer from the injection mold halves to the room temperature molding machine platens is generally not a major concern. However, the recommended mold temperatures required when molding some engineering thermoplastics or high-performance thermoplastics can be in the 100 °C to 200 °C range, leading to concerns over conductive heat transfer from the injection mold to the machine platens. This can also be a concern for the A-side of many hot runner molds. In such cases, reducing conductive heat transfer from the injection mold to the machine platens saves energy, reduces the start-up time (to reach steady state conditions), and helps to extend the life of some molding machine components. Mold insulation sheets or plates can be used to minimize the mold-to-platen heat loss, as shown in Figure 2.62. The use of mold insulation sheets can also result in a more uniform mold temperature [32].

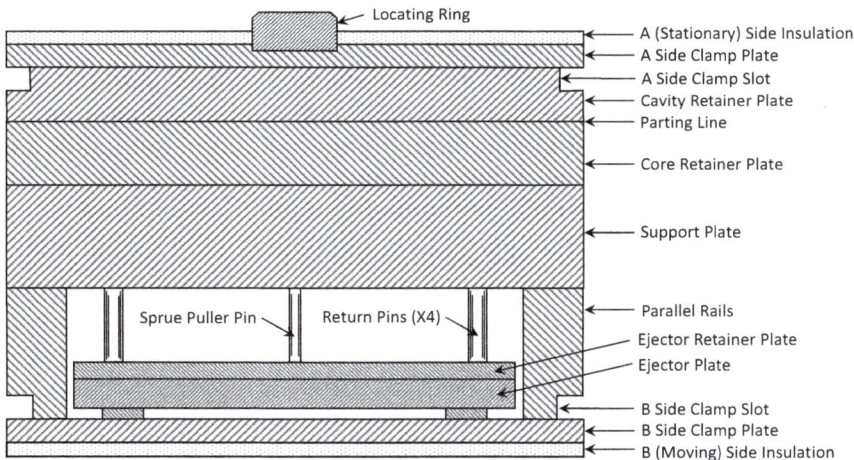

Figure 2.62 Low thermal conductivity mold insulation sheets or plates can be used to reduce conductive heat transfer from a warm mold to the cooler molding machine platens. The insulating plates are most commonly used when molding high-performance thermoplastics that require relatively high mold temperatures. Insulating sheets are also commonly used for hot runner injection molds.

Mold insulation sheets are typically bolted to the front and rear clamp plates of the injection mold. The insulation plates are typically cut from larger, rigid insulating sheets to match the overall x–y dimensions of the injection clamp plates. Mold insulation sheets are available in a wide variety of materials, ranging from fiber reinforced thermoset plastic sheets to Portland cement-based products [33]. The key common property for all mold insulation sheets is their relatively low thermal conductivity value. Unfortunately, most of these mold insulation sheet materials are relatively soft and are easily scratched or damaged (compared to steel) so they should be handled with care. The primary variable used for insulating sheet material selection is the insulating material's maximum recommended use temperature. The insulating sheet's flatness and flatness tolerances are also critical so that the mold parting line remains parallel.

Over time, the insulating sheets can compress due to the repeated compressive loading that occurs each molding cycle. This may not be a concern with directly acting hydraulic clamp units, but it can be a concern for molding machines with a toggle clamp. Toggle clamp units may require recalibration of the low-pressure mold protection settings or clamp tonnage settings for long production runs in which mold insulating sheets are utilized.

2.2.3.8 Single Minute Exchange of Die (SMED) Carts

Injection mold changes are a routine aspect of most injection molding operations. The frequency of the mold changes is limited for a dedicated injection molding process, but changes can be very frequent for many custom molding operations, particularly in molding applications involving just-in-time manufacturing or short production runs. Reducing injection mold changeover times can have a dramatic effect on profitability since the time between production runs is one form of downtime. One philosophy that has been developed and applied to reduce injection mold changeover times is known as the "Single Minute Exchange of Die" (SMED) philosophy. While the single minute descriptor is over-ambitious for injection mold changeovers, significant mold changeover time savings can be achieved if some very basic SMED guidelines are implemented [34,35].

Injection mold changeover times are influenced by a huge number of actions or elements that vary with mold size and complexity, technician training, plastic material purging, and the like. The goal of SMED is to both (i) minimize the number of required actions and (ii) simplify the required actions, all done in an effort to promote efficiency and save time. These actions or elements are further categorized as either internal elements or external elements. Internal elements are activities that occur when the machine is not in production (e. g., installation of a mold clamp). External elements are parallel activities that can be performed while the injection molding machine is still in production mode (e. g., preheating the mold to be installed). The SMED concepts are relatively simple in theory, but implementation can be quite difficult and requires very careful planning. For example, it is obvious that preheating a hot runner mold before the injection mold is installed in the press would save time compared to preheating the mold after installation. However, if the mold were preheated, and then there were some sort of mold installation malfunction that caused significant downtime, the material in the hot runner could become overheated and degraded, leading to even more significant delays. Contingencies of this type must be taken into account so that proper corrective action can be taken. Preheating a hot runner mold could also lead to a number of worker safety issues that must be taken into account. For example, additional personal protective equipment (e. g., heat-resistant gloves) is required when handling hot, electrically-heated injection molds.

Another goal of the SMED philosophy is to maximize the number of external elements that can be performed while the molding machine is in production, and to simplify or streamline the internal elements. Again, these actions are all implemented in an effort to reduce mold changeover times while maintaining safety protocols. Keys to accomplishing these goals are careful planning, workforce training, standardization, and automation in some cases. For example, the use of hydraulic mold clamps or magnetic platens will result a time savings compared to fully manual mold clamp installation, but such automation requires additional capital expenditure.

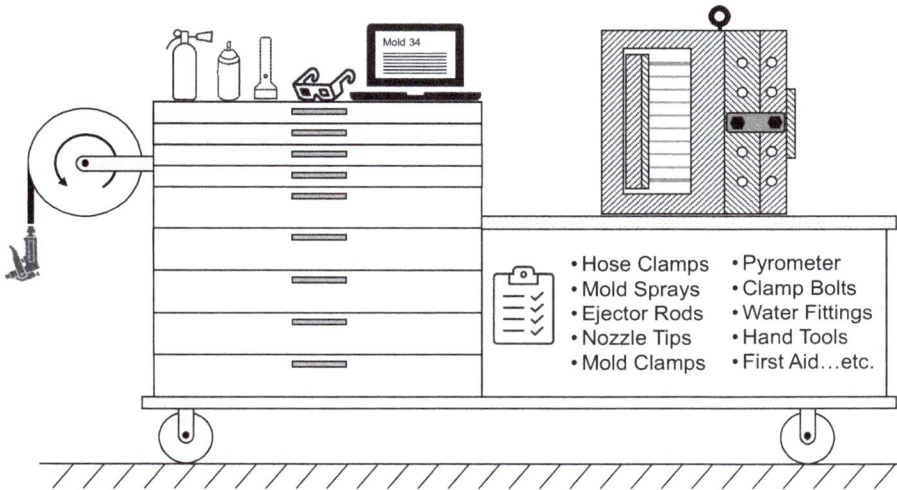

Figure 2.63 SMED carts are portable carts that contain the tools, supplies and hardware required to change an injection mold.

SMED carts can also play a very important role in reducing injection mold changeover times [34,35]. SMED carts vary in their design and construction, but the concept is to have an industrial quality portable cart that contains all of the tools, supplies and hardware required to change an injection mold. The idea is to eliminate back and forth trips to a central tool crib by physically bringing all of the necessary equipment to the molding cell. The inventory of tools and equipment required for a mold change can be minimized by implementing standardization as much as possible. Carts should be checked and re-stocked with supplies regularly since every cart item required for the mold change needs to be available on the cart. SMED carts can be a purchased item or they can be custom-built to meet the specific mold changing needs and practices of the injection molding shop. The carts should be robust with heavy duty casters and steel construction. Some SMED carts also include computer hardware so that the mold specific changeover procedures, mold change practices, and historical data, such as mold filling process simulation results, can be stored and displayed to assist with the mold change and subsequent process start-up.

2.3 Injection Molds

2.3.1 Injection Mold Functions

While many different types of injection mold are used for the production of thermo-plastic parts, all injection molds have certain features and functions in common. Figure 2.64 outlines some of the more important injection mold functions, starting with part formation and ending with ejection of the cooled and solidified molding. Injection molds must be capable of producing tight-tolerance thermoplastic parts reliably for thousands or millions of molding cycles. Injection molds must also be very robust since the temperatures, pressures and abrasive forces associated with the injection molding process are considerable. Tight mold tolerances, complex and robust mold construction, and long-term reliability all contribute to the relatively high cost of injection molds. The following sections will review some of the more important injection mold types and mold functions, but the reader is directed to other texts for more detailed information on the subjects of injection mold design and injection molded part design [36–39].

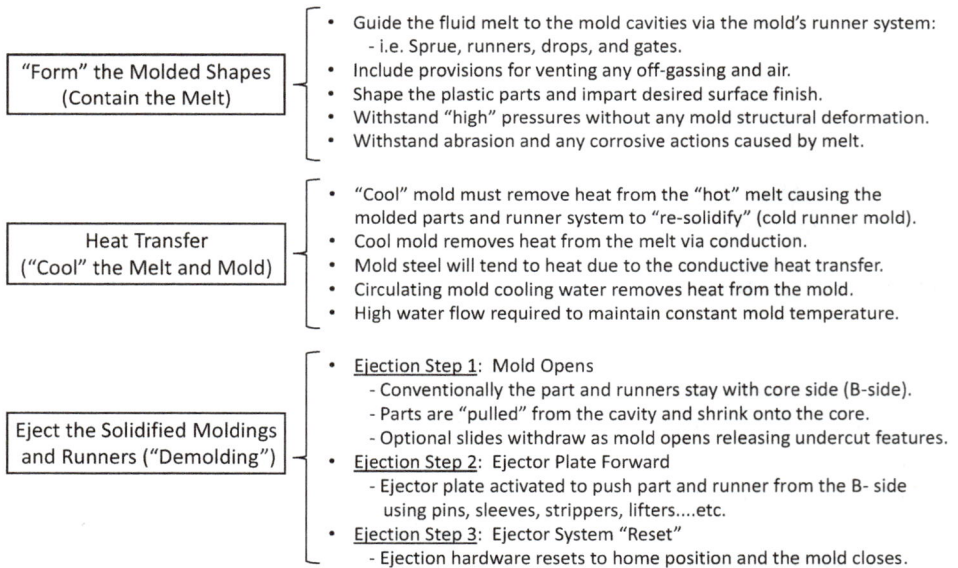

"Form" the Molded Shapes (Contain the Melt)

- Guide the fluid melt to the mold cavities via the mold's runner system:
  - i.e. Sprue, runners, drops, and gates.
- Include provisions for venting any off-gassing and air.
- Shape the plastic parts and impart desired surface finish.
- Withstand "high" pressures without any mold structural deformation.
- Withstand abrasion and any corrosive actions caused by melt.

Heat Transfer ("Cool" the Melt and Mold)

- "Cool" mold must remove heat from the "hot" melt causing the molded parts and runner system to "re-solidify" (cold runner mold).
- Cool mold removes heat from the melt via conduction.
- Mold steel will tend to heat due to the conductive heat transfer.
- Circulating mold cooling water removes heat from the mold.
- High water flow required to maintain constant mold temperature.

Eject the Solidified Moldings and Runners ("Demolding")

- Ejection Step 1: Mold Opens
  - Conventionally the part and runners stay with core side (B-side).
  - Parts are "pulled" from the cavity and shrink onto the core.
  - Optional slides withdraw as mold opens releasing undercut features.
- Ejection Step 2: Ejector Plate Forward
  - Ejector plate activated to push part and runner from the B- side using pins, sleeves, strippers, lifters....etc.
- Ejection Step 3: Ejector System "Reset"
  - Ejection hardware resets to home position and the mold closes.

Figure 2.64 Almost all thermoplastic injection molds have certain features and functions in common. Thermoplastic injection molds include features that allow for mold filling and part formation, heat transfer (i. e., mold cooling), and part ejection.

2.3.2 Cold Runner Injection Molds

Injection mold types can be characterized in variety of different ways, but it is common to describe injection molds based on the type of runner system they utilize. The injection mold depicted in Figure 2.65 is described as a multi-cavity, two-plate, cold runner injection mold. The mold's cold runner system provides the pathway through which the injected melt flows from the injection unit's nozzle to the mold cavities. With cold-runner molds, the mold's runner system cools, solidifies and is ejected along with the cooled injection molded parts. The ejected cold runner system is manufacturing scrap that is generated each molding cycle. The runners can be reground and blended with virgin material in applications that allow for the use of runner regrind. In other more critical applications, the use of regrind may not be sanctioned and the runner scrap may be discarded, sold or re-used in another injection molding application. The topic of regrind use is discussed in more detail in Section 1.7 of this text.

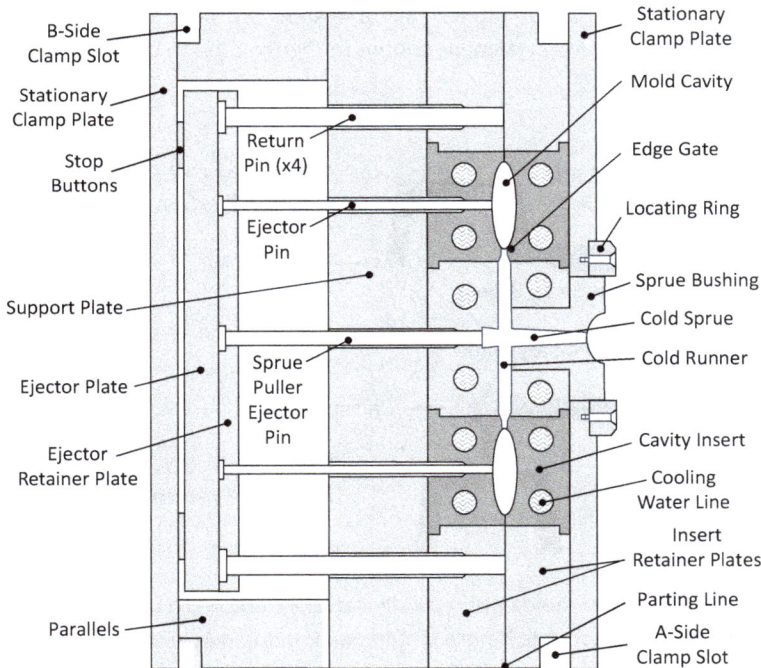

Figure 2.65 Simplified cross-sectional schematic of a multi-cavity, two-plate, cold runner injection mold. The molded parts are edge gated.

The cold-runner mold shown in Figure 2.65 is described as a two-plate cold runner mold because it has only one parting line where the stationary and moving halves

(A and B halves) of the injection mold separate. While a two-plate injection mold has more than two plates, there are only two plates (or inserts) in contact with the thermoplastic material. Two-plate molds do contain many other plates, including clamp plates, ejector plates and a support plate. Gating options with two-plate, multi-cavity molds include various forms of edge gates (edge gates, fan gates) or tunnel gates. Tunnel gates are also known as sub-gates. Two-plate molds have a single parting line through which both the solidified molded parts and the solidified runner system are ejected at the end of the injection molding cycle.

A single cavity two-plate injection mold is typically produced by machining the cavity and core geometry directly into the mold's cavity and core plates. Multi-cavity molds, such as that shown in Figure 2.65, are typically produced using tool steel cavity and core inserts. The use of inserts facilitates mold construction and allows for more options in terms of insert metallurgy and heat treating.

The cold-runner injection mold shown in Figure 2.65 also includes clamping plates, a support plate, cooling channels, vents, and a part (and runner) ejection system. The mold is typically designed such that the parts and runner will stick to the moving or B-side of the injection mold, so that the mold's ejector system can be activated by the injection molding machine's ejector system, as shown in Figure 2.43.

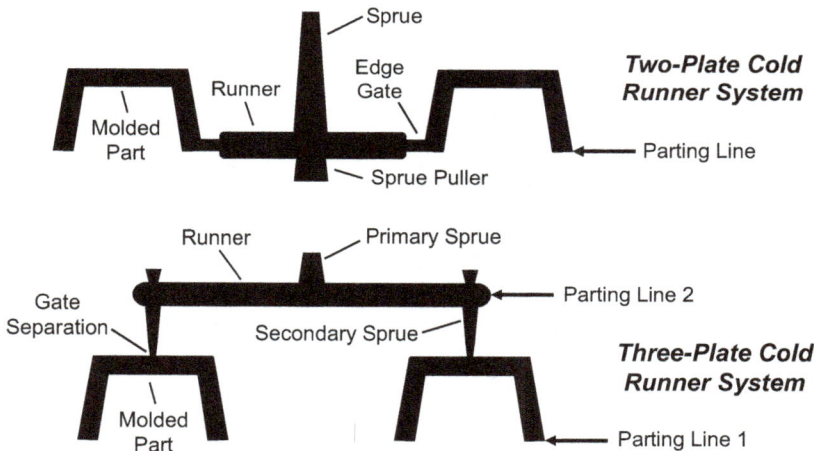

Figure 2.66 Cold runner injection molds are typically categorized based on their runner system design. Two-plate molds have a single parting line, while three-plate molds have two parting lines (one for the cold runner and one for the molded parts).

Other cold-runner injection molds are described as three-plate molds. Three-plate cold-runner injection molds have a more complex construction and are less common than two-plate cold runner injection molds. Three-plate molds have an additional mold plate and parting line that allows for top surface pin-point gating as shown in

Figure 2.66. The cold-runner system for three-a plate mold will have a primary sprue that feeds the cold runner, and reverse taper secondary sprues that transition to the pin-point gate geometry. The molded parts and the runner system separate automatically when the three-plate injection mold opens. The solidified moldings and the runner system each drop from their own parting lines when the mold opens.

2.3.2.1 Cold Runner Geometry

The cold-runner system of an injection mold serves as the conduit through which the thermoplastic melt is delivered to the injection mold cavities. Many different cold runner geometries and layouts are possible, and the decisions related to runner design will affect the injection mold's process window and molded part quality. Some of the factors that should be considered when designing a cold runner system include:

- Total runner volume - which determines the concentration of runner scrap.

- Total pressure drop to each mold cavity should not be excessive.

- Runner cooling time should be long enough to allow for sufficient holding time.

- The time required for runner ejection should be less than the molded part cooling time.

- Total pressure drop to each injection mold cavity should be equivalent.

One important cold runner variable is the cold-runner cross-sectional geometry. The three most common injection mold cold runner cross-sectional geometries are shown in Figure 2.67. These cross-sectional geometries include (i) the trapezoidal geometry, (ii) the modified trapezoidal geometry, and (iii) the full-round geometry. The full-round cold runner geometry is generally preferred for reasons that require some explanation.

A full-round cross-sectional geometry has the smallest possible ratio of circumference to cross-sectional area. This is beneficial in terms of fluid flow efficiency, but causes difficulties with heat transfer (i. e., mold cooling). Cylindrical runners have a geometry that is the most difficult geometry to cool due to their minimum surface area to volume ratio. While the cooling issue may seem like a disadvantage, it is not, since cooling of the melt as it flows through the cold-runner system is not desirable. It is best if the thermoplastic melt entering the mold cavity is at the same temperature as the melt in the molding machine barrel. Any cooling of the injected melt during mold filling will cause additional pressure drop and orientation issues. Once the melt is in the mold cavity, the runner will continue to cool, but more slowly than with the other cold runner geometries. It is worth noting that complete cooling of the runner is not required for most injection molding operations. The runner and sprue must be cool enough to be ejected, but runner distortion or warpage after ejection is typically not an issue since the runner will be reground. The full-round runner geometry is naturally drafted, which allows the runner system to be ejected while the runner is still warm. The only real disadvantage associated with full-round cold runners is that the runner system must be machined into both the A and B sides of the injection mold.

Both the trapezoidal and modified trapezoidal runner geometries shown in Figure 2.67 have greater surface area to volume ratios than a full-round runner, which results in a greater rate of cooling for these runner geometries compared to full-round runners. These geometries are simpler to machine since they are machined into only one mold plate. While full-round cold runners are preferred, the modified trapezoidal runner geometry also offers a very good balance of machinability, fluid flow characteristics, heat transfer characteristics, and ejectability.

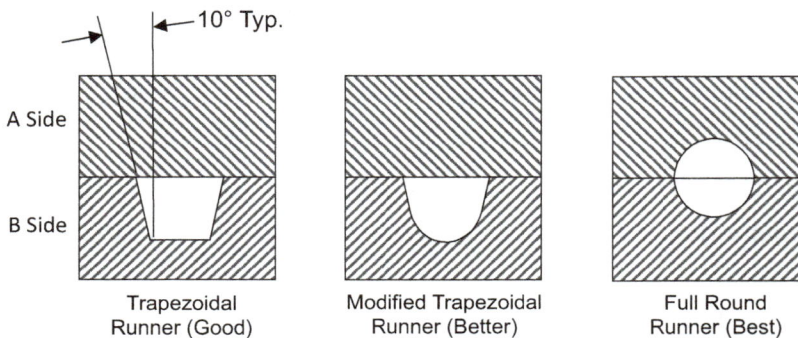

Figure 2.67 Common cold runner cross-sectional geometries include trapezoidal, modified trapezoidal and circular (full-round) geometries. These geometries are all well drafted to facilitate runner ejection. Full-round cold runners are preferred but require machining in both the A and B sides of the mold.

The magnitude of a cold runner's cross-sectional area is also an important runner design variable. Using larger cold runner cross-sectional areas will reduce the runner system's pressure drop during mold filling, but a larger cross-sectional geometry will also increase the amount of runner scrap that is generated in each injection molding cycle. These two opposing factors must be balanced when selecting the most appropriate runner cross-sectional area or runner diameter. One common cold runner design philosophy is to maximize runner diameter to reduce pressure drop and minimize runner length to reduce runner volume and runner scrap.

Cold runner cross-sectional areas should also change in proportion to the volumetric flow rate that the runner carries. Consider the geometrically balanced, full-round, H-style cold runner layout shown in Figure 2.68. The cold runner system includes a sprue (not shown), a primary runner, secondary runners, and tertiary runners. The volumetric flow rates associated with each runner section vary. For example, the primary runner has double the volumetric flow rate compared to the secondary runner. As such, the cross-sectional area of the primary runner should be greater than that of the secondary runner. As a first approximation, the cross-sectional area of the primary runner should be double that of the secondary runner. Likewise, the secondary runner should have a diameter that is greater than that of the tertiary runner. Direct proportionality is a good first approximation for the relative cross-sectional areas of the runners, but direct proportionality is an oversimplification since thermoplastics exhibit non-Newtonian flow behavior. Mold filling process simulations can be used to optimize the relative runner cross-sectional areas. Also note that the cold runner system shown in Figure 2.68 incorporates cold slug wells at the runner junctions. Cold slug wells do increase runner volume and therefore add to the amount of runner scrap, but they are beneficial in a few ways. Cold slug wells are small reservoirs or wells that are designed to trap solid or semi-solid defects that are associated with the melt flow front. Melt fronts tend to be colder and less homogeneous than the remainder of the melt flow stream. The use of cold slug wells can reduce cavity to cavity filling variations and improve molded part surface appearance.

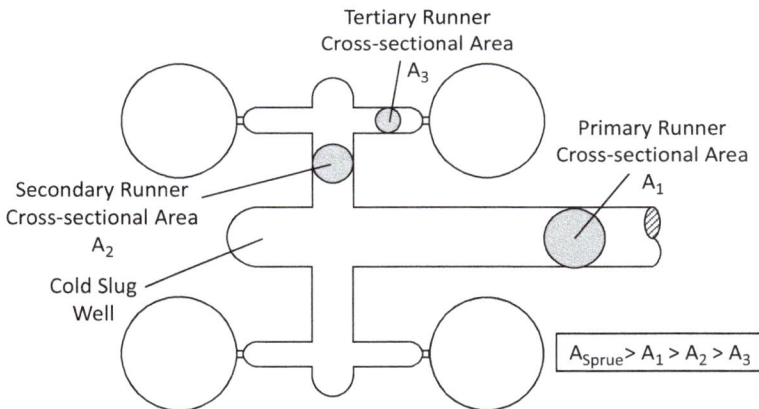

Figure 2.68 Many factors must be considered when designing the geometry of a cold runner system. These factors range from the amount of scrap generation to added mold filling pressure drop. One important consideration is to ensure that the runner cross-sectional area relates to the runner's volumetric flow rate. For example, the cross-sectional area of the primary runner shown in the figure is greater than that of the secondary runner since it carries double the volumetric flow rate compared to the secondary runner.

Injection mold runner system design is in many ways analogous to conventional plumbing system design. This applies to the runner cross sectional areas as described above and to juncture design issues. The cold runner system shown in Figure 2.68 is an H-style runner layout with a number of sharp 90° corners. These corners involve non-uniform shear effects and relatively large juncture losses in terms of pressure drop. Historically, H-style cold runner systems have been used because they are relatively easy to machine using simple X-Y machining tables. However, many other geometric runner layouts can be easily machined with today's CNC machining centers. The more streamlined primary to secondary runner transition shown in Figure 2.69 will have a reduced juncture loss in comparison to the H-style runner shown in Figure 2.68. Regardless of the cold runner design utilized, it is good practice to include a cold slug well just before the mold cavity entrance (i. e., just before the gate). A more detailed discussion of injection mold runner systems, juncture losses, and shear induced mold filling variations is given in [38].

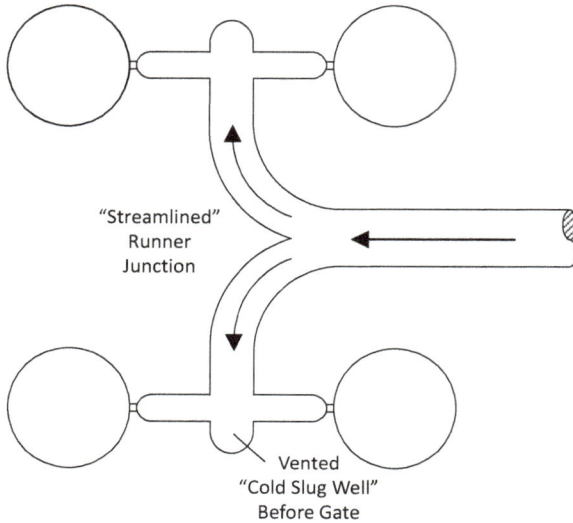

Figure 2.69
A variety of runner layouts are possible with multi-cavity, cold runner injection molds. While H-style runner layouts are common, CNC runner machining allows for the use of more streamlined runner system layouts that offer reduced juncture losses.

Injection mold runner system layouts should also be pressure balanced whenever possible. The multi-cavity mold runner system shown in Figure 2.69 is both geometrically and naturally pressure balanced. The total pressure drop associated with each mold cavity is identical since all cavities and runner branches have the same geometries. The total mold filling pressure drop for the mold is simply the sum of the individual pressure drops as indicated by Equation 2.2.

$$\Delta P_{Total} = \Delta P_{Sprue} + \Delta P_{1^\circ Runner} + \Delta P_{2^\circ Runner} + \Delta P_{3^\circ Runner} + \Delta P_{Gate} + \Delta P_{Cavity} \quad (2.2)$$

Naturally balanced multi-cavity injection molds will fill uniformly and all mold cavities should have a uniform cavity pressure history. The uniform cavity pressure history should lead to the production of molded parts having very uniform dimensions. Some other multi-cavity injection mold cold runner layouts, such as a spoke runner layout, can also be naturally balanced. While naturally balanced cold runner layouts are generally preferred in terms of part quality, balanced runner layouts typically have a greater runner volume than unbalanced runner layouts.

The injection mold runner systems shown in Figure 2.70 are *not* naturally or geometrically pressure balanced. The eight-cavity cold runner layout shown on the left in Figure 2.70 has a herringbone runner system layout. The primary runner feeding the outboard cavities has a greater length (and therefore greater pressure drop) than the primary runner feeding the inboard cavities. As a result, the total pressure drop for the outboard cavities is greater than that for the inboard cavities based on Equation 2.2. The four inboard cavities will likely fill first and have a greater cavity pressure history and less mold shrinkage than the four outboard cavities. The difference in dimension could be almost insignificant or very significant depending on the degree of imbalance and the application tolerances. While unbalanced herringbone runner designs do minimize cavity spacing requirements and runner volume (i.e., runner scrap), the molded part dimensions can vary.

The injection mold runner system layout shown to the right in Figure 2.70 is an artificially pressure balanced herringbone runner system. Injection mold runner systems can be artificially balanced by adjusting the lengths or diameters of the various runner branches. The artificially balanced runner shown in Figure 2.70 has been artificially balanced by reducing the diameter of the inboard secondary runners (compared to the outboard secondary runners). The inboard secondary runner geometry has been modified in an effort to equalize the total pressure drop for both the inboard and outboard mold cavities. The dimensional changes required to artificially runner balance a mold's runner system are normally determined using a mold filling simulation software program.

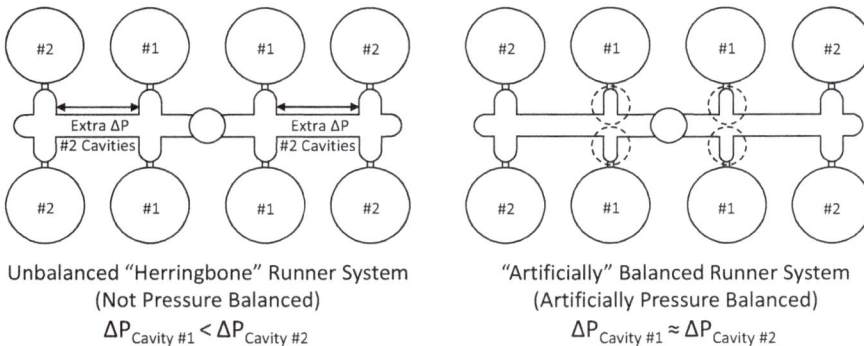

Unbalanced "Herringbone" Runner System
(Not Pressure Balanced)
$$\Delta P_{Cavity\ \#1} < \Delta P_{Cavity\ \#2}$$

"Artificially" Balanced Runner System
(Artificially Pressure Balanced)
$$\Delta P_{Cavity\ \#1} \approx \Delta P_{Cavity\ \#2}$$

Figure 2.70 Herringbone runner layouts are sometimes used for multi-cavity, cold runner injection molds since they are relatively easy to machine and minimize runner scrap compared to more traditional geometrically balanced runner system layouts. Unfortunately, herringbone runner layouts are not pressure balanced since the total runner length for the outboard mold cavities is greater than that for the inboard mold cavities. As a result, the inboard mold cavities tend to be overpacked relative to the outboard mold cavities. In such a case, the runner system can be artificially balanced by adjusting the relative diameters of the secondary runners.

While the use of naturally balanced multi-cavity injection molds is generally recommended, unbalanced and artificially balanced injection mold layouts are still common in some applications, especially when the degree of imbalance is small. While artificially pressure balanced runner systems are considered to be an improvement over unbalanced runner systems, they are not strictly equivalent to a naturally balanced runner system due to variations in shear history for the different runner branches. Artificial runner balancing is also common and strongly recommended for multi-cavity family molds that produce more than one part geometry. Family molds are neither geometrically or naturally balanced; this means that artificial runner balancing is required in order to create a more uniform fill pattern for family molds.

2.3.2.2 Injection Mold Vents

Injection molds must be vented so that the air within the mold's runner system and cavities does not impede the mold filling process. In addition, the flowing melt fronts can off-gas as the injection mold fills. If these gases are not properly vented, they can cause a variety of processing or part quality issues including;

- Internal voids
- Surface splay
- Gas traps
- Knit line issues
- Dieseling
- Short shots

There are two ways to vent injection molds. One method, known as vacuum venting, uses a vacuum system to evacuate the air from the mold runner and cavity just prior to the melt injection phase of the molding process. This method of venting is effective at removing the air from the mold runners and cavities, but the mold must be sealed to prevent air infiltration when the vacuum is drawn. Air infiltration can occur through the mold's parting line, insert gaps or ejector pin clearances. As a result, these areas of the mold must be sealed with gaskets or o-rings to prevent air infiltration. The complex construction associated with most injection molds limits the use of this mold venting technique.

Most injection molds are vented by allowing the thermoplastic melt to push the air and off-gases from the injection mold during the mold filling process. The injection mold runners and cavities incorporate shallow clearances through which the air and off-gasses can vent to the surrounding atmosphere as the mold fills. The most common injection mold vents are shallow edge vents located along the parting line of the mold as shown in Figure 2.71 and Figure 2.72. The edge vents are deep enough to allow for gas flow, but shallow enough to prevent the melt from flowing into the vent. Vent depths can be as shallow as 0.01 mm for low viscosity (high MFR) thermoplastics and as deep as 0.05 mm for very viscous thermoplastics. Edge vents are typically cut into the mold plates or inserts by surface grinding. Venting efficiency for edge vents can be improved by back grinding or machining a vent relief track that will minimize gas pressure drop as the gas flows through the vent channel as shown in Figure 2.71. Edge vents offer the added advantage of being easy to clean if they become clogged with condensed deposits from the melt front since the vents are fully accessible when the mold is open.

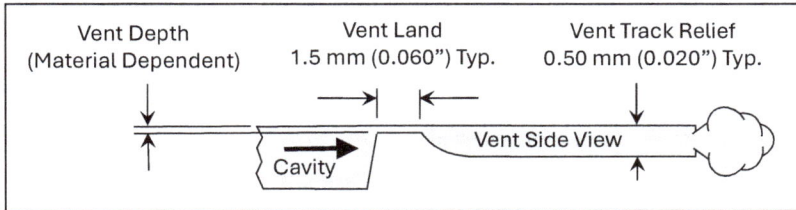

Figure 2.71 A number of different injection mold vent geometries are possible. Edge vents located along the parting line of the injection mold are most common. Edge vents are easy to machine or surface grind, and they are easily maintained (cleaned). Material suppliers will typically suggest recommended vent depths for the materials they supply, based primarily on the material's melt viscosity or melt flow rate. Vent relief tracks are added to reduce the overall vent pressure drop.

While edge vents are most common, other types of vents are also used to further enhance venting efficiency. As an example, some venting will occur naturally through the clearance gaps around an injection mold's ejector pins. Ejector pin venting efficiency can be enhanced by grinding shallow flats on the ejector pin as shown in Figure 2.72. Static venting pins or porous plugs can also be used as injection mold vents. These vents are typically located in recessed regions of the mold cavity where the gas may become trapped. Pin vents are more difficult to maintain or clean since the vent gaps are not accessible without at least some mold disassembly.

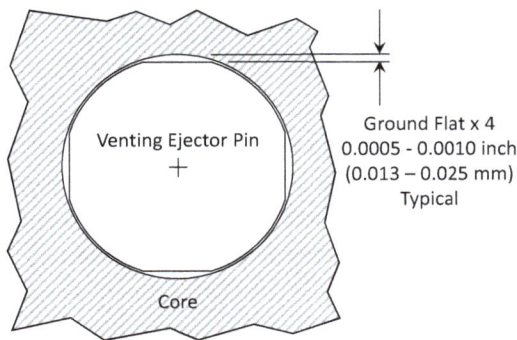

Figure 2.72 Ejector pin clearances also allow for some gas venting. Ejector pin venting efficiency can be greatly improved by grinding a few flats around the circumference of the ejector pin.

Injection molds should incorporate a sufficient number of vents to allow for unrestricted mold filling. The type, number, and locations of the vents are all important injection mold venting variables. One common rule regarding vent location is to place a vent at the end of flow (i. e., at the cavity location that is last to be filled) as shown in Figure 2.73 (left). While a vent at this location is certainly necessary, a single vent is inadequate for most injection molding applications. Vents should be located throughout the runner system and mold cavity in an effort to maximize venting efficiency and minimize mold filling resistance. Other rules of thumb suggest that up to 30% of the parting line perimeter should be vented or that there should be one edge vent for every 25–50 mm of parting line perimeter. Having multiple vents has many benefits and no real down-side other than the added machining cost. Adding vents will also result in an increase in compressive stress on the mold insert steel and should be considered when designing the amount of insert shut-off area.

Vents should also be strategically placed in specific areas of the injection mold. For example, runner system vents should be placed at cold slug well locations. Incorporating runner vents can help eliminate molded part surface splay by allowing off-gasses to be vented to the surrounding atmosphere before the melt enters the cavity. Vents should also be located near knit lines to help minimize the formation of gas traps and knit line depressions (i. e., V-notches). Mold filling simulation programs can be used to predict the locations where targeted venting may be required [39].

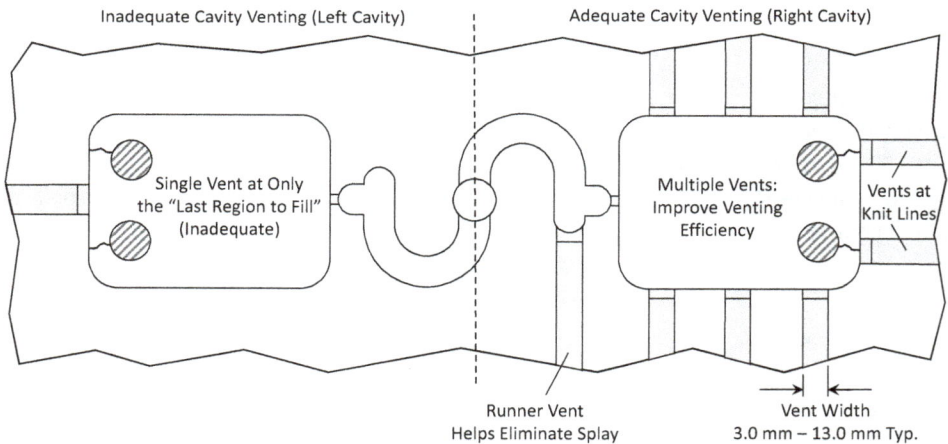

Figure 2.73 Injection mold cavities should incorporate multiple vents to minimize mold filling resistance and mold filling related defects. The mold's cold runner systems should also be vented, especially when the runner incorporates cold slug wells. Vents should also be located near areas of the molded part where knit lines or gas traps form.

2.3.2.3 **Injection Mold Cooling**

The injection mold's temperature is a very important process variable. Material suppliers will typically provide guidelines for both the minimum and maximum recommended mold temperatures for a given material grade. It is best to stay within the material supplier's recommended mold temperature range. Using a mold temperature below the minimum recommended value can result in mold filling difficulties and excessive residual molecular orientation. Using mold temperatures above the maximum recommended value will lead to an unnecessarily long cooling and cycle time.

Injection molds are typically cooled by circulating treated and temperature-controlled water through the mold's cooling channels or cooling circuits. Somewhat surprisingly, injection mold temperature is not always measured directly, although it can be. In most cases, the temperature of the mold is assumed to be the same as the temperature of the cooling water solution (or sometimes oil for higher mold temperatures) that is being circulated through the mold cooling channels. This may or may not be a good assumption. Mold cavity and core temperatures can and should always be confirmed using a portable surface pyrometer.

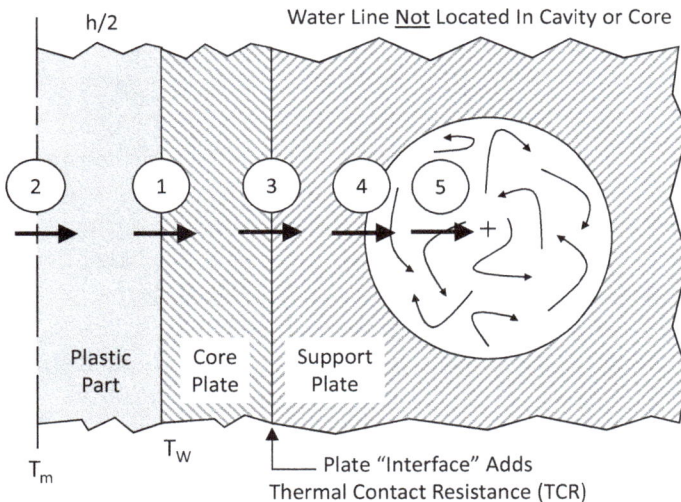

Figure 2.74 Injection mold temperatures are normally controlled by circulating treated cooling water through the injection mold's cooling channels. Whenever possible, it is best to design the mold cooling system such that the cooling channels are located within the mold cavity plates (or mold inserts) that are in contact with the thermoplastic material to be cooled. The cooling channel location shown here is not optimum, since the cooling channel is located in the support plate, not in the cavity plate. Heat transfer efficiency is reduced due to the added thermal contact resistance between the two mold plates (location 3).

There are several key factors that influence mold cooling efficiency. The first is the cooling system design or layout. Cooling channels should be located in such a way that the mold cavity surface temperature is uniform. Whenever possible, cooling channels should be located within the steel cavity and core inserts or plates that are in contact with the thermoplastic melt, as shown in Figure 2.65. This is not always possible, particularly for smaller size mold inserts. Cooling channel heat transfer efficiency is reduced if the cooling channels are not located within the cavity or core inserts (or cavity or core plates), as shown in Figure 2.74. The added thermal contact resistance between mold plates or inserts will reduce the rate of heat transfer. Mold temperatures will increase over the course of the molding cycle if the cooling fluid cannot remove the heat from the mold steel as fast as the hot melt can transfer heat to the mold steel.

The flow rate of the cooling fluid is also a critical mold cooling variable. The fluid flow within the cooling channels must be high enough to achieve turbulent flow. Turbulent flow ensures that the coolant will be well mixed (in terms of temperature mixing) so that hot fluid boundary layers do not form. A hot boundary layer will hinder the rate of heat transfer from the mold steel surrounding the cooling channel to the mold cooling channel. Volumetric flow meters are required in order to determine if the mold coolant flow rate is high enough to achieve turbulent flow.

2.3.3 Conventional Hot Runner Injection Molds

Most injection molds, such as that shown in Figure 2.65, are described as cold runner molds since the injection mold's runner system cools and solidifies along with the molded plastic parts. The cold runner system is a form of manufacturing scrap that may be reground and reused as regrind, or it may be discarded, sold or used in another injection molding application. Hot runner injection molds, such as that shown in Figure 2.75, eliminate the cold runner scrap and all of the issues associated with disposition of the cold runner scrap. Hot runner molds are most widely used in applications where production volumes are very high or for large part molding.

As the name suggests, the runner system of a hot runner injection mold is heated to a temperature that is similar to that of the molding machine's nozzle or the melt's temperature. Conceptually, a hot runner is essentially a sort of nozzle extension. Conventional hot runner systems are made up of several electrically heated components that are isolated from the cooled portions of the injection mold. The hot runner system has a heated sprue, a heated distribution manifold, and heated drops or nozzles that communicate with the mold cavities. The heated runner components are contained within the A-side of the injection mold, and isolated form the cold sections of the mold using air gaps, as shown in Figure 2.75. The heated components are also supported by low thermal conductivity stand-offs. The A-side of a hot runner injection mold usually includes an insulating plate to minimize heat loss to the stationary molding machine platen.

The most obvious advantage of a hot runner mold is the total or partial elimination of cold runner scrap. Partial elimination of runner scrap is associated with molds that have both hot and cold runner systems. In some applications, injection mold hot runner systems feed reduced-volume cold runner systems as a means of reducing the amount of cold runner scrap that is generated in each molding cycle. The added capital cost of the hot runner system and hot runner controls are typically justified for high production applications based solely on regrind elimination. While elimination of runner scrap is the primary advantage associated with a hot runner mold, hot runners offer many other secondary advantages. Unlike cold runner molds, hot runner molds allow for control of the melt's temperature all the way up to the mold cavity entrance. This typically allows for a number of process and part quality improvements. Secondary processing advantages associated with hot runners can include; gating flexibility, automatic gate separation, reduced mold filling pressures, lower clamp force requirements and reduced plastication time. Molded part quality advantages can include improved surface finish and improve knit line quality.

Hot runner molds are significantly more expensive than cold runner molds so these advantages must be balanced against the process and part quality advantages. Hot runner molds are also not appropriate for some shear-sensitive thermoplastics. In addition, hot runner molds may not be appropriate in applications where very frequent color or material changes are required.

Stringing or inconsistent gate vestiges can also be an issue with conventional hot runner injection molds as depicted in Figure 2.75. Most conventional hot runner molds have small pin gates that freeze as the molded parts cool. The molded part will separate from the hot runner when the mold opens at the end of the molding cycle. When this happens, some melt can be pulled from the hot drops (or hot nozzles) as the mold opens. These fiber-like strings are an imperfection or a defect. The fiber-like strings can cause mold parting line damage if they are caught between the mold faces when clamp tonnage is applied. The hot drop or nozzle temperatures can normally be adjusted to minimize or eliminate the stringing problem. The inconsistent gate vestige and stringing issues associated with conventional hot runner molds are eliminated when a valve-gated hot runner injection mold is utilized. An overview of valve-gated hot runner injection molds is given in the next section.

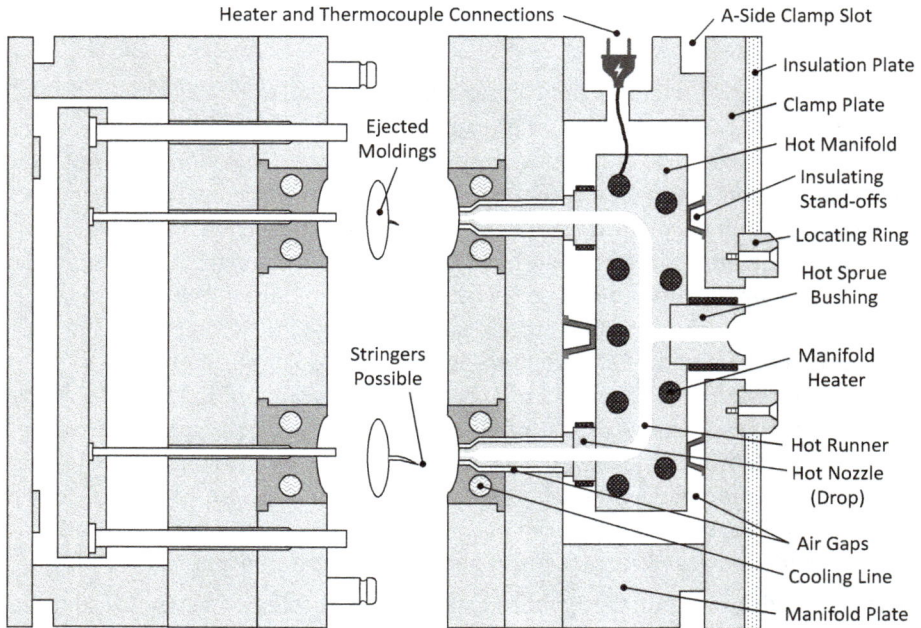

Figure 2.75 Hot runner injection molds offer many advantages over cold runner molds. Elimination of cold runner scrap is the most obvious advantage, especially for high production applications. While hot runner molds are more expensive, the heated runner system allows for improved control of the melt's temperature all the way to the mold cavity gate. This added melt temperature control capability can be used to fine-tune molded part quality (e. g., molded part surface finish, knit line quality). Hot nozzle temperature control is critical with conventional hot runner molds to avoid the formation of strings or stringers when the injection mold opens.

2.3.4 Valve-Gated Hot Runner Injection Molds

Valve-gated hot runner injection molds are growing in popularity because they offer all of the benefits of a conventional hot runner injection mold, along with a number of additional advantages over conventional hot runner molds. A simplified schematic of a typical multi-cavity, valve-gated hot runner injection mold is shown in Figure 2.76. Like the conventional hot runner mold shown in Figure 2.75, valve-gated hot runner injection molds have a heated nozzle, a heated manifold, and heated nozzle drops. However, the hot drops or hot nozzles for a valve-gated hot runner mold also incorporate a pneumatically- or hydraulically-actuated valve pin. The valve pin opens at the start of the melt injection phase of the molding process, and it closes on demand sometime later in the molding cycle (e. g., typically after some amount of holding pressure time has been applied).

One primary advantage of a valve-gated hot runner injection mold is the greatly improved gate appearance. The valve pin leaves a barely visible witness mark. The face of the valve pin actually forms a portion of the molded part surface. Valve pins are typically tapered and seat tightly within a tapered hole. Valve gates completely eliminate the stringing that can occur with conventional hot runner molds. Another advantage of valve gates is that larger gate diameters can be used. This is typically an advantage for large part molding where high volumetric melt flow rates are required (e. g., automobile bumpers). Hot runner gate diameters are typically limited to small size pin gates with conventional hot runner molds in order to limit string formation. There is no such concern with valve-gated hot runner molds. Valve-gated hot runner molds are more expensive than conventional hot runner or cold runner molds, but these gating advantages are not achievable otherwise.

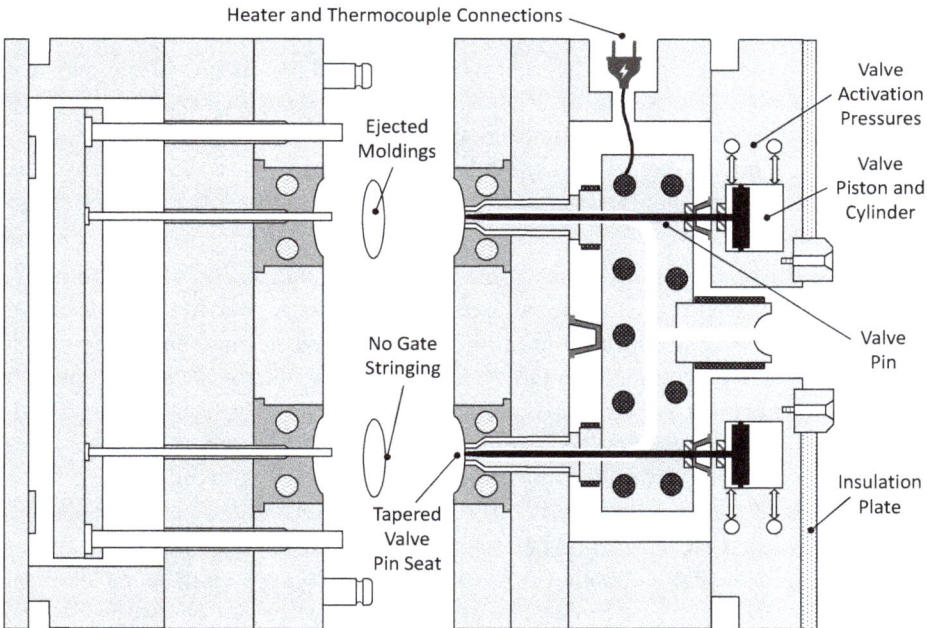

Figure 2.76 Valve-gated hot runner injection molds incorporate gate valve pins that are activated using pneumatic or hydraulic pressure control. The valve pins open at the start of the melt injection phase and close on demand. The valve gates are typically set to close once the packing and holding pressure phases of the process are complete. While valve gates add to the hot runner injection mold cost, they allow for the use of larger gate diameters, and they greatly improve gate vestige appearance when compared with conventional hot or cold runner molds. The valve gates are shown in the closed position.

Injection molding process flexibility is also greatly improved when valve-gated hot runner injection molds are used. Valve gates can be opened and closed on demand. The gates typically open at the start of the injection phase of the process, but they can be closed at any point in the molding cycle. This allows for improved control over holding time and molded part dimensions. Valve gates can also be used to eliminate knit lines for larger parts that are molded with multiple drops. The opening of the valve gates can be sequenced as the mold fills to eliminate knit line formation (i. e., sequential valve gating).

2.3.5 In-Mold Pressure Sensors

Thermoplastic melts are compressible fluids, and as such, the in-mold pressure history applied during the injection molding cycle will influence the molded part's shrinkage and dimensional characteristics. Injection molds can be instrumented with sensors so that the condition of the melt within the mold can be monitored over the course of the molding cycle. While injection molds can be instrumented with a variety of sensor types, in-mold pressure sensors are most common. In-mold pressure sensors are not a required element for the injection molding process, but their use has been shown to be very valuable as (i) a process monitoring tool or (ii) as a process control sensor.

Injection mold cavity pressure sensors are relatively expensive, however, the information that they provide is of great value to the process engineer. The use of a single in-mold pressure sensor is most common, although multiple in-mold pressure sensors are sometime used for multi-cavity molds. If an injection mold is to be instrumented with an in-mold pressure sensor, the type of pressure sensor to be used and the installation location must be determined. In-mold pressures can be determined using either (i) a self-contained, direct-reading pressure sensor, or (ii) indirectly using a force sensor and pin as shown in Figure 2.77. Direct-reading sensors are very reliable and relatively easy to install. Both strain gauge and piezoelectric in-mold pressure sensors are commercially available. Whenever possible, the sensor signal wiring and connector should be recessed into a mold plate as shown in Figure 2.77 to ensure that the sensor wiring cannot be damaged during mold changes or mold maintenance.

Figure 2.77 also shows how a force or load sensor can be used to measure in-mold pressure (indirectly). The force sensor can be placed behind the head of a functioning ejector pin, or it can be placed behind a dummy pin. Melt pressure is determined by dividing the sensor force reading by the area of the ejector (or dummy) pin face. Sensor cable channels and recessed sensor connections are also recommended for indirect pressure sensors to minimize the potential for sensor wire damage.

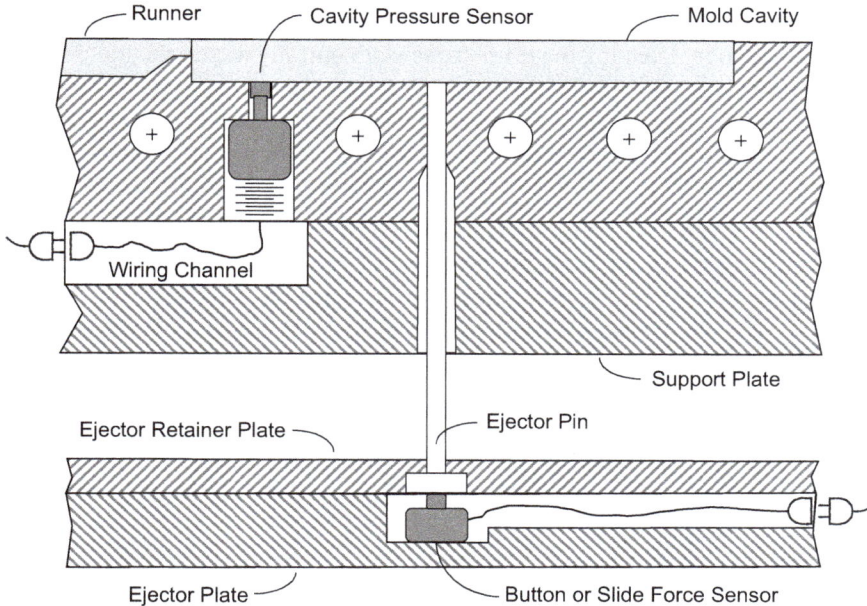

Figure 2.77 Injection molds are sometimes instrumented with a piezoelectric or a strain gauge in-mold pressure sensor. The in-mold pressure sensor signal can be used for either process monitoring or packing pressure control. The in-mold pressure can be measured using either a direct-reading pressure sensor, or a force sensor located behind an ejector pin (or dummy pin). In-mold pressure sensors are normally located within a mold cavity. Wiring channels and a recessed sensor signal connector are recommended when installing an in-mold pressure sensor to avoid damage to the sensor signal wiring.

In-mold pressure sensor placement is also an important decision. Sensors can be placed in either the injection mold's runner system or within a mold cavity. Sensing pressure within a mold cavity is the most common practice. In-cavity pressure sensors are typically located within the cavity but close to the gate. Locating the sensor about 25% into the flow length is typical for a general-purpose molding application. Locating the in-mold pressure sensor close to the gate maximizes the amount of information that can be obtained as the mold cavity fills, since the sensor provides a pressure reading only after the melt front crosses the sensor. While not very common, a cavity pressure sensor can be placed towards the end of the flow length for an injection mold cavity that is hard to fill in order to confirm that the fill is complete.

The decision regarding cavity pressure sensor location is relatively easy for a single cavity injection mold. With a single cavity mold, the primary decision is where the cavity pressure sensor will be located within the single cavity. Pressure sensor installation location is not such an easy decision for multi-cavity injection molds. Some feel that if the multi-cavity mold is naturally balanced, then any of the cavities would be an appropriate location. Locating the pressure sensor centrally within the mold's runner system is another option for multi-cavity injection molds. However, pressure sensors placed within the mold's runner system will not provide as much process information as those located within a mold cavity (e. g., no information related to gate solidification).

In-mold pressure sensors can be used to monitor the injection molding process or as a means of process control for some molding processes. Cavity pressure signals can be used to initiate velocity to pressure transfer, or for pack pressure to hold pressure transfer (i. e., peak pack pressure control). A typical cavity pressure vs. time trace for the injection molding process is shown in Figure 2.78.

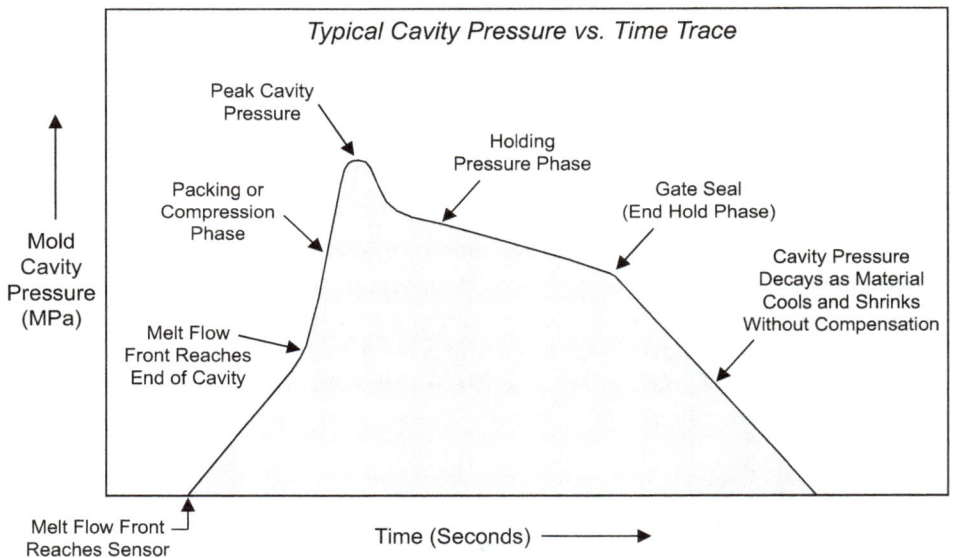

Figure 2.78 The figure shows a cavity pressure vs. time trace for a typical injection molding process. Cavity pressure traces are particularly useful because they provide a direct indication of the condition of the thermoplastic material in the mold cavity. For example, the area under a cavity pressure vs. time trace is indicative of the total cavity pressure history, which relates to molded part shrinkage and part dimensions.

References

[1] Osswald, T., Turng, T., and Gramann, P., "Injection Molding Handbook", Hanser Publications, Munich (2022).

[2] Johannaber, F., "Injection Molding Machines", Hanser Publications, Munich (2008).

[3] Malloy, R. and Chabot, J., "A History of Injection Molding", *Journal of Injection Molding Technology*, 1, (1), 1 (1997).

[4] RJG Corporation, 3111 Park Drive, Traverse City, MI 49686, https://rjginc.com.

[5] Gattshall, R., "A Molder's Plea: Let's Standardize Controllers", *Plastics Technology*, September 27 (2016).

[6] https://www.ptonline.com/knowledgecenter/blending/blending-basics/how-blenders-work.

[7] https://www.ptonline.com/knowledgecenter/plastics-feeding/feed-rate-basics.

[8] Mehos, G. et.al., "Designing Hoppers, Bins, and Silos for Reliable Flow", *AIChE Publication*, April (2018).

[9] Orbetron, 317 Willow Street, Gastonia, NC 28054, https://orbetron.com.

[10] Sotos, N., "Liquid Color for Plastics: Friend of Foe", *Plastics Business Magazine*, June 2 (2017).

[11] Bozzelli, J., "Injection Molding: Get Control Over Barrel Zone Temperature Override", *Plastics Technology*, December 21 (2015).

[12] Technical Bulletin, "General Purpose Screw Designs: Disfunctional and Expensive", Dray Mfg. Inc., Hamilton, Texas (2017).

[13] Technical Bulletin, "Barrier Screws", Dray Mfg. Inc., Hamilton, Texas (2017).

[14] Lang, A., "Screw Recovery Problems Call for Special Screw Design", *Plastics Technology*, June 22 (2020).

[15] Sun, X. et. al., "Design Optimization of Maddock Mixers for Single Screw Extrusion Using Numerical Simulation", SPE Annual Technical Conference, Anaheim, CA (2017).

[16] Bozzelli, J., "Injection Molding: Sliding vs. Locking Ring–Which Non-Return Valve Is Right for You", *Plastics Technology*, October 1 (2018).

[17] https://www.sumitomo-shi-demag.eu/activelock.

[18] Bozzelli, J., "The Importance of Hot Nozzle-Tip Mating Against a Cold Sprue Bushing", *Plastics Technology*, June 14 (2023).

[19] Azzopardi, C., "Know Your Options in Injection Machine Nozzles", *Plastics Technology*, May 20 (2023).

[20] Bozzelli, J., "Screw Speed Versus Recovery Time", *Plastics Technology*, March 21 (2014).

[21] Anonymous, "Why Control Melt Decompression? Just Shut it Off", *Plastics Today*, September 8 (2011).

[22] Bozzelli, J., "Mold-clamping Details for Profit", *Plastics Technology*, March 19 (2019).

[23] Schiller, G., "A Practical Approach to Scientific Molding", Hanser Publications, Munich (2018).

[24] Malloy, R. and Schott, N., Polymer Processing Chapter for "Applied Polymer Science Handbook", 2nd Edition, American Chemical Society (1985).

[25] Losek, J., Technical Literature, "Clamp Tonnage and Injection Pressure", ICOMold.com, November (2022).

[26] Technical Bulletin, "Clamp Tonnage Calculations", Kruger Industries, September 7 (2023).

[27] Williams, J., "Clamp Tonnage: More is Better, Right?", *Plastics Technology*, February 2 (2017).

[28] Bozzelli, J. and Fattori, J., "How to Mount an Injection Mold", *Plastics Technology*, October 2 (2022).

[29] Technical Bulletin, "SmartBolts® Promote Safety for Plastics Manufacturers", Industrial Indicators, Frederick, MD 21703.

[30] PSC Company, 34500 Doreka Drive, Fraser, MI 48026.

[31] "Mold Clamping", Routsis Training and Technical Blog, Routsis Training LLC, Nashua, NH 03063.

[32] Hartwig, K., Using Mold Insulation to Reduce Costs", *Plastics Today*, January 5 (2002).

[33] Technical Bulletin, Jaco Products, Middlefield, OH 44062.

[34] Brodine, D., "Designing the Perfect SMED Cart", Plastics Business, March 5, (2015).

[35] Currence, B., "Achieving Continuous Improvement Using SMED Programs", *Plastics Technology*, June 18 (2020).

[36] Kazmer, D., "Injection Mold Design Engineering", Hanser Publications, Munich (2022).

[37] Mennig, G., "Mold-Making Handbook", Hanser Publications, Munich (2013).

[38] Beaumont, J., "Runner and Gating Design Handbook", Hanser Publications, Munich (2019).

[39] Malloy, R., "Plastics Part Design for Injection Molding", Hanser Publications, Munich (2011).

3

Injection Molding Process Variables and Their Effect on Part Quality

3.1 Pre-Production Process Planning

The injection molding process is a very complex manufacturing process with many different process variables. The goal of this chapter is to review these variables and show how they can impact a molded part's performance, dimensions, physical properties, and appearance. While there are no universally applicable procedures for setting these different process variables, there are guidelines and principles that apply to most injection molding processes. For example, the principles and techniques associated with scientific molding are quite well-established and these proven procedures help process engineers develop more robust injection molding processes. Several texts have been written on the subject [1,2].

The goal here is to review the primary injection molding process variables that are associated with the injection molding of thermoplastics. The basic premise associated with this chapter is illustrated in Figure 3.1. When injection molding process engineers are starting up a new injection molding process, they are given three fundamental inputs, namely;

- A new or unfamiliar injection mold.

- A new or unfamiliar thermoplastic material.

- Molded part specifications (e. g., dimensional, cosmetic, performance).

Given these inputs, the injection molding process engineer (or molder) must then (i) select the most appropriate manufacturing equipment (i. e., the injection molding machine and auxiliary equipment) and (ii) determine the injection molding process variables that result in both a repeatable robust molding process and high-quality, in-specification injection molded parts.

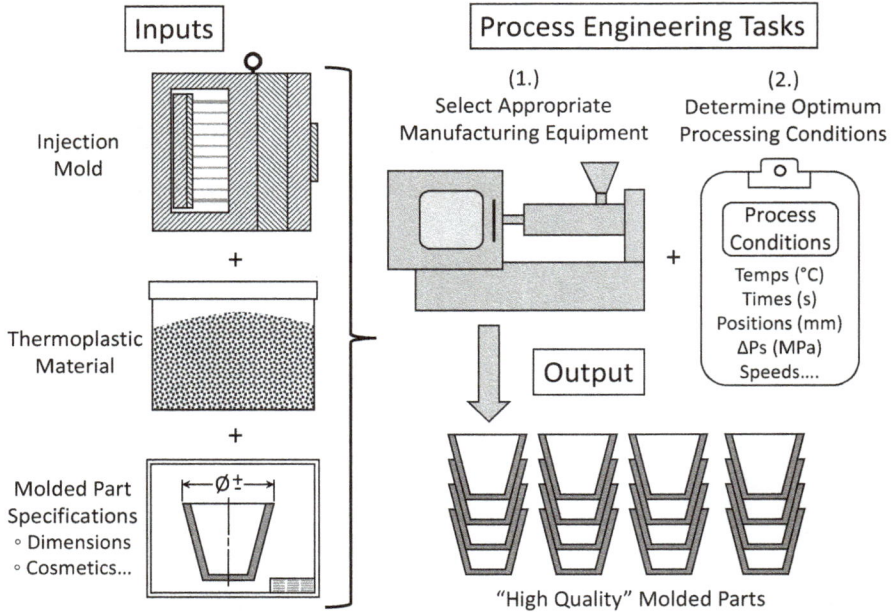

Figure 3.1 Process engineers starting-up a new injection molding process must make decisions regarding (i) the injection molding process machinery and (ii) the injection molding process conditions that lead to the production of high-quality, in-specification thermoplastic parts.

The difficulties in laying out the road map for establishing the optimum injection molding process (given a new or unfamiliar injection mold and new or unfamiliar thermoplastic material) stems from the fact that the injection molding process is an extremely versatile process that can be used to process a very wide variety of plastic materials. Best practices that apply to thin-wall molded thermoplastic packaging may have little or no correlation with those associated with molding thick-wall optical lenses. The best practices associated with molding a high melt flow rate semi-crystalline polypropylene may have little or no correlation with those associated with molding a high-performance amorphous polyetherimide. The processing concerns associated with these applications are in fact very different. While there is no *one size fits all* solution to optimizing injection molding process conditions, having an understanding of the function of each injection molding process variable is essential. Having an understanding of these injection molding process variables is a pre-requisite to establishing an appropriate and optimized injection molding process for a new application.

3.1.1 Customer Communications and Molded Part Target Specifications

The thermoplastic parts that are produced by the injection molding process all have requirements in terms of their performance, dimensional and cosmetic specifications. These requirements or specifications can be very loose or they can be very tight depending on the application. In a new application, they may even be overly or unrealistically tight. It is also important that these molded part target specifications are documented in a quantitative way, and that there is a very clear understanding of these part specifications by both the customer and the molder. The molding operation could be in-house or the molder could be an off-site custom molder. Either way, both parties should have a very clear understanding of all the target part specifications so that a suitable molding process can be established.

Dimensional or geometric specifications are by nature quantitative and, as such, they are usually unambiguous. Geometric specifications include attributes such as critical dimensions, general and specific dimensional tolerances, and flatness requirements. Even sink marks or knit lines can be considered as geometric features having target specifications. For example, the maximum allowable depth for a sink mark or the knit line V-notch depth can be specified.

Other, non-dimensional target part specifications are fundamentally more qualitative in nature. As an example, it can be very difficult to quantify cosmetic issues for molded thermoplastic parts. Cosmetic defects such as surface gloss or gate blush are more qualitative than quantitative. One published guide assists plastic part molders and customers in objectively evaluating the cosmetic attributes and surface quality for injection molded thermoplastic parts [3]. This publication also describes methods for inspection and measurement of a molded part's surface finish. Some cosmetic issues have become even more difficult to define in recent times due to the vast array of lighting technologies that are now available. Lighting conditions can have a very significant effect on a molded part's surface appearance.

A customer's position on the use of regrind should also be clearly defined. The issues associated with the use of regrind can be quite complex, and many customers do not give these issues the consideration they deserve. First, it must be clear as to whether the use of regrind is sanctioned or not authorized. If regrind use is sanctioned or authorized, all of the practices and procedures associated with the regrind use should also be clearly defined (e.g., target regrind concentration, fines removal) as shown in Figure 3.2. More details regarding regrind use are given in Section 1.7 of this text.

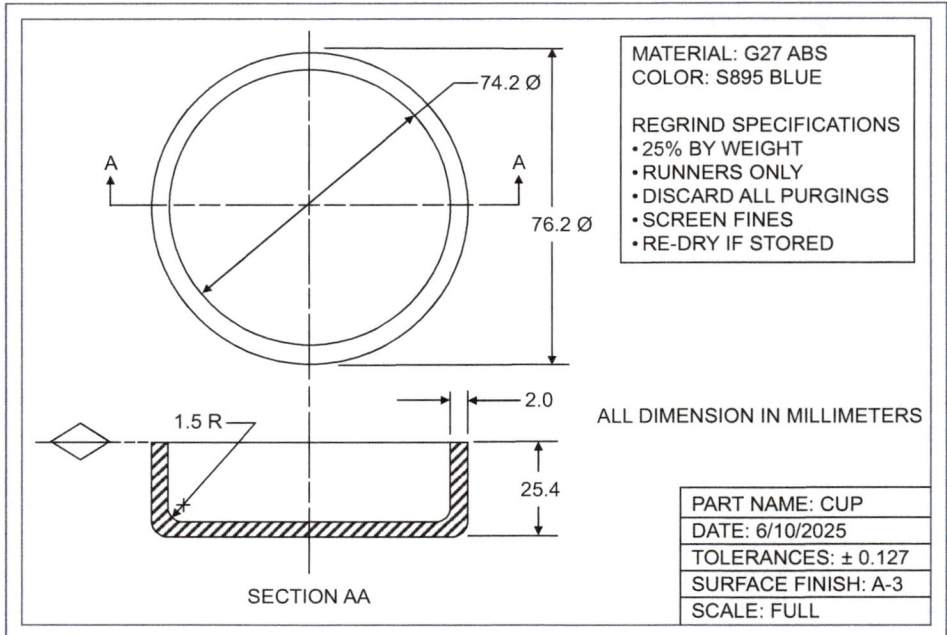

MATERIAL: G27 ABS
COLOR: S895 BLUE

REGRIND SPECIFICATIONS
• 25% BY WEIGHT
• RUNNERS ONLY
• DISCARD ALL PURGINGS
• SCREEN FINES
• RE-DRY IF STORED

74.2 Ø
76.2 Ø
2.0
1.5 R
25.4

ALL DIMENSION IN MILLIMETERS

SECTION AA

PART NAME: CUP
DATE: 6/10/2025
TOLERANCES: ± 0.127
SURFACE FINISH: A-3
SCALE: FULL

Figure 3.2 The dimension specifications and tolerance requirements for injection molded parts should be well defined. Molded part specifications should also indicate if the use of regrind is sanctioned. If so, the practices associated with regrind use should be clearly defined.

3.1.2 Material Supplier Molding Process Recommendations

Thermoplastic material suppliers will generally publish both (i) material property data sheets and (ii) recommended injection molding practices for the thermoplastic materials they supply, as discussed in Section 1.12.1. While most material supplier data sheets have a similar (but unfortunately not a standardized) format, their molding process recommendations can vary. The material supplier process guides can have process information that can range from very limited to very detailed depending on the specific material supplier. Nevertheless, molding process engineers should attempt to gain as much processing information as possible from the material supplier. When the material supplier molding process information is limited, process engineers may also consider process recommendations from other sources, including those published in textbooks [1] or those for a very similar material. For example, the nozzle tip design recommended for one material supplier's neat nylon 66 material grade is expected to be quite similar to that for a neat nylon 66 available from another material supplier.

At a minimum, material suppliers generally provide recommendations for at least the following process variables:

- Material drying requirements for hygroscopic plastics (e. g., maximum allowable moisture content).

- Mold temperature range (minimum and maximum recommended mold temperature).

- Melt (and hot runner) temperature range (minimum and maximum recommended melt temperature).

Each of these topics is discussed in more detail later in this chapter. Thermoplastic material suppliers also provide mold shrinkage and vent depth data that is used primarily for mold design. They may also provide recommendations for other process variables such as injection velocity or back pressure, but they really cannot do so with any certainty since most molding process variables are also influenced by the mold's design, molded part geometry and injection molding machinery (and not just by the material being molded). For example, material suppliers cannot recommend a specific holding pressure value with any certainty since this variable's setpoint is typically based on critical part dimensions (not based on the thermoplastic material itself). In practice, most injection molding process variables are set based on a combination of theoretical considerations and experimental procedures associated with molding process start-up.

Some material suppliers also provide guidance associated with the use of regrind. For example, they may suggest a maximum recommended regrind concentration. Alternatively, they may publish property data from a regrind study showing how certain physical properties are influenced by the regrind concentration. Keeping regrind free from contamination, and also dryness and concentration uniformity are all critical issues when regrind is to be used.

3.1.3 Molding Machinery Selection for Captive Injection Molding

The scheduling and technical issues associated with injection molding machinery selection can vary significantly depending on whether the molding is to be done in-house or by a custom molder [4]. The economic issues associated with in-house vs. custom injection molding also vary significantly based on production run volumes. At one extreme, in-house injection molding is common for very high-volume applications such as 24–7 injection molding operations. For example, high-volume in-house injection molding is very popular for plastic packaging. On the other hand, lean molding practices that include just-in-time manufacturing are also common for some captive molding operations. In-house molding can speed up production by improving fulfillment times and eliminating external supplier shipping times (and shipping costs). Having all molding related services under one roof is beneficial in that it can result in a smoother design to manufacturing path which includes complete control over quality issues. While parts are typically designed in-house, and molding may be done in-house, tool design and tool building is most frequently subbed out to a tooling vendor. It is difficult to justify the labor cost for skilled toolmakers and the costs of the specialized tool making equipment for a limited number of tool builds.

With high volume 24–7 captive molding, it is possible that the molding machinery is literally purchased when a new mold is put into production. The long payback times associated with molding equipment capital investment costs are more acceptable for high production volume applications. This is a best-case scenario in terms of matching the machinery characteristics with those of the material and the mold. In such a case, the molding machine can be optimized in terms of the machine's clamp specifications and tonnage, shot capacity, screw design, plasticating capacity, metallurgy, and the like. The same holds for auxiliary equipment.

In other captive molding applications, existing in-house injection molding machinery is used for production. In such cases, molds are designed such that they are compatible with the existing injection molding machinery. In this case, the technical issues associated with injection molding machine selection are essentially the same as they are for a custom molder (as discussed in the next section). The goal is to select the most appropriate molding machinery (based on the material and mold) from the machinery available in-house. This can present a challenge since most captive molding operations have limitations associated with the number and type of injection molding machines that are available. In addition, balancing production scheduling can be difficult due to the peaks and valleys associated with product demand, although in-house molders do have complete control over production scheduling. Machinery downtime due to breakdowns can also be problematic for captive production if there is no redundancy in terms of injection molding machinery. These are some of the many reasons that commercial custom injection molding or contract plastic manufacturing businesses are very popular in the plastics industry.

3.1.4 Molding Machinery Selection for Custom Injection Molding

Custom injection molders are essentially contract manufacturers that offer injection molding and other related services. At a minimum, custom molders sell injection molding machine capacity. However, many custom molders offer a far wider array of services that include part design, prototyping, tool design, tool building, and secondary operations such as ultrasonic welding or part decorating. When custom molders are utilized for molded part production, the capital equipment costs, labor costs and overhead costs associated with in-house molding are transferred to the custom molder. The custom molding industry is very large and very well-established. Due to the large number of custom molders, it is relatively easy to identify a custom molder that has the process capabilities that you (as a customer) require. The custom injection molding industry is also very competitive and therefore cost effective for most applications (provided production volumes are sufficient to justify the injection mold cost).

Developing a strong relationship between the customer and the custom molder is critical. It is also important that the custom molding partner has expertise, skill sets, and molding process capabilities that correlate closely will the customer's application. For example, some custom molders specialize in the molding of optical parts, while others specialize in molding high-performance plastics or molding for food contact applications. In addition, different industries require certifications, which the molder must comply with. The technical qualification process used for custom injection molding vendor selection typically involves a review of material handling capabilities, molding process capabilities and operational procedures related to quality control. Having complete customer confidence in the custom molder is important since molded part quality becomes more of a potential issue when molded part production is conducted off-site or off-shore.

The new mold and material processes development task depicted in Figure 3.1 involves two main steps. The first step is to select appropriate molding machinery for the application. Injection molding machine hardware topics have been discussed in Chapter 2 of this text. The second step of this task is to select molding process conditions that lead to the production of in-specification parts. The bulk of this chapter deals with this second step. Ideally, the equipment and process conditions used for molded part production should be consistent over time. Ideally, the same injection molding machinery is used for production over the life of the plastic part. This can be difficult in custom molding applications since the same molding machinery is typically used for multiple molds and materials over time. Equipment modifications over time and machinery wear can lead to part quality differences over the long run. The amount of flexibility that custom molders have in terms of process conditions or injection molding machinery changes is customer-dependent. However, generally speaking, the injection molding equipment and injection molding process parameters should correlate as closely as possible to those that are optimum for the new material and mold. Some of the more important issues related to injection molding machinery selection are discussed below. The goal is to select molding machinery that will allow for the development of a robust injection molding process that is stable and has a wide process window. Ideally, the wide process window will allow the molding process to be less sensitive to ambient temperature variations, barrel temperature cycling or thermoplastic material lot variations.

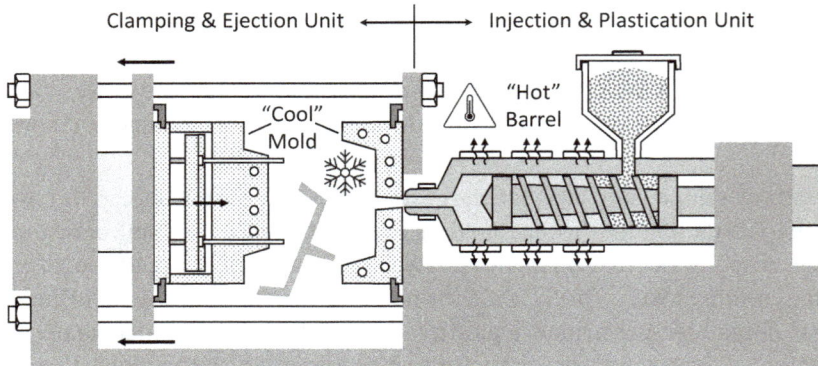

Figure 3.3 Ideally, the injection molding machine selected for a new production application (i. e., a new mold and new material) should have characteristics that are very similar to a molding machine that would be purchased for the application if a new injection molding machine purchase was authorized for the production run.

Material dryer: If material pre-drying is required, the material dryer should be sized such that the dryer capacity and throughput rate is at least equal to that of the molding process output. The expected molding process production rate can be determined theoretically using molding simulations or cycle time calculators, while recommended drying times are normally published by the material supplier.

Injection unit metallurgy: The metallurgy used for the injection unit's plasticating screw, non-return valve, barrel, and nozzle should be reviewed when molding thermoplastic materials that are either abrasive or corrosive. For example, chemically resistant molding machine components are required when molding fluoropolymers or chlorinated polymers. Some flame-retardant additives can also be corrosive. Other materials, such as a glass fiber reinforced thermoplastic or heavily filled thermoplastics, can be very abrasive. The plasticating screws and barrels used when injection molding these materials should have surface harness values that reduce the rate of abrasive wear. Free flow check valves are also more appropriate for glass fiber reinforced materials in order to minimize glass fiber length reduction.

Injection unit shot capacity: All injection units have a maximum shot size or shot capacity. This value is typically specified in terms of the maximum shot volume (cm^3) that the injection unit can inject. The maximum shot size can also be specified based on the maximum weight (ounces or grams) of general-purpose polystyrene that can be injected. The total volume or weight of the shot that will be molded (i. e., the molded parts and runner in the case of a cold runner mold) should be less than the maximum injection unit shot capacity. Many sources suggest that the shot size utilization factor should be somewhere between 40–80% of the injection unit's maximum shot capacity. This is a bit of an oversimplification but it is a common rule of thumb or first approximation.

It is acceptable to venture outside of this recommended shot size utilization range, but the consequences of doing so should be considered. For example, there are two main issues when a shot utilization factor is less than 40%. These smaller shot sizes often result in long barrel residence times and possible material degradation. This may or may not be an issue depending on the length of the molding cycle time. The material's residence time in the barrel is influenced by both the shot size utilization factor and the molding process cycle time. The material residence time can be estimated theoretically or it can be determined experimentally (e. g., using a color concentrate). In cases where the material's residence time is excessive, a starve feeder can be used to reduce the fill factor for the plasticating screw and therefore reduce residence time. Some molding machine manufacturers also offer quick change barrel assemblies of different shot capacity that can be used for the same molding machine.

A second issue associated with a small shot size is that the process variability that occurs when the plasticating screw's check valve closes during injection. This process variability is magnified when the shot size is small. The variability, as a percentage of the shot size, becomes greater as the shot size decreases. Injection units equipped with a more positive self-locking check ring offer improved shot consistency which is especially important when the shot size utilization factor is low. There are also issues when the shot size utilization is greater than 80%. Some of these issues are related to the fact that the screw reciprocates during the plastication phase of the molding process. The melt quality (e. g., melt temperature uniformity) tends to be more variable as the shot utilization factor increases.

Injection unit plasticating capacity and screw geometry: Molding machine injection units also have limitations in terms of melting efficiency and plasticating capacity. The injection unit's plasticating screw design must be appropriate for the material to be molded. Many injection molding machines utilize a general-purpose plasticating screw that is suitable (but not optimum) for plasticating a variety of different thermoplastic materials. Injection molding screws have many attributes but the screw's compression ratio is one of the more important variables. Material supplier molding process guides will typically list the optimum compression ratio for a given material. A more complete review of injection molding plasticating screws is given in section 2.2.2.3 of this text.

Injection rate and injection pressure: The injection pressure and injection rate capabilities of an injection molding machine can vary significantly from machine to machine. Historically, injection molding machines had a maximum melt injection pressure of around 145 MPa. This injection pressure was suitable for molding most of the conventional wall thickness thermoplastic parts of the day. However, over the years, trends towards both thin-walling and larger part molding led to the need for molding machines with greater injection pressure capabilities. Molding machinery manufacturers now offer molding machines having injection units capable of 300 MPa or more injection pressure capability. The higher pressures are typically achieved by using smaller diameter screws and barrels which facilitate the pressure generation but limit shot capacity. Pre-production mold filling simulations can be used to esti-mate the injection pressure requirements and optimum injection rate for a new mold and material combination. Very high injection rates that may be required for very thin-wall parts can are best achieved using accumulator assist injection molding ma-chines.

Nozzle type and tip geometry: It would be uncommon to select a molding machine for a new application based only on the molding machine's nozzle geometry. Injection molding nozzles are relatively inexpensive interchangeable components that can be replaced as required. It is especially important that the molding machine nozzle or at least the nozzle tip (for a two-piece nozzle) is fully compatible with the injection mold to be run. The optimum nozzle tip orifice diameter and the nozzle tip radius are dic-tated by the dimensions of the injection mold's sprue bushing dimensions as de-scribed in Section 2.2.2.5.

Clamp requirements: There are many molding machine clamp specifications that will determine if the clamp is suitable for an injection mold. Many of the clamp variables are related to the physical size of the clamps (e. g., platen size, tie bar spacing, min. and max. die height, ejector rod pattern). The optimum relationship between mold and clamp size is discussed in Section 2.2.3. Along with physical size, molding ma-chine clamps must provide sufficient clamp force to counteract the injection force and prevent flash. Issues related to the required clamp force are discussed in Section 2.2.3.2. Clamp force is adjustable, which means that molding machines with a higher-than-required-clamp capacity can be utilized for lower clamp force molding applica-tions by dialing down the applied clamp force.

3.2 Molding Material Preparation

3.2.1 Hygroscopic Material Pre-Drying

One of first process considerations when starting up a new injection molding process is the thermoplastic material's pre-drying (or drying) requirements, since the time required for the material drying process can be as long as hours in some cases. Hygro-scopic thermoplastics, such as acrylic or ABS, need to be dried prior to molding to ensure that there is no moisture present when the material is melted. While most thermoplastics do not absorb large quantities of water, even small concentrations of residual moisture will lead to molding defects that include splay or silver streaks. Drying is even more critical for condensation polymers, such as polycarbonate or ny-lon 66, since moisture can also cause a reduction in the average molecular weight and loss of toughness for these materials as discussed in Section 1.4.2.3. Other plastics, such as HDPE or polypropylene are hydrophobic in nature (i. e., non-hygroscopic), have no tendency to absorb moisture from the surrounding atmosphere as such, and therefore do not require drying. While materials such as HDPE have little or no tendency to absorb moisture, there can be material storage conditions (e. g., cold silos) that can lead to condensation or surface moisture. Finding ways to eliminate or prevent this surface moisture condensation is preferable to drying for these applications.

The recommended drying requirements for a given thermoplastic material are normally provided by the material supplier. The material supplier will normally recommend a maximum acceptable moisture content (or a range of acceptable moisture contents typically described as the moisture content window). For example, a supplier of a neat polycarbonate recommends that the maximum moisture content should be ≤0.02% before molding [5]. As a result, it is important that molders have the capability to accurately measure moisture content (prior to molding) using a moisture analyzer. Gravimetric moisture analyzers based on total weight loss are relatively simple and cost effective, but more sophisticated analytical chemical techniques can provide for greater accuracy.

Typically, material suppliers will also suggest recommended drying conditions, although the breadth or detail associated with this guidance varies considerably from material supplier to material supplier. At a minimum, material suppliers will typically recommend a drying temperature and a drying time (or range of drying times). For example, the drying condition recommendation for the neat polycarbonate referenced above is given as 2–4 hours at a drying temperature of 120 °C. While such drying conditions are also impacted by the type of drying equipment that is being utilized, it is best to stay within the recommended range of drying conditions to avoid under-drying or over-drying. It is therefore important to coordinate the dryer residence time with the throughput of injection molding process (as much as possible). For example, over-drying is likely to occur if the dryer capacity is too large based upon the injection molding machine production rate. Over-drying can lead to excessive material oxidation, yellowing, or volatile additive loss, all of which are undesirable outcomes. Over-drying can also be the result of a molding production delay (e. g., due to machine maintenance). If an unscheduled production delay does occur, the dryer temperature can be temporarily reduced in order to minimize the potential for over-drying issues.

There are many types of material dryers on the market ranging from small tray dryers, desiccant hopper dryers of various capacity, or compressed air dryers. Most dryers function by circulating dry, heated air, through the dryer hopper at specific air flow rates. Since most thermoplastics are supplied in pellet form, the pellets have a bulk density that allows air to pass though the mass of pellets in the hopper. Dryer hoppers are typically designed to promote a uniform air flow rate. The dryer process variables should be routinely monitored to ensure the dryer is operating efficiently and consistently. A typical desiccant dryer has two (or more) alternating desiccant beds that are used to keep the circulating air dry. While one desiccant bed (i. e., the active bed) is used to remove moisture from the moist return air, the second (now saturated) desiccant bed is heated off-line to remove the previously absorbed moisture (i. e. the desiccant bed regenerates). Ideally the desiccant beds keep the dew point of the dry circulating air to about –40 °C. The delivery air temperature, the return air temperature, the drying air dew point, the air flow rate and the hopper material level are all critical drying variables. These variables should be monitored as a means of ensuring that the material is being dried properly. Ensuring that these variables are within an acceptable tolerance range can help minimize the frequency of moisture content verification testing. Aftercoolers are typically required if the dryer return air temperatures exceed 65 °C in order to optimize the desiccant efficiency. Dryers also contain delivery and return air filters that remove dust and fines from the circulating dry air. These filters should be cleaned or replaced periodically in order to maintain dryer air flow rate consistency [6].

3.2.2 Vented Barrel Processing

It should also be noted here that vented-barrel injection molding machines can be used as an alternative to material pre-drying for some hygroscopic thermoplastics. The concepts, advantages and limitations of vented-barrel devolatilization have been discussed earlier in Section 2.2.2.3.5. The main advantage of vented-barrel processing is that moisture removal and melting occur simultaneously during plastication. This eliminates the need for material pre-drying altogether if this method of melt processing is suitable for the material being molded. Vented-barrel processing seems to be losing favor in recent years, most likely due to advancements in material dryer technology. In addition, vented-barrel processing is not appropriate for all hygroscopic thermoplastics. Vented-barrel processing is best suited for polymers produced by the addition polymerization process (such as ABS). Vented-barrel processing is not common with condensation polymers, such as polycarbonate or polyamides, since they can react with water at high temperatures, causing them to partially depolymerize. As a result, material pre-drying is more appropriate when injection molding a condensation polymer.

3.2.3 Moisture Re-absorption

There are several issues that can complicate the material drying process. Moisture re-absorption is one complicating issue. Properly dried plastics will reabsorb moisture quickly if they are exposed to wet or moist ambient air. The rate of moisture re-absorption depends on the material type, the relative humidity of the ambient air, the pellet surface area to volume ratio, and other variables. *Any* exposure to ambient air is unacceptable with very moisture-sensitive materials, such as PET. The key is to keep the drying system sealed (to the extent possible) such that the dried pellets are not exposed to ambient air after drying. The molding machine should have a sealed material delivery system and a sealed or gasketed hopper design. Moisture re-absorption can also be an issue when utilizing hygroscopic thermoplastic regrind (e. g., regrind from cold runner scrap). The cold runners should be dry when they are ejected from the injection mold, if the thermoplastic molding material was dried properly before molding. It may be possible to utilize reground runners without re-drying the regrind, if the material is reground and reused immediately, as in the case of a closed loop beside the press granulation process. However, any time delay before the thermoplastic regrind is re-introduced into the injection molding process may result in an unacceptable moisture concentration (due to moisture re-absorption).

In some cases, particularly for small part or low production volume applications, materials are supplied as a pre-dried material. Pre-dried materials are typically supplied in foil lined or other moisture resistant barrier packaging. In theory, such packaging should eliminate the need for the molder to re-dry the material prior to molding. However, in practice, there is a chance the barrier packing may be compromised, which exposes the pre-dried material to the ambient air allowing for moisture re-absorption. In such a case, the material packaging integrity should be confirmed by performing an analytical moisture analysis on the pre-dried material, or the material should be re-dried if the packaging integrity is in any way suspect.

3.2.4 Multi-Material Feed Streams

Injection molding operations that involve multi-material feed streams, such as a thermoplastic and masterbatch pigment, can also introduce drying complications. The drying requirements for each component material must be determined prior to process start-up. If a masterbatch pigment and the base thermoplastic material both require drying, *and* the drying conditions are similar for each component material, it may be best to pre-blend the component materials prior to drying. The type and grade of carrier thermoplastic used to bind masterbatch additives is often unknown and its importance is often dismissed since the carrier concentration is relatively low. However, it is best to check with the masterbatch supplier to determine both the masterbatch compatibility with the thermoplastic being molded and for any drying condition guidelines (if drying is required). On the other hand, if the component material drying requirements (e. g., drying temperatures) are different, the materials will most likely require separate drying operations to avoid softening and/or over-drying of the lower drying temperature material. In addition, if separate material feeders are utilized for a multi-material molding process, it is important to use sealed material handling systems that minimize ambient air infiltration for the dried material(s) to minimize moisture reabsorption.

3.3 Injection Mold Temperature Considerations

The temperature of an injection mold will have a very significant impact on both the quality of a molded thermoplastic part and the time required to cool the part so that it can be ejected without distortion. Unfortunately, these two factors are at odds with one another, making the choice of the mold temperature setpoint a very difficult but important decision when starting up a new injection molding process. In general, higher mold temperatures tend to produce higher quality parts, however, higher mold temperatures require longer cooling times and therefore longer cycle times. Some guidelines regarding mold temperature setpoint considerations are given in the following section. Note that the variable mold temperature refers to the temperature of a cold runner mold or the cold portions of a hot runner mold. Hot runner temperature settings are more closely related to the molding machine's nozzle temperature setpoint than mold temperature.

3.3.1 Mold Temperature Range

First and foremost, the recommended mold temperature setpoints are material dependent. Thermoplastic material suppliers will generally indicate the range of mold temperatures suitable for the materials that they supply. When molding thermoplastics, the mold temperature is always cool relative to the melt temperature. While molds are cool relative to the material's melt temperature, the temperature of the mold can be well above room temperature. For example, a material supplier for a neat general-purpose polycarbonate indicates that the mold temperature for this material should be in the range from 80–120°C [5]. It is best to stay within this recommended range for conventional injection molding applications. While the material supplier can provide the recommended mold temperature range, the selection of the most appropriate value within this range is the more difficult decision. This decision is further complicated by the fact that most injection molds *do not* incorporate mold temperature sensors for mold temperature control. The temperature of an injection mold is usually controlled by managing the temperature and flow rate of the cooling fluid being circulated through the mold's cooling or heat transfer channels. The conventional injection molding process involves the continuous circulation of fluid through the mold cooling channels, as depicted in Figure 3.4. While it is the actual mold cavity and core temperatures that are most important, the temperature of the cooling fluid being circulated through the mold is usually the mold temperature control variable. Ideally, the steel and coolant temperatures are similar, but this correlation is highly dependent on a large number of variables that include the design of the injection mold's cooling system, cavity and core steel type, coolant flow rate, scale formation, and other factors. It is therefore very important to have the ability to measure the actual cavity and core temperatures. At a minimum, a fast-response, hand-held surface pyrometer can be a very useful tool for the start-up of a new injection molding process. Instrumented coolant manifolds that indicate coolant flow rate, and allow for coolant flow rate control, are equally important.

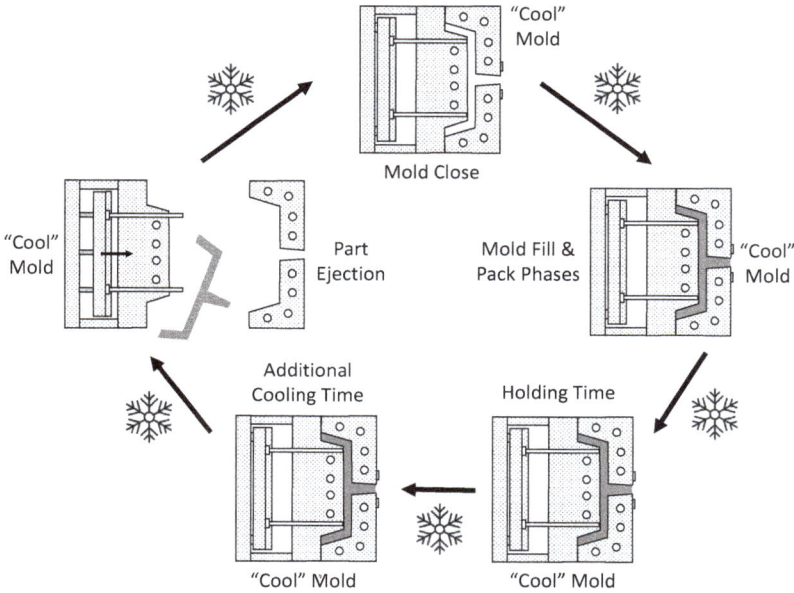

Figure 3.4 Injection mold temperatures are controlled by managing the temperature of the fluid circulating through the injection mold. In conventional injection mold processing, the coolant delivery temperature is maintained at a constant set value. While the coolant temperature increases as it passes through the mold's cooling channels, it is brought back to the set temperature prior to delivery.

Once the mold has been installed, and all clamp and ejector parameters have been set, cooling system connections can be made. After a mold has been built, there is no easy way to view the cooling channel layout so it is very important that all cooling connections are properly identified. The cooling circuit designations are typically stamped or engraved into the mold plates (e.g., Zone 1 IN, Zone 1 OUT). Properly marked mold cooling connections allow set-up technicians to install cooling lines without the need to review mold CAD models or prints during the start-up of a new mold. The use of color-coded hoses for IN and OUT water connections can also help avoid confusion. Once cooling connections are made, it is common to close or nearly close the mold (without tonnage) so that the injection mold's leader pins (and other A plate and B plate alignment features) engage, allowing the mold halves to expand more uniformly as the mold heats to the set temperature. Note that clamp settings, especially low-pressure mold protection, should be checked or reset once the mold reaches operating temperature. Also note that injection molds are typically installed while at the ambient temperature, but larger size molds are sometimes preheated prior to installation in an effort to reduce the mold pre-heat and changeover time.

There are many factors to consider when setting the injection mold temperature for a new process. As stated above, mold temperature is usually controlled by managing the temperature of the heat transfer fluid that is circulated through the mold. The actual mold cavity and core temperatures are typically not instrumented with temperature sensors, and the mold temperature is assumed to be the same as the temperature of the heat transfer fluid. This assumption is not always valid. The actual injection mold cavity and core temperatures should always be confirmed using a fast response surface pyrometer. It is also common to set the mold temperatures for the stationary and moving mold halves to the same value, at least initially. The cavity and core mold temperature setpoints may be fine-tuned based on part quality issues (e. g., amount of warpage) as the process start-up progresses.

Mold temperature selection should begin by reviewing the material supplier recommendations for the mold temperature range. Material suppliers will recommend a minimum mold temperature and a maximum mold temperature. In some cases, this range is very narrow, while in others the range is very broad. Regardless, it is best to stay within this recommended range. It is possible to mold parts outside of the recommended mold temperature range, but the outcomes may not be optimum. In general, molded part quality tends to be better when mold temperatures near the upper end of the recommended range are used. On the other hand, mold temperatures closer to the lower end of the recommended range will allow for a faster cooling time and shorter overall cycle time. Both part quality and cycle time should be considered when selecting the optimum mold temperature. It is very evident that a lower mold temperature will result in a shorter cycle time and greater production rate. However, the effect of a lower mold temperature on part quality is not always so obvious. The effects that mold temperature can have on molded part quality, cooling time and other process variables are discussed in the following sections.

3.3.2 Skin Layer Development

Consider the effect of mold temperature on the mold filling phase of the process. Mold filling is described as a non-isothermal process, since a relatively hot melt is injected into a relatively cool injection mold. For example, the recommended melt temperature for a neat polycarbonate is about 280 °C, while a mold temperature of 90 °C is within the recommended mold temperature range. Because the injection mold is cooler than the melt, the portion of the melt that is in contact with the relatively cool mold cavity walls will solidify as the injection mold fills as depicted in Figure 3.5.

While injection mold fill times are relatively short, it does take a finite amount of time to fill the mold, during which the solid layers will form. These solid skins or frozen layers tend to have micro-structures that differ from the core regions of the molded part. Injection molded parts are often described as having a skin-core-skin or layered morphology. The laminar flow velocity profile also results in viscous heating as the mold fills, especially in the melt layers that are adjacent to the frozen layers.

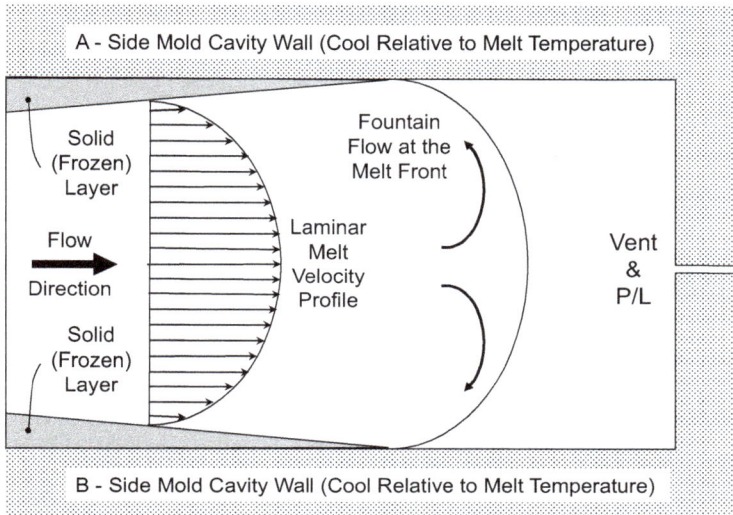

Figure 3.5 The mold filling process is a non-isothermal laminar flow process since (i) the mold is set to a far lower temperature than the melt, and (ii) it takes time to fill the mold. Thin solid layers (i. e., skins) will form at the surfaces of the plastic part as the mold fills, while the bulk of the part thickness (i. e., the core) remains molten. The thickness of the solid layers depends on many factors including melt and mold temperatures, mold and melt thermal properties, flow length, and injection rate. The melt adjacent to the solid layers also tends to heat due to viscous dissipation.

The absolute thickness of a molded part's solid layer is influenced by many variables, including mold temperature. The thickness of the solid layers increases as mold temperature decreases. Mold filling also becomes more difficult at lower mold temperatures since the effective thickness of the mold cavity (through which the melt flows) decreases as the solid layer grows. The molded part surface quality is also influenced by mold temperature since a cooler more viscous material will not replicate surface features as well as a warmer, less viscous material. It should also be noted that solid layer formation is generally more of an issue for thin wall applications since the solid layers represent a greater percentage of the overall part thickness.

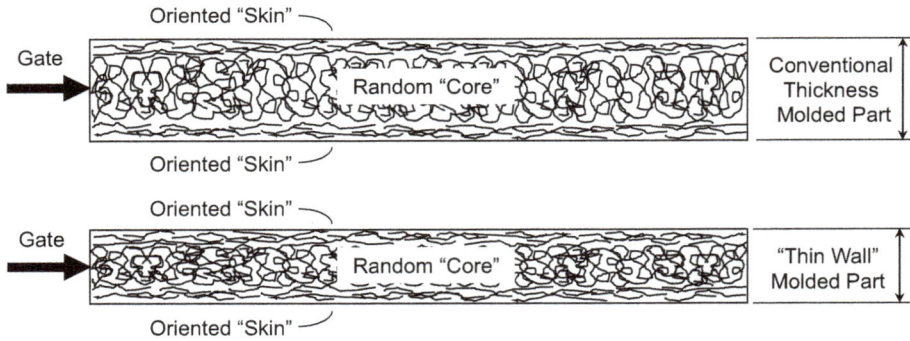

Figure 3.6 Solid skin layer formation tends to be more significant issue for thinner wall injection molded parts since the solid skin layers represent a greater proportion or percentage of the overall molded part wall thickness.

3.3.3 Frozen-in Molecular Orientation

When thermoplastics are in the melt state, the molecules have a random (or random coil) configuration. This is the case for the plasticated melt sitting in the injection molding machine's shot chamber before it is injected into the injection mold cavity. These polymer chains tend to become aligned or oriented as the melted shot is injected into the mold due to the laminar flow profile that develops during injection. In the case of an amorphous plastic, such as polycarbonate, these oriented chain-like molecules will relax (or re-randomize) almost instantaneously (i. e., within milliseconds) if the melt is hot enough to allow for segmental movement. This is the case for molecules located within the hot core regions of the molded parts. However, oriented polymer molecules located within the thin frozen solid skins are unable to relax, causing the molecular orientation to be frozen-in. This frozen-in orientation is undesirable in most applications since it can lead to anisotropic mechanical behavior and mold shrinkage.

Frozen-in orientation is also one cause of molded-in stress as shown in Figure 3.7. The images on the left in Figure 3.7 depict a part that is molded at a relatively high mold temperature where molecular orientation is able to relax. The images on the right in Figure 3.7 show a part molded at a relatively low mold temperature, where there is a significant degree of frozen-in molecular orientation. Oriented molecules are in many ways analogous to stretched springs. The springs radiate from the gate causing that area of the part to have a high level of internal stress. The internal stress associated with the frozen-in orientation can cause room temperature warpage in the case of more flexible plastics or distortion at elevated temperatures for more rigid plastics. These internal stresses will superimpose on other stresses caused by mechanical loads, reducing the overall mechanical performance for the plastic part having frozen-in molecular orientation. Injection molded parts that have frozen-in molecular orientation will tend to exhibit anisotropic mechanical properties and performance. Internal stresses caused by frozen-in molecular orientation can also reduce the chemical resistance of a plastic part due to the phenomenon known as environmental stress cracking or crazing (ESCC).

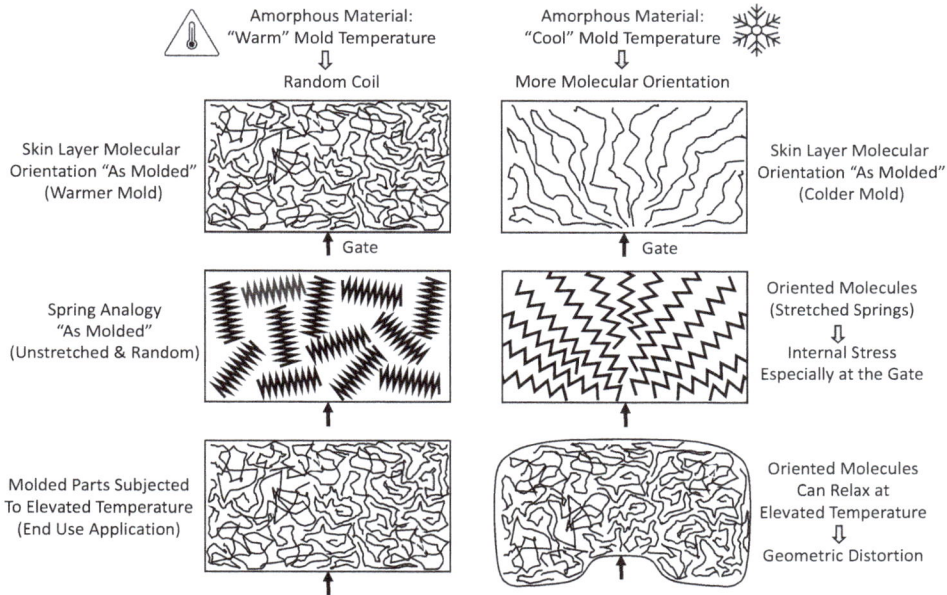

Figure 3.7 The long chain-like polymer molecules of a thermoplastic material tend to align along the flow direction during the laminar flow mold filling phase of the injection molding process. The oriented chains within the solid skins are described as "frozen-in" while oriented chains within the core quickly relax since the core is still molten at the end of the mold filling phase. Frozen-in molecular orientation is one cause of "molded-in stress". The molded-in stress can cause molded part deformation (particularly at elevated temperatures), reduced chemical resistance, and anisotropic mechanical performance

In summary, selecting a mold temperature toward the higher end of the material supplier's recommended mold temperature range will help minimize the amount of frozen-in molecular orientation, especially when molding amorphous thermoplastic materials. Reducing the amount of frozen-in molecular orientation is expected to improve overall part performance and promote more isotropic behavior. However, higher mold temperatures also increase the cooling time required and therefore the overall cycle time. This can be a tough sell since the performance of a plastic part over the long term is difficult to predict in the short term. The degree of orientation is also difficult to quantify. One simple way to gain some understanding as to the level of frozen-in molecular orientation is to perform a de-molding test where plastic parts are warmed up to a temperature close to material's softening temperature or heat deflection temperature (HDT). The degree of dimensional deformation that occurs provides some indication as to the level of molded-in stress which is typically greatest in the region of the part near the gate as shown in both Figure 3.7 and Figure 3.8.

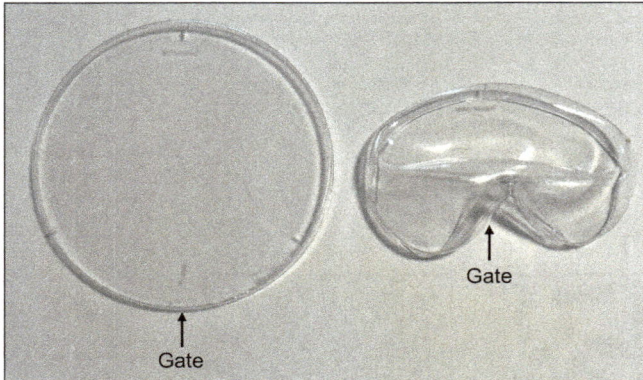

Figure 3.8 The thin-wall molded polystyrene on the left is fully functional, but it does have a relatively high degree of frozen-in molecular orientation and internal stress. The stress level is not high enough to cause deformation at room temperature. The duplicate part on the right is the result of a de-molding test by placing it in an oven heated to the polystyrene's heat deflection temperature (about 95 °C). This de-molding test can be used to help determine how various molding conditions influence the level of molded-in stress or frozen-in molecular orientation. Note that the maximum deformation occurs in the region of the part near the gate.

There are some applications where controlled molecular orientation is actually desirable, as in the case of a cable tie, where the orientation direction can be coincident with the direction of applied stress in the end application (i. e., uniaxial stress). The issues associated with frozen-in molecular orientation tend to be more severe for higher viscosity (lower MFR) material grades, and they are more significant for thin-wall moldings, since the skin layers represent a greater percentage of the overall part thickness.

3.3.4 Degree of Crystallinity

Mold temperature selection when molding a semi-crystalline plastic is important for the reasons discussed above, *and* because mold temperature will influence the degree of crystallinity for the molded part. The injection mold temperature will affect the solid skin layer thicknesses and the amount of frozen-in orientation for a semi-crystalline material in much the same way as it would for an amorphous material. This is because semi-crystalline plastics are partially amorphous and partially crystalline. In addition, the degree of crystallinity that develops as a semi-crystalline thermoplastic cools is dependent on the rate at which the melt cools and solidifies. A slower cooling rate gives the molecules more time to find themselves in spatial positions that favor the formation of crystals. At one extreme, a very thin-wall semi-crystalline thermoplastic part molded in a very cool mold cavity will solidify very quickly, limiting the time available for crystal formation. At the other extreme, a thick-wall semi-crystalline part molded in a relatively hot mold cavity will cool more slowly and will tend to exhibit a far greater degree of crystallinity. This relationship between melt cooling rate and the degree of crystallinity holds for most semi-crystalline materials, but its significance varies from material type to material type. For example, Figure 1.14 shows that very hot mold temperatures are required when molding polyphenylene sulfide (PPS). There is a critical minimum mold temperature that is required when molding this material in order to achieve the desired degree of crystallinity for optimum mechanical performance. This material, like most semi-crystalline plastics, depends upon its crystalline structure (or degree of crystallinity) to achieve optimum performance.

Mold temperature also has a significant effect on the mold shrinkage for a semi-crystalline plastic. The observed mold shrinkage will generally increase as mold temperature increases. This increase in shrinkage is associated with the increase in the degree of crystallinity that is achieved at the higher mold temperatures and slower cooling rate. The crystalline regions of a semi-crystalline plastic are more tightly packed than the amorphous regions, resulting in greater mold shrinkage when the degree of crystallinity increases. However, there can also be other confounding factors (other than degree of crystallinity) that influence the mold shrinkage of a thermoplastic. For example, increasing mold temperature typically results in a greater degree of crystallinity and mold shrinkage for a semi-crystalline thermoplastic, but a warmer mold temperature also allows for a wider process window. As an example, a warmer mold temperature will extend the gate seal time, allowing for a greater control over cavity pressure history and mold shrinkage compensation.

3.3.5 Mold Temperature Controllers

The auxiliary equipment that is used for mold temperature control can vary depending on the production facility infrastructure, mold size, mold temperature range and a number of other variables. It is common practice to use one (or more) mold temperature controllers (MTC) to cool or heat the mold to the desired mold temperature. MTCs are used to circulate the heat transfer fluid through the heat transfer channels (or cooling channels) of the mold cavities and cores. Treated water or a water glycol mixture are the most widely used heat transfer fluids. MTCs are electrically heated, semi-closed pumping systems. The pumping capacity of the MTC should be high enough to achieve the required coolant flow rate. The MTC units have an electrical heater that brings the coolant to the desired coolant (and mold) temperature. These systems also have temperature valves that will discharge hot water and introduce cooler water in order to maintain the set coolant temperature.

As stated earlier, most injection molds are not instrumented with a mold temperature sensor. Mold temperature is usually controlled by controlling the temperature of the mold coolant. Most MTC units have an internal temperature sensor that is used as the control thermocouple for coolant temperature control. Ideally, the MTCs have both delivery and return coolant temperature sensors for additional insights into the mold cooling process. Some MTCs also include provisions for the use of a remote temperature sensor. This option allows for the coolant temperature control thermocouple to be placed within the injection mold steel or a mold cooling channel.

Typical low-pressure water based MTCs can operate at temperatures ranging from the plant water temperature to around the boiling temperature of the cooling fluid [7]. This range of coolant temperatures is suitable for the majority of thermoplastics, but not all thermoplastics. Some engineering or high performance plastics (e. g., polyphenylene sulfide or polyether imide) require higher mold temperatures. In such cases, either (i) hot oil MTCs, or (ii) high-pressure water MTCs can be used to achieve the higher mold coolant and mold temperatures [8]. It should be noted here that there are a number of safety concerns associated with mold cooling, especially for high mold coolant temperatures. For example, neoprene/nitrile coolant hose with reinforcing braid may be used for conventional coolant temperatures, while hose having a fluoropolymer core with stainless steel over-braid (or equivalent) are required for applications involving higher coolant temperatures. Some high temperature stainless over-braid hoses are also available with a colored (blue or red) silicone covering that is easily identified and more easily cleaned [9]. All of the mold coolant hoses and hose connections should be checked for leaks, to eliminate the potential for injury, corrosion, electrical or slip hazards.

Use of Multiple MTCs or a Multi-Zone MTC

When MTCs are utilized for injection mold cooling, it is best to use multiple MTCs or multi-zone MTCs as opposed to using a single unit. When a single-zone MTC is used for mold cooling, the coolant must be looped from one half of the mold to the other as shown in Figure 3.9. The mold coolant fluid in Figure 3.9 first flows through the stationary injection mold half where it picks up heat energy and then flows through the moving mold half. In such a case, the coolant temperature entering the B-side mold half will be greater than that entering the A-side mold half. The temperature difference can be minimized using high coolant flow rates, but there will always be a ΔT between the mold halves if a single MTC is used. As a result, one surface of the molded part will tend to be warmer than the other surface of the part due to the non-symmetrical cooling rate. In such a case, the molded part will tend to exhibit more shrinkage on the hot side particularly in the case of a semi-crystalline material. The hot side of the part will also exhibit more post ejection shrinkage (for virtually any material) due to a greater ΔT for the hot side as the part cools to room temperature. This differential shrinkage through the thickness of the molded part can lead to both internal shrinkage stress and molded part warpage. There is really no way for a process technician to eliminate the non-uniform cooling when a single MTC is used.

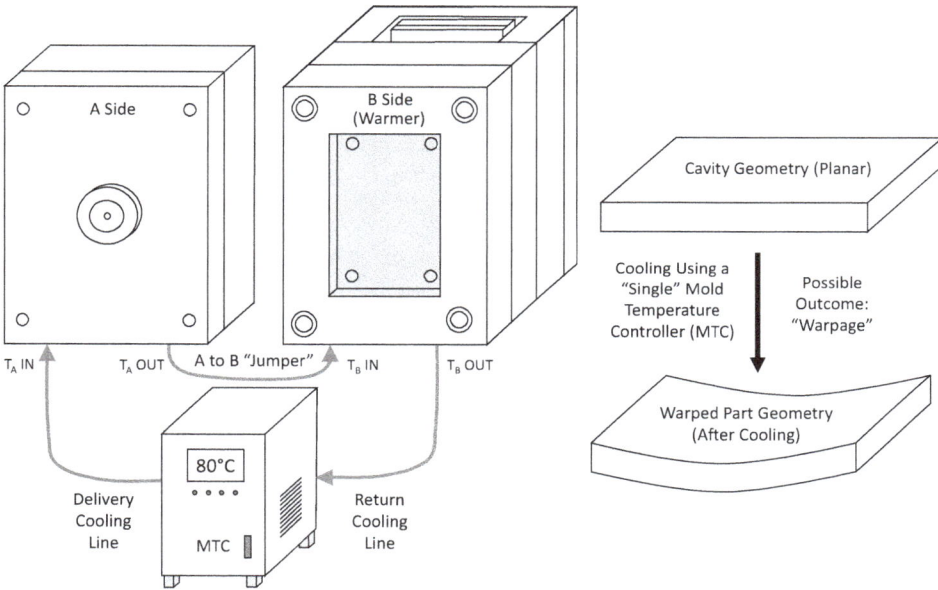

Figure 3.9 Using a single mold temperature controller to control both the A-side and B-side mold temperatures is *not* recommended. In this case, the coolant flowing through the B-side of the injection mold will tend to be warmer than that flowing through the A-side, creating an unbalanced temperature distribution through the thickness of the molded part, which can lead to warpage and/or internal shrinkage stresses.

The use of multiple (two or more) MTCs or a multi-zone MTC will greatly improve control over mold cooling and molded part warpage issues as shown in Figure 3.10. The use of a multi-zone mold temperature control (one MTC circuit for each mold half) minimizes the coolant travel distance and residence time, which limits the coolant temperature rise ($T_{OUT} - T_{IN}$). Many sources suggest the ΔT between the coolant inlet and the coolant outlet should never exceed 2 °C. The multi-zone approach to injection mold cooling also allows the process technician to balance the cooling so that both molded part surfaces cool at the same rate. Injection mold and part geometries can be very complex and it is virtually impossible to design an injection mold that has an equally effective cooling circuit for both the cavity and the core sides of the tool, even when cooling simulations are utilized. Having the ability to independently control the temperature of each mold half provides for significantly greater control over the mold cooling process and the uniformity of mold shrinkage through the thickness of the part to minimize the potential for molded part warpage.

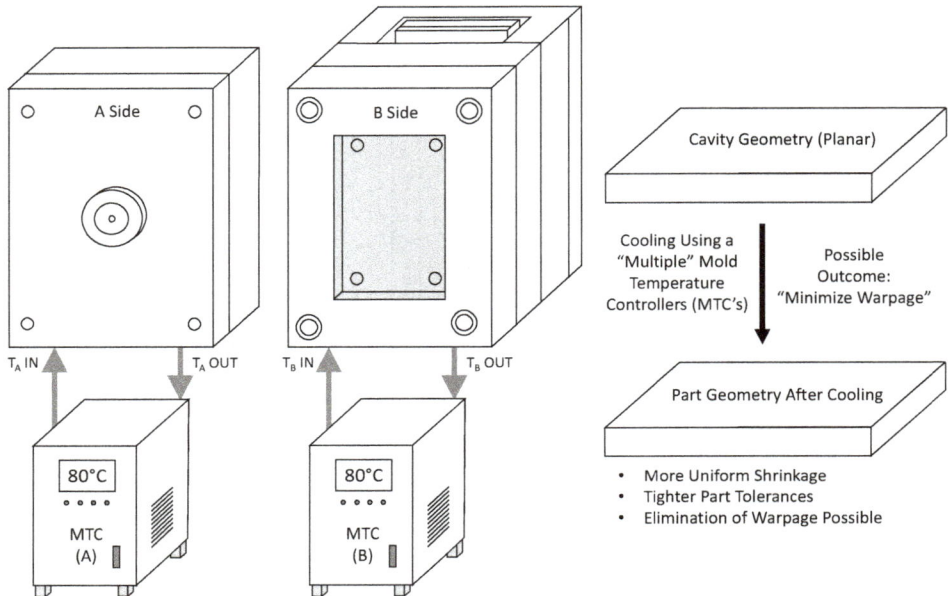

Figure 3.10 The use of multiple mold temperature controllers (MTCs), or a multi-zone MTC for the A-side and B-side mold halves offers a number of benefits. This can lead to more uniform mold temperatures by reducing the residence time for the coolant due to shortening the flow distance between the coolant fluid inlet and outlet. This also allows process technicians to adjust the A and B side coolant temperatures independently in order to balance the cooling rate through the part thickness to help minimize the potential for molded part warpage.

3.3.6 Mold Cooling Water Flow Rate

While mold coolant temperature is one critical variable associated with injection mold cooling, the cooling fluid flow rate is equally important. Mold temperature is a variable that is relatively easy to confirm or measure using a hand-held surface pyrometer (typically just after part ejection). Coolant flow rates cannot be determined unless the molding cell is equipped with flow meters. Most MTCs are not equipped with positive displacement pumps, so their flow rate will vary with the cooling circuit pressure drop. Coolant flow rates are often unknown or overlooked as important molding process variables.

The use of multi-port coolant distribution and collection manifolds offers a number advantages in terms of efficient injection mold cooling. They can improve neatness and housekeeping, reduce mold change-over times and hose management as shown in Figure 3.11. Coolant manifolds equipped with electronic, mechanical vane, or float type flow meters can be used to provide the mold coolant flow rate data. Without flow meters, there is no way to ensure that flow within the mold coolant channels is both adequate and repeatable for future production runs [11].

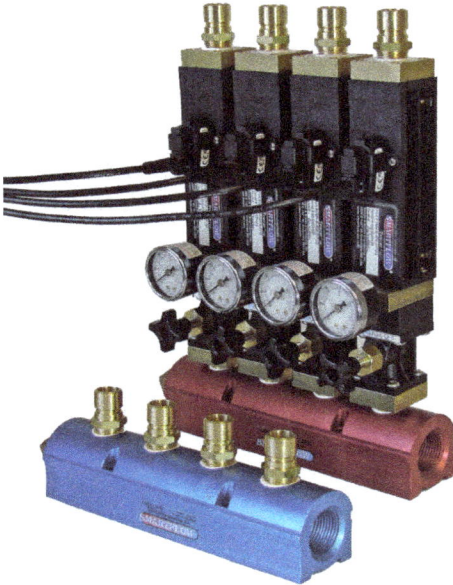

Figure 3.11
Multi-port, color coded mold coolant distribution and collection manifolds offer a number of advantages in terms of housekeeping and reducing mold changeover times. Flow meters should also be used to ensure that the mold coolant flow is both adequate and repeatable (Image provided courtesy of Burger & Brown Engineering, Inc., 4500 E 142nd Street, Grandview, MO 64030, www.burgereng.com).

3.3.6.1 Turbulent Coolant Flow

The most critical issue associated with mold coolant flow rate is ensuring that the flow rate in each cooling channel is high enough to achieve turbulent flow. At low volumetric flow rates, fluids such as treated cooling water, exhibit laminar or layered flow. It is very important that the coolant flow rate within the cooling channels is high enough to result in turbulent rather than laminar flow. Turbulent flow involves chaotic motion which provides for good coolant temperature mixing and improved heat transfer at the boundary between the coolant and the mold steel as shown in Figure 3.12. There is little or no temperature mixing with laminar flow, which causes the boundary layer to heat up, thereby reducing ΔT and heat transfer from the injection mold steel to the coolant.

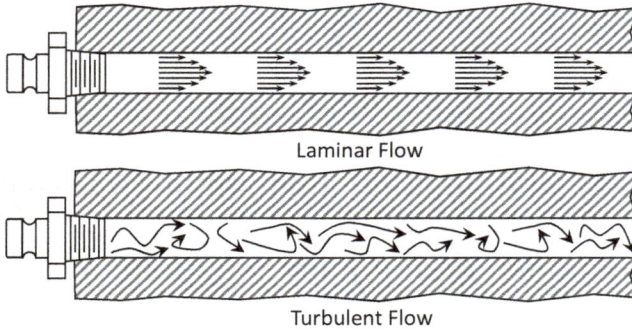

Turbulent Flow

Figure 3.12
The flow rates used for the injection mold coolant should be high enough to result in turbulent flow within the mold's cooling channels for efficient heat transfer.

The injection mold cooling process is actually a very complex heat transfer process involving several modes of heat transfer, as shown for the mold cavity cross sections in Figure 3.13. At the end of the mold filling phase of the process, the molded part's midplane temperature (2) is essentially the initial melt temperature. At the instant of fill, the cooler mold cavity surface temperature (1) begins to heat as the hot melt transfers heat to the cooler mold via conductive heat transfer. As the mold surface heats, the mold cavity steel (3) temperature increases and heat then transfers to the coolant (5) through the coolant boundary layer (4). An effective and efficient injection mold cooling circuit will remove heat from the mold steel as fast as the plastic part can transfer heat to the mold cavity steel. If this is not the case, the mold steel temperature will increase during the initial portions of the molding cycle, thereby reducing the rate of heat transfer between the plastic part and the mold steel.

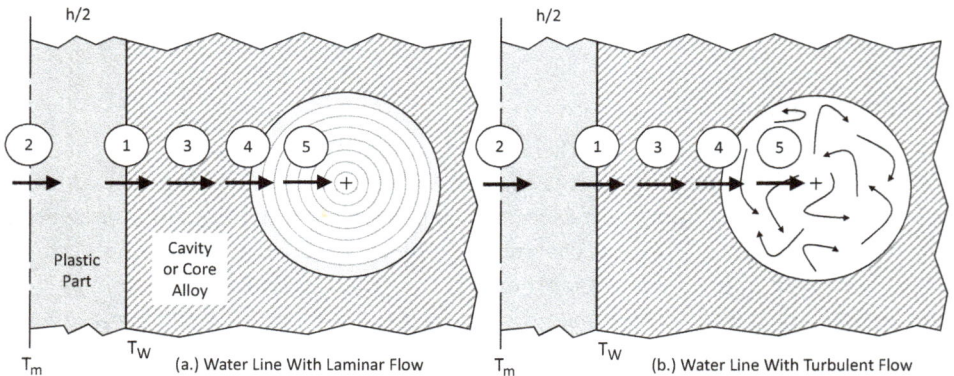

Figure 3.13 Low coolant flow rates that result in laminar flow reduce heat transfer from the mold steel to the coolant due to heating of the coolant boundary layer (labeled 4). Heat transfer is greatly improved by increasing the coolant flow rate so that turbulent flow is achieved. Turbulent flow improves heat transfer by providing for chaotic motion and coolant temperature mixing, maximizing the ΔT between the coolant boundary layer and the mold steel.

Mold cooling water has a relatively high heat capacity but a relatively low thermal conductivity. If the coolant flow is laminar, the coolant boundary layer heats up, stays hot, and thus reduces the rate of heat transfer between the mold cavity steel and the coolant as shown in Figure 3.13 (left). However, if turbulent flow is achieved as in Figure 3.13 (right), the coolant boundary layer is constantly replenished due to the chaotic mixing action associated with the turbulent flow. This constant replenishment at the boundary layer maximizes the molded part cooling rate. The optimum coolant rate for a mold cooling circuit will (i) result in turbulent flow and (ii) be high enough to keep the average coolant fluid temperature rise ($T_{OUT} - T_{IN}$) to no more than 2 °C.

It should also be noted here that the mold cooling channels should be located within the cavity and core mold steel that is in contact with the plastic melt (if their physical size permits). Cooling channels that are located in the support plate, or in cavity and core retainer plates, are less effective, as depicted in Figure 2.74. The partial mold cross section shown in Figure 2.74 has a plate interface between the molded part and the cooling channel that adds contact resistance, reducing the overall rate of heat transfer.

The coolant flow rates required to achieve turbulent flow can be determined using charts, equations, or on-line calculators [12]. The Reynolds number provides a theoretical indication of whether flow is laminar or turbulent. The Reynolds number is a dimensionless number that is the ratio of the inertial force divided by the viscous force for a flowing fluid. The transition from laminar flow to turbulent flow occurs at a Reynolds number of about 4000. Reynolds numbers greater than 4000 are suitable for injection mold cooling circuits but there is a point of diminishing returns since pumping energy also increases with increasing Reynolds number. Details regarding the cooling circuit layout (series vs. parallel layout), cooling channel geometry (shape, diameter), coolant fluid composition, and temperature are required for Reynolds number calculations.

$$N_{Re} = \sigma V \oslash / \mu \qquad\qquad\qquad (3.1)$$

Where:

N_{Re} = dimensionless Reynolds number

σ = coolant fluid density (at the coolant temperature)

V = average coolant velocity

$\oslash$ = diameter of the coolant channel

μ = dynamic viscosity of the coolant fluid (at the coolant temperature)

3.3.6.2 Cooling Channel Scaling

Scaling or fouling of the cooling channels can also be an important variable related to mold cooling. Cooling channel scale can cause a reduction in the coolant flow rate and boundary layer heat transfer. Some estimates suggest heat transfer rates can be reduced by as much as 20% if the cooling channels are not maintained properly. Coolant additives that help prevent corrosion and mineral deposits are the most common way to help prevent or minimize the scaling problem. The use of a coolant fluid having a neutral or slightly alkaline ph will help minimize cooling channel corrosion. The mold cooling circuits can also be maintained and cleaned by mechanical scouring or by closed loop cleaning solution circulation when the mold is off-line. Stainless steel cores and cavities are less prone to scaling or corrosion compared to other alloys. In some cases, mold cooling channels have been plated using an electroless nickel plating process as another means of minimizing cooling channel corrosion [13]. It is also beneficial to remove any residual coolant from the mold cooling circuit prior to mold storage. Once the coolant has been evacuated from the cooling circuit, a corrosion preventative can be applied before the mold is stored.

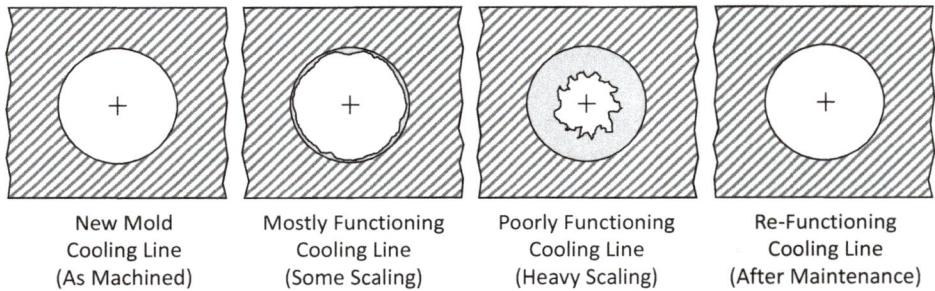

New Mold
Cooling Line
(As Machined)   Mostly Functioning
Cooling Line
(Some Scaling)   Poorly Functioning
Cooling Line
(Heavy Scaling)   Re-Functioning
Cooling Line
(After Maintenance)

Figure 3.14 Fouling or scaling of cooling channels due to corrosion or mineral deposits will also reduce mold cooling efficiency, leading to longer cycle times. Scaling can be minimized for water or water/glycol cooling systems using the appropriate anti-fouling additives. Periodic preventative maintenance is recommended to remove any deposits. This can be accomplished off-line using a pump and cleaning solution or by mechanical means [14].

3.3.7 Rapid Heat Cycle Molding Processes

The various issue associated with the solid (frozen) skin formation that occurs with the conventional injection molding process can be eliminated or minimized using one of the many rapid heat cycle molding (RHCM) processes [15–17]. There are many RHCM variants, but they all have common process features. Fundamentally, all RHCM processes involve rapid heating of the injection mold's cavity and core surfaces prior to mold filling. The cavity and core surface temperatures are heated to a temperature that is significantly greater than the mold temperatures used for the conventional injection molding process shown in Figure 3.4. The hotter-than-conventional mold temperatures allow for mold filling without solid or frozen layer formation. Once the injection mold is filled and packed using the RHCM process, the cavity and core surface temperatures are then rapidly cooled back to a more conventional mold temperature. This cyclic mold heating and cooling repeats each RHCM molding cycle, as depicted in Figure 3.15.

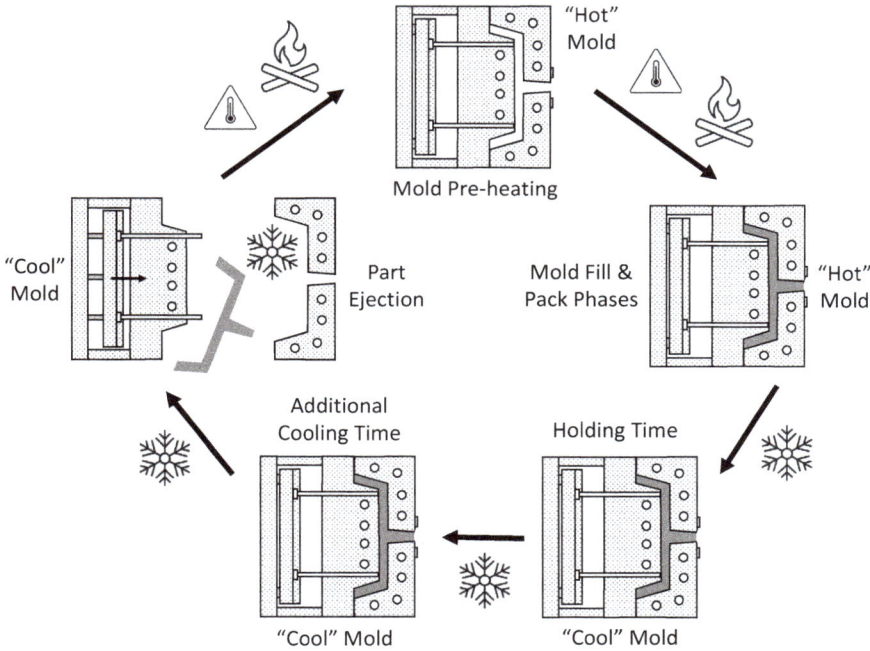

Figure 3.15 Rapid heat cycle molding (RHCM) processes involve rapid heating of the mold cavity and core surfaces prior to melt injection. This is done to eliminate or minimize solid skin layer formation during mold filling. Once the mold is filled and packed, the mold is then cooled as it would be in conventional injection molding. The rapid heating and cooling process repeats each molding cycle.

RHCM processes offer a number of advantages over conventional injection molding. Heating the mold cavity and core surfaces to a greater-than-conventional mold temperature helps prevent solid or frozen skin layer formation during mold filling, as shown in Figure 3.16. There are many advantages associated with injecting the thermoplastic melt into a hotter-than-normal mold cavity. These include improved part surface esthetics (or improved surface replication), improved knit line quality, greater flow length (especially for thin-wall molding applications), reduced pressure drop, lower molded-in stress, and better dimensional stability for the molded parts. Figure 3.16 also shows that the effective cavity thickness in which melt can flow is greater for the RHCM process, leading to a lower pressure drop since the mold filling pressure drop is inversely proportional to the cavity thickness cubed (at a given flow rate). RHCM process techniques have been used successfully in many applications, especially in applications with stringent cosmetic demands. While the processes are more energy intensive, they can eliminate the need for painting in applications that range from injection molded television bezels to molded automotive interior parts.

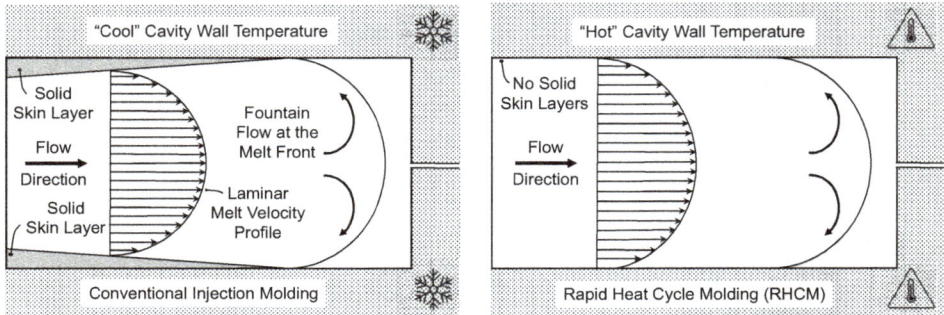

Figure 3.16 Mold temperatures used with the conventional injection molding process are cool relative to the melt temperature. This causes a solid skin (i. e., frozen layer) to form as the injection mold fills (left). Rapid heat cycle molding (RHCM) processes involve heating the mold cavity (or cavity surface) to a higher-than-conventional mold temperature in order to eliminate or minimize solid skin layer formation (right).

One team of researchers has studied the RHCM process for a neat polypropylene homopolymer [17]. The results of their study showed that the frozen skin layer thickness decreased as the temperature of the mold cavity surfaces (during melt injection) increased. The frozen layer was no longer detectable once at mold surface temperatures of 90 °C or higher. As a reference, the mold temperatures used for conventional injection molding with neat polypropylene are typically less than 50 °C [1].

While RHCM molding processes have clear advantages in terms of injection-molded part quality, RHCM processes are not as popular as conventional injection molding processes. The RHCM processes are more energy intensive and require additional equipment and process variables. This all stems from the fact that the mold cavity and core surface temperatures must be heated to greater-than-conventional temperatures prior to melt injection. They must then be rapidly cooled immediately after the mold is filled and packed. It is only the cavity and core surfaces that need to be heated for such a process to be effective but that is technically difficult. The vast majority of the RHCM processes heat more than just the surface of the core and cavity and may require heating and cooling of the entirety of the cavity and core mass. In the simplest cases, this can be done by alternating the circulation of hot and cold fluids within one set of mold heat transfer channels (e. g., steam during injection and water after injection). Other processes use two sets of heat transfer channels, one set for heating and one set for cooling. Keeping these heating and cooling channels close to the mold cavity and core surfaces improves efficiency. The development of cavity and core manufacturing techniques that allow for conformal heat transfer channels should be very beneficial for the development of RHCM processes. Channel design flexibility is greatly increased with conformal heat transfer channels.

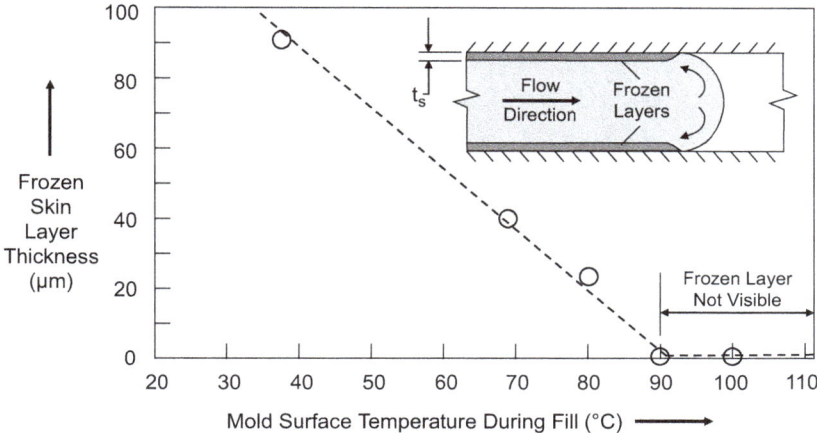

Figure 3.17 One rapid heat cycle molding process study was conducted with a neat polypropylene homopolymer at a variety of higher-than-conventional mold cavity surface temperatures (during mold filling only). The study results show that the frozen layer thickness that forms during mold filling decreases as mold surface temperature increases. The frozen layers were no longer visible for parts molded above a critical mold filling surface temperature of 90 °C in this case [17].

Other RHCM processes use a combination of electrically powered heaters (e. g., induction heaters, infrared heaters or resistance heaters) and conventional water cooling to achieve the rapid heat/cool molding process each cycle. The heaters can be integral to the injection mold or they can be robotically inserted into the open mold after part ejection.

3.3.8 Post Mold Annealing

Plastic parts are sometimes annealed in a secondary operation after they are produced by the injection molding process [18,19]. Annealing is a secondary operation that can be used to provide added control over a molded part's internal stress level or crystalline structure development (in the case of a semi-crystalline material). Annealing is a time-consuming secondary operation that adds cost and can be avoided in most injection molding applications. However, annealing is common in some injection molding applications where part performance is critical. Proper tool design and proper injection molding process condition selection are both key to eliminating the need for annealing.

Annealing is a thermal process where molded plastic parts are exposed to hot air, hot water, hot oil or infrared energy following very specific, material-dependent annealing protocols (times, temperatures). Annealing can result in molded part deformation and fixturing is required in some cases, further complicating this secondary operation. There are two primary reasons for annealing injection-molded thermoplastic parts. The first is to help relieve internal stress caused by either frozen orientation or differential cooling which causes differential shrinkage. Annealing can improve the chemical resistance and mechanical performance of the plastic part, particularly for amorphous plastics that are subject to environmental stress cracking and crazing (ESCC). Material suppliers can recommend annealing protocols that (i) minimize molded part deformation and (ii) minimize internal stress formation as the annealed part cools. A second common reason for annealing plastic parts is increasing the degree of crystallinity for semi-crystalline plastics. Molded parts that are annealed do exhibit improved performance and improved long term dimensional stability, but issues with deformation and higher molded part cost limit this practice to only the most critical applications.

A possible alternative to injection molded part annealing is to mold parts in such a way that the internal stresses are minimized and the desired degree of crystallinity (for semi-crystalline plastics) is achieved without the need for this secondary operation. Ideally, mold filling and cooling simulations can be used to optimize the tool design with regard to these issues. In addition, injection molding process conditions that minimize internal stress and promote crystallinity (for semi-crystalline plastics) help eliminate the need for annealing. Process variables related to mold filling, packing and cooling are all important. As an example, mold temperature is one of these variables. Material suppliers will always recommend a range of mold temperatures for a given material. It is common to select a mold temperature towards the lower end of this range to reduce cycle time, which is typically dominated by the time it takes to cool the part. Selecting a mold temperature at the higher end of this range will help reduce orientation related stress and promote greater crystallinity for semi-crystalline plastics. While it is true that a higher mold temperature will increase the molding cycle time, a longer molding cycle time is likely to be acceptable if the need for an annealing operation can be eliminated.

3.4 Barrel, Nozzle and Melt Temperature Considerations

The barrel and nozzle of an injection molding machine take time to heat; these temperatures therefore need to be set well before molding can begin. In addition, it is best to add a barrel heat soak time once the barrel is up to the set temperature to ensure that any thermoplastic material left in the barrel from the previous production run is fully melted. However, it is also important that the heat soak time is not excessive, as long heat soak times or cycle delays can lead to plastic material degradation. This is another example of why production planning is so important. Further, it is important that the required material pre-drying time (for hygroscopic thermoplastics) be coordinated with the time required for the barrel heating and heat soak time. The material to be molded should be fully dry by the time the barrel heat soak time has expired. It is also important to know what material was used for last production run, or as the final purge material for the last shut down, since some of that material will remain in the barrel. Hopefully that material is thermally stable at the new production temperature. If not, it may need to be purged with a more stable purge material (e. g., polystyrene or a stable purge compound). If, for some reason, the start of production is going to be delayed, barrel temperatures should be reduced during the delay time in order to minimize the potential for material degradation. This applies to both start up and production delays or interruptions.

The starting point in setting barrel temperatures is to review material supplier data sheets or processing guides. These can be very variable in their content but at a minimum, they should provide guidance as to the acceptable range of melt temperatures for a given thermoplastic material. The first step in setting the barrel temperature, or barrel temperature profile, is to select a target value for the melt temperature that falls within the range given by the material supplier. For example, the supplier literature for *Lexan*® 101, a general-purpose polycarbonate, indicates that the melt temperature should be in the range from 290 °–320 °C [5]. The material may not be fully melted at temperatures below 290 °C and may be degraded above 320 °C.

Selecting a melt temperature towards the lower end of this range will result in a shorter cooling time (all other factors being equal). However, the material will be more viscous at the lower melt temperature which could lead to mold filling issues (e. g., short shots, weak knit lines), especially for thinner-wall molded parts. Selecting a melt temperature towards the higher end of the range will result in a viscosity reduction, which in turn provides for lower fill pressures, improved surface finish, stronger knit lines and the like. Cooling time will be longer at the higher melt temperature, but part quality is generally improved. There is also a greater potential for material degradation at the higher melt temperature since the material heat history (total time at that melt temperature) will be greater. This is a particular concern in applications where the shot size utilization is low (i. e., there is a small shot), or when cooling time is very long, or both as a worst case. In such cases, starve feeding may also provide benefit by reducing residence time. In recent years, there has been a trend towards downsizing barrel shot capacity by molding machinery suppliers to help minimize the potential for material degradation especially for thin-wall applications. The smaller shot sizes are achieved by reducing the injection unit's plasticating screw diameter which also offers the added advantage of greater injection intensification ratios (i. e., higher melt pressure capabilities).

Once a target melt temperature has been selected, the next step is to determine what barrel temperature profile will provide for the target melt temperature. Some material suppliers will suggest barrel temperature settings as very rough guidelines, since so many other molding machine and process variables also effect the material's melt temperature. Variables such as screw type, screw compression ratio, L/D ratio, screw rotational speed, back pressure value, and cooling time all affect the material's melt temperature. In fact, most of the melting energy is associated with viscous dissipation during screw rotation (and even injection). It is also important to note here that melt temperature can be very difficult to measure, or measure in a repeatable manner, as discussed in Section 3.5.

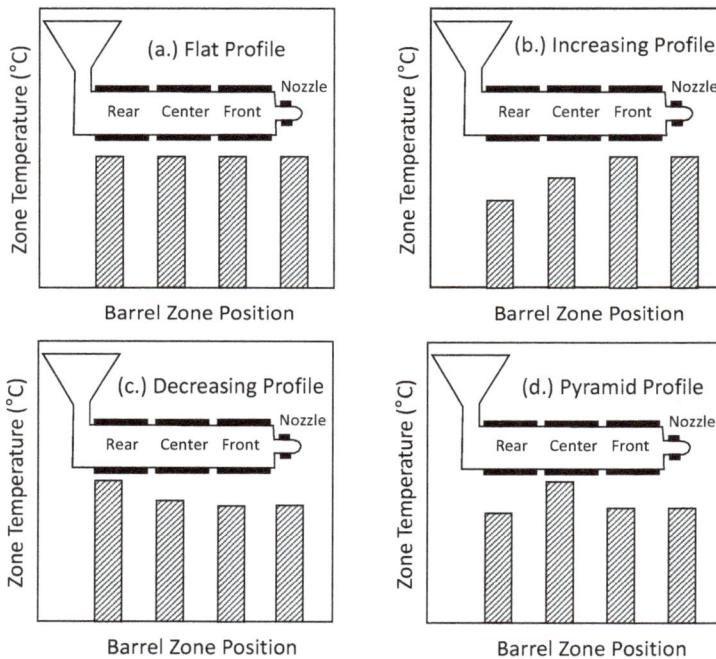

Figure 3.18 Barrel temperatures are typically set in an effort to achieve a target shot melt temperature (i. e., the average temperature of the plasticated shot). There are no simple rules in terms of setting barrel temperatures since many other variables including screw design, screw rotational speed, and back pressure also affect melt temperature. An increasing barrel temperature profile (b. above) is common in many injection molding applications. The increasing profile helps avoid premature melting that can cause bridging, air entrapment and other issues. Decreasing profiles or pyramid profiles are sometimes used, particularly for highly crystalline materials that are more difficult to melt. They also tend to reduce screw torque and compensate for plasticating screws that have a shorter than optimum L/D ratio.

Most injection molding machines have at least three barrel temperature zones and at least one nozzle temperature zone. The three barrel temperature zones loosely correspond to the feed section (zone 1) transition section (zone 2) and metering section (zone 3) of a typical plasticating screw. The nozzle temperature zone will typically control the temperature of the nozzle body. Some molding machines, particularly larger molding machines, can have up to six or more barrel temperature zones and it is helpful to understand where each zone begins and ends relative to the molding machine's screw geometry. Injection molding machines also have a feed throat temperature which is controlled using open- or closed-loop water cooling at the feed throat. A number of different strategies have been used to establish the optimum barrel temperature profile to achieve a specific melt temperature [20]. A few of the more common barrel temperature profiles are shown in Figure 3.18.

The most common barrel temperature profile is an increasing profile such as that shown in Figure 3.18b. One strategy in setting the increasing barrel temperatures is to set the temperature of *both* the last barrel zone (Zone 3 in Figure 3.18) *and* the nozzle temperature to the same temperature as the target melt temperature. The remaining barrel temperatures are then reduced by 10°–20°C per zone as you move backwards along the barrel towards the feed throat. The first barrel heating zone is then at a significantly lower temperature so that premature melting is less likely to occur. Premature melting can cause pellet feeding problems (like bridging) or air entrapment into the melt. Air must be able to escape through the feed throat as the pellets are compressed and melted while they travel along the plasticating screw. Premature melting can cause a melt seal that can result in air entrapment. Premature pellet melting in the feed throat area is more likely to occur if a flat temperature profile is set. An increasing barrel temperature profile is also common for applications where a volatile additive, such as a chemical blowing agent (CBA). This is common for structural foam molding, as discussed in section 4.3.

Some materials that are very difficult to melt, such as a highly crystalline material like nylon 66 or polybutylene terephthalate may benefit from a reverse barrel temperature profile or a pyramid temperature profile. Highly crystalline thermoplastic materials can be more difficult to melt than amorphous plastics and require greater heat input. They are also less likely to melt prematurely in the feed throat area. The reverse and pyramid barrel temperature profiles also tend to reduce screw torque and compensate for a shorter than optimum L/D screw.

It should also be confirmed that actual barrel temperatures are equivalent to the barrel set temperatures. Because most of the energy input during plastication is from viscous heating, barrel temperature zones, particularly those surrounding the compression zone of the screw, can override the temperature set point. This is most common in applications where the melt is very viscous and the screw compression ratio is too high. Unlike extruders, most injection molding machines do not have barrel cooling capabilities (i. e., two-mode temperature control). Temperature override may necessitate a screw change to a lower compression ratio screw or a reduction in plastication rate or screw rotational speed. Regardless of the strategy used to set the initial barrel temperature profile, the next step should be to confirm that the barrel temperature profile actually does result in a melt temperature that is close to the desired target value.

3.5 Evaluation of Melt Quality and Melt Temperature

Once the injection molding machine barrel and nozzle temperatures have reached the set points, additional heat soak time should be added to ensure that any residual thermoplastic in the barrel has been fully softened or melted. Once the heat soak is complete, material can be loaded into the hopper so that melt preparation or plastication can be initiated. The first shots that are generated are typically used to purge the residual material from the last production run and to ensure that the melt quality for the current material is acceptable in terms of consistency, mixedness, purity and bulk melt temperature. Barrel and nozzle purging is normally accomplished using an air shot technique with the injection unit in the retracted position, although in-mold purging is also possible. Air shot purging is generally effective at removing any residual material or contamination associated with the previous production run, but there are times where mechanical cleaning may be necessary. In such cases, the heated nozzle, screw, check valve and barrel are cleaned using copper gauze, brass tools and a barrel brush. These machine components are typically massive enough so that they can be cleaned more safely using residual heat immediately after the barrel and nozzle heaters are shut down.

3.5.1 Safety Considerations When Purging

Air shots can represent a significant safety hazard and as such, proper purge safety precautions should be followed. Both the melt pressures and melt temperatures associated with the purging process can be dangerously high. As a result, effective personal protective equipment (PPE) is required for the operator purging the barrel. Molding machines are typically equipped with an interlocked purge guard mounted to the stationary platen that will prevent molding machine operation if the guard is not in the fully closed position. An open purge guard should prevent melt injection, but an overheated or chemically unstable shot of melt could unexpectedly blow-out through the retracted nozzle. In such a case, the PPE is then the last line of protection for the operator. The PPE will help prevent possible injury, especially burns from the high temperature molten thermoplastic. Such unexpected blow outs can occur for a variety of reasons, but a long cycle delay that increases residence time is a very common trigger for such an unexpected event.

At a minimum, PPE for the purge operator should include; heat-resistant long gloves that include forearm protection (like welding gloves), full face protection (not just safety glasses with side shields), and a good quality (non-disposable) lab coat as shown in Figure 3.19.

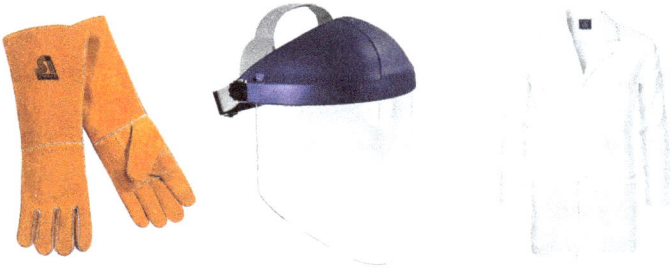

Figure 3.19 Purging the barrel of an injection molding machine can be dangerous due to the high pressure and high temperatures associated with injecting thermoplastic melt into the purge area. The purge operator should have the right personal protective equipment (PPE) to help prevent possible injury, especially burns from molten plastic. The purge PPE should include at least long heat-resistant gloves that provide hand and forearm protection (e. g., *Kevlar®* or welding gloves), full face protection, and a protective lab coat [21].

Proper ventilation of the purge area can also be an important safety consideration. Typically, there is more off-gassing when thermoplastic melts are purged (i. e., air shots) than during normal molding. During normal molding, the melted shot is injected into a relatively cool, closed mold where it cools fairly quickly. During purging, the molten air shots remain hot for a much longer period of time and more off-gassing tends to occur than during the normal production run. Molding machines may not be supplied with purge area ventilation so this is typically a facilities issue for the molding plant. The concept of local purge area ventilation is shown in Figure 3.21. Purge area ventilation ducts can incorporate a sliding gate that is opened during purging and fully or partially closed during production molding so that the airflow does not interfere with barrel or nozzle temperature control. Self-contained, portable fume extractors that can be brought to the molding machine during purging are also available [22]. Purgings, especially those from larger shot size barrels, can stay dangerously hot for long periods of time unless they are cooled in water filled purge pails.

3.5.2 Visual Observations When Purging

While there are many variations on the techniques used to evaluate purge melt quality, they generally involve examining air shot purging once all of the previously molded material or contamination has been purged from the barrel. The air shots are ideally generated at molding conditions that are close to those expected in production. Visual observations provide an initial evaluation to ensure that there are no major issues with melt quality and a follow-up evaluation is recommend once process conditions have been dialed in. Several plastication process variables must be set to allow for visual observation of the purge shots. These include screw rotational speed and back pressure.

Screw rotational speed: Up to this point in time, only the drying conditions, mold temperatures and barrel temperatures have been set. Other process variables, such as the shot size, screw rotational speed, and back pressure, have yet to be determined and optimized, but preliminary settings or estimated values for these variables will need to be set so that purge shots can be generated. For example, setting the screw rotational speed to a medium value will allow for plastication and purging, but this value may be increased or decreased for the actual production run. Because screw rotation speed does affect melt temperature and melt quality, this melt quality evaluation process is best described as the initial melt quality evaluation process. A second melt quality evaluation is recommended once the final process variables to be used in production are dialed in. One source recommends that, prior to plastication, a momentary injection or decompression of the screw is initiated to ensure the screw moves freely and to clear the check valve before initiation of screw rotation [23].

Back pressure: The back pressure value used during air shot purging is usually set very low or to zero, depending on the desired method of purging. Back pressure adds resistance to screw reciprocation during plastication as discussed earlier (see Section 2.2.2.4.2). Adding back pressure when the injection unit or carriage is retracted will generally cause the injection unit to simply extrude material through the open nozzle during screw rotation and not allow for shot build-up (unless a shut-off nozzle is used). Adding back pressure so that the melt extrudes through the nozzle is, in fact, one method of purging; this is often a multi-mode process involving extrusion followed by small shot purging. Reducing back pressure minimizes the likelihood of extrusion and allows for shot generation. In some cases, particularly for very low viscosity melts (i. e., high MFR materials), decompression prior to plastication (i. e., screw suck back to a value near the shot size) can also be used to minimize melt extrusion and help with shot generation when the injection unit is retracted. Shot build-up is less of a problem with viscous melts (i. e., low MFR materials) or when a shut-off nozzle is utilized.

Visual observations: Visual observations can be helpful when evaluating melt quality, but not all purgings are expected to have the same appearance. Evidence of contamination (e. g., black specs) would indicate that additional purging is required with either the production material, purge material or a commercial purge compound [24, 25]. Any indication of foaming indicates that there may be an issue with material pre-drying or possible air entrapment. Observations or conclusions regarding color uniformity can be difficult to make at this point since plastication conditions (especially the back pressure value) will be different than those used during production. One important note here is that air shot purgings often appear to show a brown discoloration, particularly if the material is modified with butadiene rubber (e. g., ABS). This discoloration is most likely due to oxidation of the hot purging since the purging is exposed to oxygen for an extended period of time. While such discoloration could indicate that the melt is too hot, this type of oxidation may not occur when the melt is injected into the relatively cold mold during production.

The overall shape of the purging will also vary depending on the melt viscosity of the material. Purgings all begin as individual strands of material that have been extruded or injected through the molding machine's nozzle orifice. These strands are hot and tend to merge into one mass as shown in Figure 3.20. Merging of the strands tends to be incomplete if the material is very viscous. This may be an indication that the melt temperature is lower than optimum, but the melt's bulk temperature should be measured independently as confirmation. Some more viscous material's exhibit this behavior even when the melt temperature is within the recommended range.

Purgings for lower viscosity materials tend to form more of a pool of melted material. Very low viscosity, hot or overheated melts can even splash during purging validating the importance of the machine purge guards and personal protective equipment. It is best to avoid very fast injection velocities with such materials when purging to avoid the possibility of splash-back. Very low viscosity melts are sometimes so fluid that they can drip through purge pan gaps, then solidify, fusing them in place. It is for this reason that spraying a mold release agent onto the stationary platen and purge pan surfaces is recommended prior to purging. Removing solidified plastic from the purge area is far easier if a release agent has been used. This can be avoided by purging into a heat-resistant container, ideally an insulated container that both contains the melt, and helps keep it warm so that a more accurate melt temperature can be measured.

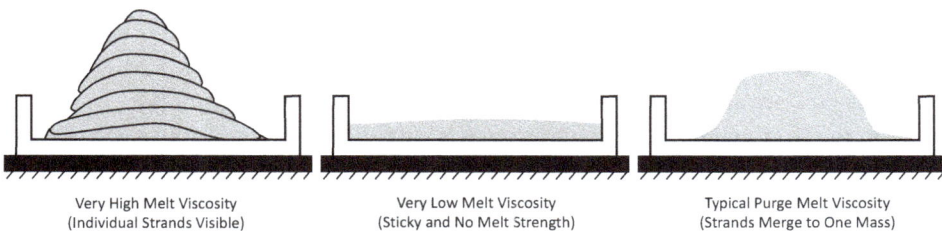

Very High Melt Viscosity
(Individual Strands Visible)

Very Low Melt Viscosity
(Sticky and No Melt Strength)

Typical Purge Melt Viscosity
(Strands Merge to One Mass)

Figure 3.20 All purgings start as strands that are extruded or injected through the molding machine's nozzle. The individual strands tend to merge into a more uniform molten mass as purging continues. The strands may only partially merge in the case of very viscous melts while very fluid melts form more pool-like homogeneous purgings. Very low viscosity melts may be fluid enough to drip through purge pan gaps and fuse in place on cooling. This problem can be avoided by purging into a heat-resistant container. Spraying the purge pan area with mold release will minimize purge adhesion and assist with clean-up of the solidified purgings.

Hot runner purging: Hot runner systems also require cleaning or purging if they become contaminated or there is to be a molding material change. Physical cleaning of hot runners is time consuming as it requires disassembly of the hot runner system, but is sometimes required. However, hot runner systems are more commonly cleaned by purging, either with a purging compound or the new molding material. If commercial purge compounds are to be used, it is best to consult with the purge compound supplier to ensure the compound grade is suitable for the application [24,25]. It is also good practice to consult with the hot runner supplier for specific purging recommendations

Hot runners can be purged using either an open mold technique or a closed mold technique [26,27]. Regardless of the technique to be used, it is best to purge the molding machine barrel and nozzle before purging the hot runner. Mechanical or abrasive purge compounds that contain scouring additives, such as glass fiber, should not be used if the hot runner gates are very small. One technique for closed mold purging is to inject shots with no packing or holding time. The shots should be short but full enough to allow for part ejection (e. g., for drafted stripper plate part ejection). The use of mold release is suggested for closed mold purging, if the application allows this, in order to minimize ejection forces. Hot runner temperatures (especially hot nozzle tip temperatures) are typically set on the higher side of the acceptable melt temperature range when purging to minimize material viscosity and pressure drop. This technique is said to be most appropriate for high cavitation hot runner molds. It is recommended that the specific hot runner supplier be consulted for their hot runner purge procedure recommendations.

3.5.3 Measuring Bulk Melt Temperature

Once purging of the contaminated melt is complete, it is good practice to measure the actual bulk temperature of the melt itself (i. e., the melt temperature). As stated earlier, barrel and nozzle temperature zones are set based upon a target melt temperature that falls within the range deemed appropriate by the thermoplastic material supplier (or based on previous experience with the same material). On-line melt temperature measurement is common for many plastics processes such as extrusion. However, on-line melt temperature measurements are popular for injection molding in large part due to the high material velocities that occur during melt injection.

Shear heating and melt stagnation issues both limit the use of intrusive melt thermocouples for most injection molding applications. While intrusive (in-stream) nozzle melt thermocouples are available for injection molding nozzle bodies, they are not widely used. Non-intrusive melt thermocouples (e. g., infrared temperature sensors, flush mount melt thermocouples) eliminate the shear heating and stagnation issues associated with intrusive thermocouples, but they have not been shown to provide a reliable indication of average bulk melt temperature. Flush-mount thermocouples provide little or no information as to the average bulk melt temperature. Infrared temperature sensors measure total infrared energy emissions, not temperature. The methodologies for correlating the total infrared emission with average bult temperature can be problematic without a detailed understanding of the thermoplastic material's infrared emissivity. As a result of these issues, the most common way to measure the average bulk melt temperature of the plasticated shot is to measure the temperature of air shot purgings as depicted in Figure 3.21 [24, 28].

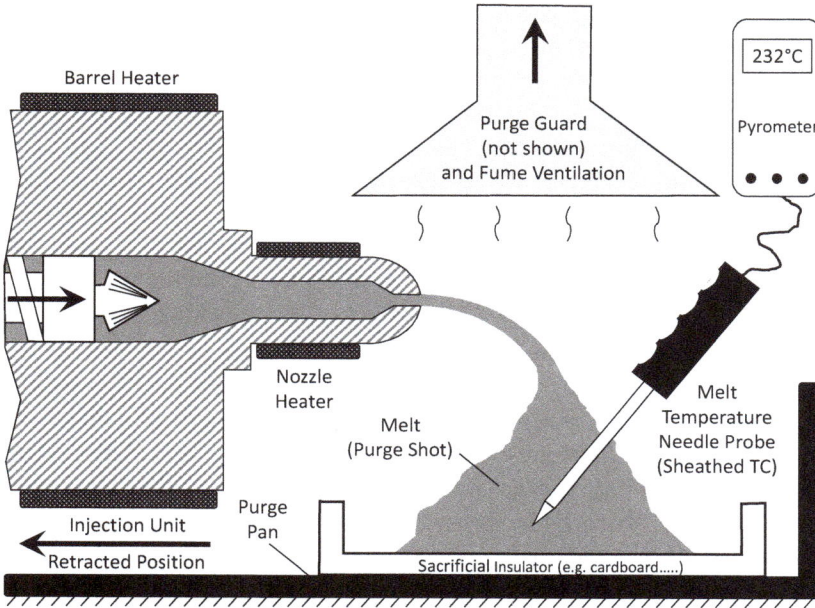

Figure 3.21 The average bulk temperature of the purged plastic melt is most commonly measured using a hand-held melt pyrometer and needle thermocouple probe. Ideally, the temperature probe of the pyrometer that is inserted into the purged melt should be a low mass, sheathed needle thermocouple that has a fast response time and is easy to clean. The probe is repeatedly inserted into multiple purge shots until the melt temperature reading plateaus or stabilizes. Purging into an insulated container helps keep the melt from cooling, especially for smaller shot size purgings. The purge area should also be ventilated to capture any fumes or off-gassing from the air shot.

It is important to note here that the temperature of the melt within a shot can be very variable. Studies have shown that the temperature variation within a shot can be 20 °C or more depending on screw design, shot capacity, plastication conditions, and cycle time [29, 30]. For example, slower screw rotation speeds tend to produce a more uniform melt temperature when compared to higher screw speeds. This is why the plastication conditions used for this initial melt temperature evaluation should be similar to those expected during production. It can be very difficult to obtain a reliable or repeatable indication of melt temperature and, as such, the tools and techniques associated with the measurement should be documented and standardized to the extent possible. While there are many variations associated with melt temperature measurement procedures, the basic melt temperature measurement concept follows.

Melt temperature measurements are most difficult for small shot sizes because the small purgings cool quickly. Purging into an insulated container can help minimize heat loss especially for smaller shot sizes. The bulk melt temperature value is obtained by inserting the fast-response melt thermocouple into the air shot. Exposed junction thermocouples tend to have the fastest response times, but they are more difficult to clean once the melt solidifies. Fast response sheathed needle type melt thermocouples probes are recommended because they are more easily cleaned than exposed junction thermocouples. Heavier gage thermocouples are more durable or robust, but have slower response times so these factors (i. e., response time vs. probe durability) must be balanced. Mold release can also be sprayed onto the probe to reduce the adhesion between the melt and the probe to facilitate cleaning.

The sheathed melt thermocouple should be inserted into the purging such that the tip of the probe is near the center mass of the purging. Multiple purge shots are typically required before an accurate and stable melt temperature reading can be obtained. Melt temperature probes are sometime pre-heated to minimize heat transfer from the melt to the needle probe but this must be done in a very controlled manner for consistent results. Inserting the probe into repeated purge shots is a more common approach that minimizes the need for melt probe pre-heating. A brass spatula placed on the purging adjacent to the probe is helpful when inserting and removing the probe from the purging so that the operator's protective gloves never come in contact with the hot mass of purged thermoplastic. It can also be difficult to maneuver the melt probe *and* observe the pyrometer reading simultaneously so a pyrometer with a peak-hold capability is helpful. The experimentally measured melt temperature value is typically reported as the temperature plateau or the highest repeatably observed bulk temperature reading.

One manufacturer, MTMS, has developed a more repeatable melt temperature measurement system that is designed to minimize operator variability and improve both accuracy and repeatability [31]. The system uses a magnetically mounted, insulated purge puck that fits within the stationary platen opening directly opposite the molding machine nozzle. A disposable insulated cup is first placed into the puck. A sheathed melt thermocouple probe is then positioned precisely at the center of the insulated cup. Melt is then injected into the cup and the peak temperature is recorded as the thermoplastic's melt temperature. The insulated puck helps minimize heat losses, thereby improving accuracy, especially for smaller shot sizes. The precise thermocouple placement helps to improve repeatability and minimize operator related errors.

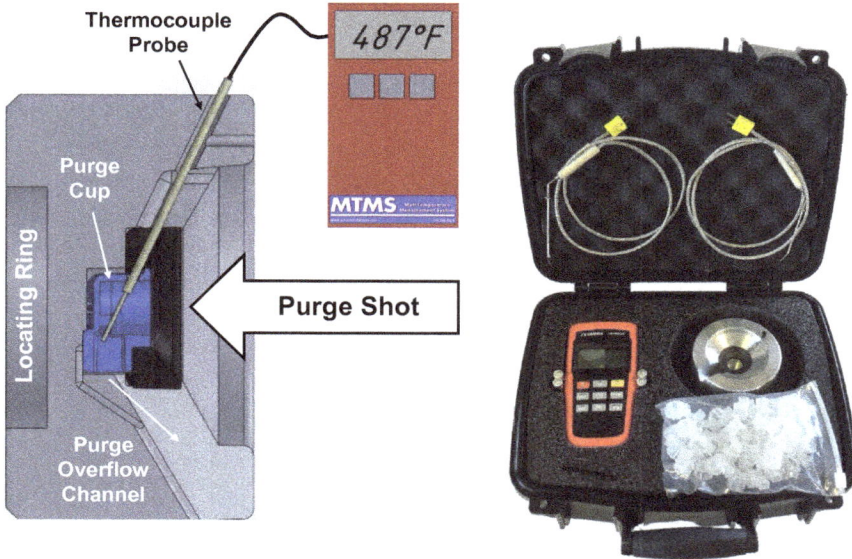

Figure 3.22 The purge puck melt temperature measurement system uses an insulated cup and a precisely located thermocouple to improve both the accuracy and repeatability of melt temperature measurements. [Courtesy MTMS, LLC, Janesville, WI 53546, *www.PlasticMeltTemp.com*].

3.6 Phases of the Injection Molding Process

The injection molding process is a complicated, cyclic process involving a large number of both sequential and concurrent process phases and process variables. Many of these process phases are mandatory, while other phases of the molding process are optional. For example, the mold filling phase is a mandatory process phase common to all injection molding applications, while the use of sprue break is an optional process phase, used only in some applications. Both basic and more detailed descriptions of each process phase are given in the following sections. In addition, each process phase has variations or options that may be utilized. For example, the injection velocity used during mold filling can be constant or it can be variable or profiled. These variants, and the methodologies or philosophies associated with setting the process variables associated with each process phase are also discussed.

While it is impossible to fully describe the injection molding process using a single figure, a Gantt chart showing the process phases, such as that shown in Figure 3.23, is helpful in describing the sequencing of the molding process phases. The injection molding cycle depicted in Figure 3.23 shows the relative times associated with a typical injection molding process. The overall cycle time required to produce a plastic part is generally dictated by the time it takes for the part to cool and is therefore largely dependent upon the nominal wall thickness of the plastics part(s). The overall cycle time for an injection molding process can range from a few seconds for very thin-wall molded parts to multiple minutes for very thick-wall molded parts. As a reference, parts having a nominal wall thickness of say 2.0 mm may have an overall cycle time of about 20 seconds (+/-). Other variables, such as melt temperature, mold temperature, material type, and mold design also have an impact on the cooling time that is required to produce the thermoplastic part. While the range of possible overall cycle times can be quite variable, the relative times associated with each process phase for a typical injection molded part are similar to those shown in Figure 3.23.

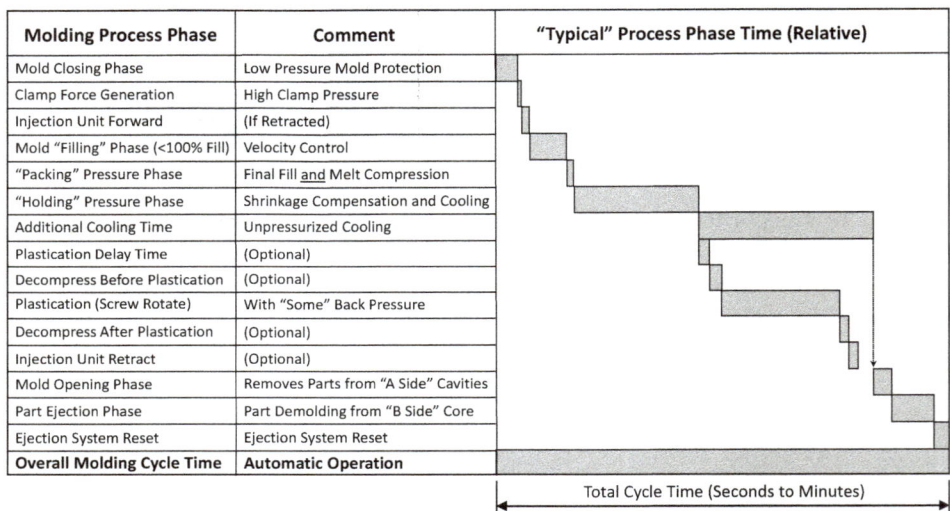

Molding Process Phase	Comment	"Typical" Process Phase Time (Relative)
Mold Closing Phase	Low Pressure Mold Protection	
Clamp Force Generation	High Clamp Pressure	
Injection Unit Forward	(If Retracted)	
Mold "Filling" Phase (<100% Fill)	Velocity Control	
"Packing" Pressure Phase	Final Fill and Melt Compression	
"Holding" Pressure Phase	Shrinkage Compensation and Cooling	
Additional Cooling Time	Unpressurized Cooling	
Plastication Delay Time	(Optional)	
Decompress Before Plastication	(Optional)	
Plastication (Screw Rotate)	With "Some" Back Pressure	
Decompress After Plastication	(Optional)	
Injection Unit Retract	(Optional)	
Mold Opening Phase	Removes Parts from "A Side" Cavities	
Part Ejection Phase	Part Demolding from "B Side" Core	
Ejection System Reset	Ejection System Reset	
Overall Molding Cycle Time	**Automatic Operation**	

Total Cycle Time (Seconds to Minutes)

Figure 3.23 The injection molding process is a cyclic process involving a large number of both sequential and concurrent process phases. Many process phases are mandatory, while others are optional. Some process phases have a very short time duration while others are more time consuming. The relative times associated with these process parameters vary depending on the specific application. The process sequence and relative timeframes for a typical injection molding process are shown above. The overall cycle time can range from several seconds to several minutes depending primarily on the molded part's nominal wall thickness.

A brief description of each injection molding process phase is given below. More detailed descriptions of each injection molding process phase are provided later in this chapter. The more detailed descriptions also include different methodologies or philosophies that can be used to help determine the optimum process variable associated with each phase of the molding process.

Mold closing phase: The molding cycle begins by initiating the mold closing process. The mold close speeds are typically set to minimize mold close time while mold closing pressures are set low enough to prevent possible tool damage from an incomplete ejector system reset or imperfect or incomplete molded part ejection.

Clamp force generation: Once the faces of the injection mold halves touch, the machine clamp goes into a high-force mode in which the required clamp tonnage is generated. The clamp force value can be adjusted based on the total molding projected area and peak packing pressure. Excessive clamp force can damage both the injection mold and the molding machine clamp. Insufficient clamp force results in flash. The clamp force is typically set to be just greater than that required to prevent flash.

Injection unit forward: The nozzle of the injection unit must be brought forward so that it mates with the injection mold's sprue bushing. If sprue break is not to be used, the nozzle will remain in the forward (home) position for the duration of the production run. If the optional sprue break mode is used, the injection unit retracts a short distance (typically after plastication) and moves to the forward position prior to the injection phase each molding cycle. The optional sprue break mode of operation can be used for a variety of reasons. As an example, sprue break is common in cold runner molding applications where nozzle freeze-off is encountered. The use of sprue break minimizes the conductive heat loss from the nozzle tip to the cold sprue bushing.

Mold filling phase: Once the clamp force has been generated, the melted shot can be injected into the injection mold. The plasticated shot must contain enough material to fill the mold and allow for post fill compression or pressurization. Most current generation injection units utilize a velocity-controlled mold filling phase. As a result, the mold filling phase is a short-shot injection process where the mold is typically filled 95–99% of the way. The final fill and melt compression phases are accomplished after transfer from velocity control to the packing pressure phase.

Packing pressure phase: Unlike the velocity-controlled mold filling phase, the packing phase is pressure controlled. The packing phase is used to both (i) complete the final mold fill, and (ii) cram or pack more of the compressible thermoplastic melt into the mold. Part quality attributes such as surface replication, knit line quality, dimensions and the like are dependent upon both the magnitude of the packing pressure and packing pressure time. The pack pressure must be high enough to complete mold filling, but low enough to avoid flash. Pack time is typically very short relative to the overall mold filling time (e. g., about 1/10 of the mold fill time as a ballpark value).

Holding pressure phase: The holding pressure (or follow-on pressure) phase of the injection molding process is essentially a continuation of the pack pressure phase, but usually at a lower pressure and for a longer time. The purpose of holding pressure is to keep the compressed melt in the cavity. This helps compensate for mold shrinkage and provide for fine control over the molded part dimensions. Hold pressure can also cram additional material into the mold cavity as the part shrinks. The hold pressure is typically set based on target part dimensions and/or to help eliminate shrinkage defects such as sink marks or voids. The hold time is typically set to be equal to or less than the time required for the gate to freeze or solidify.

Additional cooling time: As stated above, hold time is typically set based on gate freeze time. Gates cool faster than the molded parts themselves due to their normally thinner cross section. While the part begins cooling as soon as it is injected, it will not be solidified enough for ejection at the end of the holding time. Additional cooling time (commonly known as cooling time) is generally required before the part has solidified to the point where the mold can be opened and the parts can be ejected without distortion. This additional cooling time is also used as the time to generate the shot for the next molding cycle (concurrently). The injection mold is typically set to open shortly after the additional cooling timer expires.

Plastication delay time: Plastication delay time is an optional time setting that may or may not be required. A short delay time (e. g., sub-second) is sometimes used to allow time for residual holding pressure to decay. Residual holding pressure can increase check ring seat wear at the start of screw rotation. A longer plastication delay time is sometimes used when the required screw recovery time (plastication time) is significantly shorter than the additional cooling time. The thought is that thermal gradients can develop in the full shot if the shot remains in the barrel for an extended period of time. It is typical to set plastication conditions such that the mold opens shortly after the plastication process is complete.

Decompress before plastication: Screw decompression or suck back (axial screw retraction without rotation) can occur before or after plastication. Decompression before plastication (a.k.a. decompression forward) is also an optional injection molding process step. A short decompression stroke is very effective at relieving any residual holding pressure and can reduce check ring seat wear. Using a very large decompression stroke (slightly less than the shot size setting) allows the screw to plasticate the shot with virtually no backpressure since the screw is pumping into a void space. This may be an advantage in applications where (i) plastication energy input needs to be minimized or (ii) for applications where screw recovery rates are excessively slow (e. g., using a screw with an incorrect compression ratio or using a badly warn screw). It should be noted that air can be sucked into the shot chamber (e. g., through the nozzle/sprue seat) when using screw decompression before plastication.

Plastication: Plastication or charging the shot occurs as the screw rotates (and reciprocates). The screw literally pushes itself backwards towards a pre-set shot size limit as melt is pumped forward by the metering section of the screw. The screw will rotate at the set rotational speed until the shot size set point (axial distance) is reached. The time required for plastication is determined by the screw rotational speed, back pressure setting, and pumping efficiency. The screw rotational speed is typically set such that the screw reaches the shot limit shortly before the injection mold opens. Studies have shown that slower screw rotational speeds tend to produce a more uniform melt temperature.

Decompress after plastication: Decompress or suck-back after plastication (a.k.a. decompress rear) is also an optional process variable, but it is very commonly utilized. A primary reason for using suck-back is to relieve residual back pressure which is likely to cause drool from the nozzle tip orifice when the mold opens. The nozzle tip is blocked by the sprue (from the previous shot) but nozzle drool (or hot runner tip drool) can occur as soon as the mold opens if the residual back pressure is significant. Nozzle drool is even more problematic when sprue break is used. Decompress after plastication increases the shot chamber volume, thereby eliminating any residual back pressure. Decompress after plastication can eliminate the need for a shut-off nozzle in some but not all applications.

Injection unit retract: If sprue break is being used, the injection unit will retract a short distance either before or after plastication and decompression (most commonly decompress after plastication). Decompressing the shot will help eliminate drool for free flow nozzles. A shut off nozzle is typically required if sprue break before plastication is to be used. Either way, the use of sprue break helps minimize conductive heat loss from the hot nozzle tip to the sprue bushing for cold sprue applications.

Mold opening phase: The injection mold will open once the additional cooling timer has expired. This additional cooling time must be long enough to ensure that (i) the molded parts are cool enough to be ejected with any permanent deformation, and (ii) plastication of the shot for the next molding cycle is complete. Mold opening velocities are typically profiled to include a slow breakaway speed, followed by a fast traverse speed and then deceleration.

Part ejection: Part ejection is typically accomplished by connecting the injection molding machine's ejector plate to the injection mold's ejector system. Ejection is initiated once the injection mold is fully open or while it is opening. Injection molds containing angle pin actuated slides or split molds are example of molds where the mold opening movement assists with part ejection.

Ejector system reset: Once the molded parts have been ejected from the injection mold, the ejector system must be reset to the home ejector position. This can be accomplished by resetting the molding machine ejector plate for applications where the mold's ejector plate is rigidly connected to the machine ejector plate. Ejector springs can also be used for the initial ejector system reset. The mold's "return pins can be used to reset the mold's ejector system but their main function is to ensure the mold ejector plate is accurately returned the home position when the mold is fully clamped. Some ejection mechanisms, such as angle pin activated slides, are reset as the injection mold closes.

3.7 Injection Unit Carriage Position Options

3.7.1 Operation Without Sprue Break

The injection unit of an injection molding machine is sometimes referred to as the carriage or sled. The carriage or sled movement is controlled using either (i) hydraulic cylinders (pull-in cylinders) or (ii) an electronic servo motor. These actuators are used to move the sled forward and backward and provide the nozzle contact force. The nozzle contact force, or touch force, holds the nozzle tip against the mold's sprue bushing to prevent melt leakage during the pressure phases of the molding process. During purging, the carriage is moved to the rear position so that melt can be purged into the molding machine's purge pan area. After purging and initial shot plastication, the injection unit is moved to the forward position where the nozzle seats against the injection mold's radiused sprue bushing. Having a good seal between the nozzle and the sprue bushing is required in order to prevent leakage of melt during the pressure phases of the molding process. A good seal can also minimize air entrapment during the decompression or suck-back phases of the molding process. The nozzle contact force is also required to prevent blow-back of the injection unit due to the applied packing and holding pressures. The nozzle contact force can have a magnitude of about 5–10% of the molding machine's clamp tonnage [32]. This force is normally sufficient to prevent blow back or seal leakage if the nozzle tip and sprue bushing radii are matched correctly, and the nozzle tip and sprue bushing faces are in good condition.

When sprue break is not utilized, the hot nozzle tip remains in contact with the injection mold's hot or cold sprue bushing for the entire injection molding cycle. However, there are some applications where the nozzle tip may freeze off and prevent melt injection for the next molding cycle. This is most likely to happen when the injection mold has a cold sprue bushing, and the molding cycle time is relatively long. Nozzle freeze-off is also more likely to happen when molding semi-crystalline thermoplastics that have a very narrow melting temperature range or a sharp melting point (e. g., nylon 66). Nozzle body heater bands typically have a sufficiently high watt density to counteract the effects of conductive heat loss to the sprue bushing, but nozzle freeze-off can still occur in some situations. The use of nylon style nozzle tips that have a reverse taper design helps to facilitate removal of frozen material from the nozzle tip if freeze-off does occur (see Section 2.2.2.5.2). The possibility for nozzle freeze-off is also related to the fact that the nozzle tip itself is normally not heated and relies upon heat conduction from the electrically heated nozzle body. A typical two-piece nozzle usually has a heated nozzle body, and a nozzle tip that relies on heat conduction from the nozzle body. Extended length nozzle tips that are commonly used for molds having a recessed sprue bushing are typically heated to help prevent nozzle freeze-off.

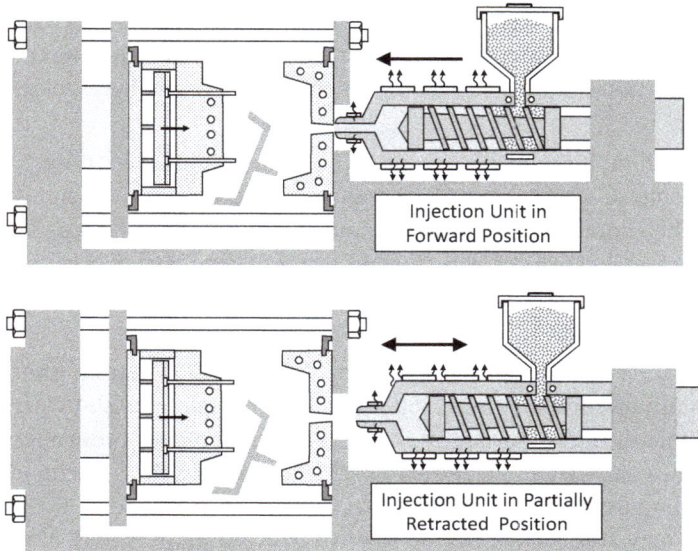

Figure 3.24 In most cases, the injection unit of the molding machine and its nozzle tip remain in contact with the mold's sprue bushing for the entire injection molding cycle (top figure). This mode of operation can lead to nozzle tip freeze-off, particularly in applications where the mold has a cold sprue bushing, and a relatively long molding cycle time. Nozzle tip freeze-off is also more common when molding semi-crystalline thermoplastics that have a narrow melting temperature range. In such cases, the injection unit can be retracted a short distance for portions of the molding cycle to minimize nozzle tip heat loss (bottom figure). The injection unit can be retracted after plastication or before plastication if a shut-off nozzle is used. Injection unit retraction, or sprue break as it is known, helps minimize heat transfer from the hot nozzle tip to the cold mold sprue bushing. The injection unit moves forward just prior to injection of the next shot.

3.7.2 Operation With Sprue Break

It is also possible to retract the sled from the mold during portions of the molding cycle where nozzle tip contact with the sprue bushing is not required. The nozzle tip must be in contact with the sprue during the mold filling, packing and holding phases of the injection molding process. During these portions of the cycle, the injection unit pull-in cylinders are energized to prevent leakage of melt between the nozzle tip and sprue bushing. However, it is not absolutely necessary to maintain contact between the nozzle tip and the sprue bushing for other portions of the molding cycle. If the injection unit is retracted sometime after the holding phase is complete, the process is described as using sprue break. If sprue break is used, the injection unit will retract a very short distance, thereby eliminating conductive heat transfer from the nozzle tip to the cold sprue bushing. The use of sprue break should allow the nozzle tip temperature to recover and remelt the solidified thermoplastic in applications where nozzle freeze-off occurs.

Sprue break before plastication: If sprue break is used, it can be initiated any time after the holding pressure phase of the process is complete. Using sprue break before plastication minimizes the contact time in which conductive nozzle tip heat loss occurs, but it can result in nozzle orifice drool during plastication, even if back pressure is set very low. If sprue break before plastication is to be used, a shut-off nozzle is generally required to prevent nozzle tip drool.

Sprue break after plastication: While sprue break prior to plastication can be used, sprue break is most commonly initiated after plastication is complete as shown in Figure 3.24 (bottom). While initiating sprue break after plastication does reduce the sprue break dwell time (i. e., injection unit retraction time), it reduces the severity of the nozzle drool issue compared to the use of sprue break before plastication. Some amount of nozzle drool may still occur if the nozzle is retracted immediately after screw recovery, due primarily to residual back pressure. Using decompress rear (i. e., suck-back) after plastication will eliminate the residual back pressure and hopefully prevent nozzle drool. As a result, sprue break is commonly initiated after both the plastication and decompression phases of the molding process are complete. This effectively eliminates the need for a shut-off nozzle, although a shut-off nozzle can still be used. The injection unit will then pull-in to the forward home position against the mold's sprue bushing once the part has been ejected, while the mold closes for the next cycle.

3.8 Shot Size (Dose)

All injection molding machine injection units have a maximum shot capacity that is given in either (i) maximum volume that can be injected or (ii) in the weight of general-purpose polystyrene that can be injected (as the molding industry's standard reference material). Ideally, the required shot volume (or dose) for an injection mold should be somewhere between 20% and 80% of the injection unit's capacity. Shot size utilization factors less than 20% may lead to an excessive residence time and poor shot-to-shot repeatability. Shot size utilization factors greater than 80% can result in a non-uniform melt quality.

The shot volume required for a given mold can be determined either theoretically from mold CAD files or experimentally by conducting molding trials. Having a good understanding of the theoretical shot volume requirement for a given mold is very helpful when setting up the process conditions for a new injection mold. Alternatively, the required shot volume for an injection mold can be determined experimentally by conducting a series of molding trials where shot size is gradually increased as shown in Figure 3.25. In such a molding trial, the shot size is increased gradually in each molding cycle to produce a progression of short shots until the cavity is full, or nearly full, as shown in Figure 3.25. Once the volume of thermoplastic material that is required to fill the mold has been determined, an additional volume of material, known as the cushion is added (see Figure 2.7). The cushion will serve as a volume of melt that will remain in the shot chamber after injection so that the melt within the mold can be pressurized during the packing and holding phases of the molding process.

It should be noted that some injection molds are not receptive to short-shot processing. For example, it is common to eject drafted thin-wall plastic cups or containers with a stripper plate. The stripper plate will engage the molded cup if the part is full, but the stripper plate may not function properly for short shots. In such a case it is advantageous to have a good theoretical understanding of the required shot size for a full part so that ejection issues do not arise during the process start up.

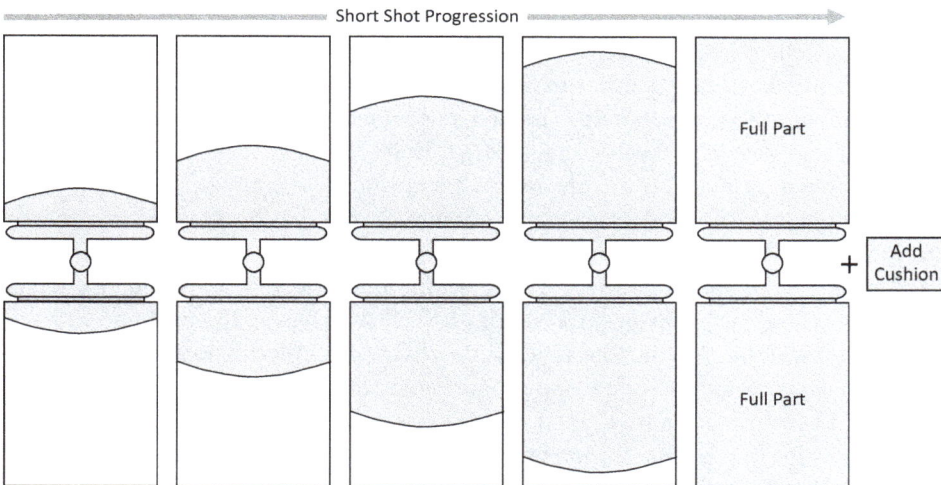

Figure 3.25 The shot size required for an injection mold can be determined theoretically from CAD files, or it can be determined experimentally. The required shot size for a new mold is typically determined experimentally by gradually increasing shot size during the molding start-up process. The correct shot size is equal to the volume of thermoplastic required to fill the mold plus an added volume for the melt cushion.

3.9 Mold Filling Injection Velocity

Most modern reciprocating screw injection molding machines control the mold filling phase of the process using closed loop velocity control. This is the case for both contemporary servo-hydraulic and all-electric injection units. Velocity controlled injection units offer a number of advantages over previous-generation open loop injection units, which are far more basic in terms of process control. A short discussion of open loop injection unit control follows since many of these molding machines are still in service.

3.9.1 Open Loop Injection Units

Previous-generation reciprocating screw injection molding machines commonly have open loop mold filling process control systems. Open loop injection units have two pressure settings, typically described as first-stage pressure and second-stage pressure. On the previous generation molding machines, the first-stage injection pressure is used to *both* fill and pack the molded parts. The second-stage pressure (a.k.a., hold pressure or follow-up pressure), is used to hold the pressurized melt within the mold cavities as the parts cool in order to help control the amount of molded part shrinkage, as shown in Figure 3.26 (left). The first-stage injection pressure needs to be high enough to completely fill the mold cavities (i. e., 100% volumetric fill) and high enough to pack the mold cavities (i. e., squeeze a bit more of the compressible melt into the mold). On the other hand, the first-stage pressure must also be low enough to prevent the mold from flashing (i. e., low enough to prevent the pressurized melt cushion from forcing mold half separation). The injection rates possible with open loop injection units can be fast, slow or anywhere in between. However, the injection rate (or fill velocity) whether fast or slow, tends to decrease as the fill progresses since the fill process is controlled by an open loop proportional valve. This is especially true if the fill process becomes pressure limited. In addition, open loop injection units normally do not allow for the use of injection velocity profiling as described in Sec 3.9.2.5.

With open loop control, transfer from first-stage pressure to second-stage pressure can be based on time, screw position, hydraulic pressure, or melt pressure in the molding machine nozzle or the injection mold, if such sensors have been installed [33]. A pressure transfer controller retrofit may also be required for pressure-based transfer depending on the machine's process control capabilities. Transfer based on cavity pressure has been shown to be the most consistent method in terms of repeatability, but transfer based on position offers a good balance of simplicity and repeatability. Transfer based on time tends to be the least consistent method of transfer. With open loop control, the actual time required to fill the cavity will vary if there is even a small change in the thermoplastic's melt viscosity. When time-based transfer is utilized, a change in fill time will then also effect pack time since the first stage time is constant using this mode of control. A small change in pack time has a very large effect on the peak pack pressure due to the steepness of the pack pressure – time curve as shown in Figure 3.26. Pressure-based transfer is far more consistent since transfer will always occur at the same peak pack pressure, even when there is some variation in the thermoplastic's melt viscosity (caused by barrel temperature cycling).

With open loop control, the holding pressure is applied immediately after transfer, as shown in Figure 3.26 (left). Hold pressure is set based on the molded part's dimensional targets or cosmetic specifications (e. g., sink mark depth) in essentially the same way it is set with today's more capable velocity-controlled injection units. Figure 3.26 shows that once transfer has occurred, the open loop and closed loop processes are essentially equivalent when a constant holding pressure is used. The current generation of closed loop injection units does have greater process flexibility in terms of hold pressure and back pressure profiling.

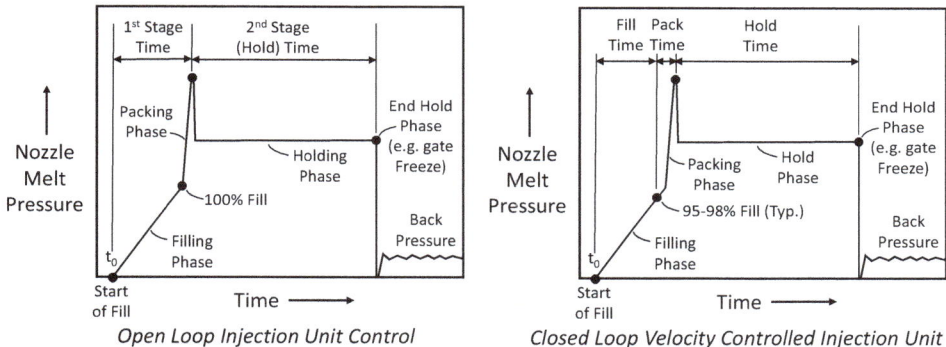

Open Loop Injection Unit Control   *Closed Loop Velocity Controlled Injection Unit*

Figure 3.26 Typical melt pressure vs. time curves for both open loop and velocity-controlled injection units. Previous generation reciprocating screw injection molding machines typically use open loop controls for mold filling and packing. What is known as first-stage or injection pressure is used to *both* fill and pack the mold cavity. This is followed by a second-stage or holding pressure, which helps control part dimensions, sinks and the like. Modern day injection molding machines have a closed loop velocity-controlled injection unit. This allows for improved control over mold filling and part quality by separating the mold filling, mold packing, and holding phases of the molding process (depicted above right).

Open loop mold filling controls have several major drawbacks. The first is that the mold filling phase of the process and the packing phase of the process are both controlled by a single variable. This lack of individual variable control results in less-than-optimum part quality, particularly in applications where part dimensional tolerances are tight. Compensation for issues such as melt viscosity variations, due to barrel temperature cycling or thermoplastic material lot changes, can be problematic with such limited control (unless cavity pressure control is utilized). Open loop injection units also have relatively crude control over injection velocity. With open loop control, the injection velocity is dependent upon the resistance to flow, which tends to increase as the mold filling progresses.

3.9.2 Closed Loop Velocity Controlled Injection Units

Most modern injection molding machines have closed loop velocity-controlled injection units. Closed loop injection units can maintain a constant injection velocity (or profiled velocity if desired) regardless of the resistance to flow caused by mold filling progression or viscosity variations, unless the mold fill process becomes pressure limited. Improved molded part quality and consistency are generally achievable when using this improved level of injection process control. While open loop injection units fill and pack the mold cavity using first-stage pressure, velocity controlled injection units separate the mold filling function from the mold packing function. Only the mold filling related variables are discussed in this section. The mold packing variables are discussed later in Section 3.10.

Once the appropriate shot size has been determined (i. e., enough material to fill the mold plus the melt cushion), the rate of mold filling or injection velocity must be set. In the simplest mode of control, the mold filling injection velocity is set to a constant value. Velocity-controlled injection units have the ability to vary injection velocity from very slow rates to very fast rates up to a some maximum value. Molding machines that are equipped with accumulator assist can inject at extremely fast rates and are commonly used for very thin-wall molding applications (e. g., molded thin-wall packaging).

Regardless of the rate of injection, velocity control does not allow for complete filling of the mold cavity. When mold filling is initiated, the screw begins to move forward at the set velocity. As mold filling progresses, the resistance to flow increases, and the pressure required to maintain the set injection velocity increases. The closed loop feedback from the injection velocity sensor (e. g., LVDT) makes that happen, unless the process becomes pressure limited. If the process becomes pressure limited, constant velocity cannot be maintained. As the fill progresses further, more and more pressure is required to maintain the velocity as shown in Figure 3.36 (right). If the mold was completely filled under velocity control (i. e., 100% fill), the screw would slow to a near zero velocity and the velocity sensor would keep calling for more pressure, which would most likely cause flashing of the mold. As a result, molds are usually filled to somewhere between 95% and 99% of the injection mold volume (i. e., a 1% to 5% short shot) when closed loop velocity control is utilized, as shown in Figure 3.27. Once that degree of fill is achieved, the process transfers to pressure control for the combined final fill and packing phase of the process, as shown in Figure 3.26 (right).

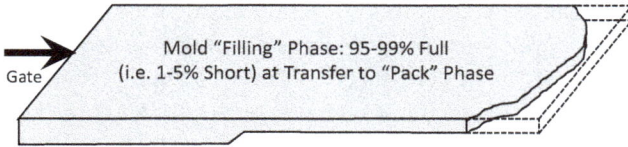

Mold "Filling" Phase: 95-99% Full
(i.e. 1-5% Short) at Transfer to "Pack" Phase

Gate

Figure 3.27 The mold filling phase of a velocity-controlled injection molding process involves injecting 95–99% of the shot under velocity control (i. e., a 1–5% short shot). The process then transfers to a pressure-controlled packing phase for final mold filling and melt compression.

3.9.2.1 Setting Injection Velocity

While velocity-controlled injection molding machines have been available for more than fifty years, there continues to be debate as to how the optimum injection velocity should be determined. Some sources indicate that very slow injection velocities are beneficial in terms of molded part quality. Other sources suggest that very fast injection velocities are beneficial in terms of part quality. Yet other sources suggest that a medium injection velocity provides for the best balance of injection mold filling related variables. It is the opinion of this author that material type, nominal part wall thickness, gating scheme, flow length, and many other part, injection mold, or molding machine variables need to be considered when determining the optimum injection velocity. There are many different ways to cook a turkey depending upon the desired outcome. The same is true in terms of mold filling velocity. Different fill rates have different outcomes.

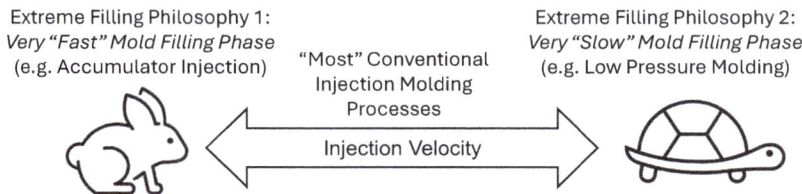

Extreme Filling Philosophy 1:
Very "Fast" Mold Filling Phase
(e.g. Accumulator Injection)

"Most" Conventional
Injection Molding
Processes

Injection Velocity

Extreme Filling Philosophy 2:
Very "Slow" Mold Filling Phase
(e.g. Low Pressure Molding)

Figure 3.28 There continues to be significant debate as to selecting the optimum rate associated with the mold filling phase of the injection molding process. One philosophy is to inject as quickly as possible, leading to the development of accumulator assist injection units that can inject at super-fast mold filling rates. On the other hand, extremely, low (constant) pressure molding processes preach the reported advantages of very slow mold filling rates. Mold filling rates for most conventional injection molding process lie somewhere between these two extremes.

3.9.2.2 Considerations Associated with Fast Mold Filling Rates

That said, it is also the opinion of this author that faster rates of injection are generally preferred over slow injection rates (more often than not). There are many reasons for this as outlined below. The shear thinning or pseudoplastic flow behavior exhibited by most thermoplastics is very beneficial in terms of using a fast injection or mold filling rate. As an example, if you double the rate of injection with a thermoplastic melt that exhibits pseudoplastic behavior, it would take less than double the force to maintain that velocity. Another fact supporting the use of fast injection rates is that there is a trend where molding machinery manufacturers continue to increase the maximum injection rates possible with their machinery. Customers seem very willing to pay the additional cost associated with high injection rate accumulator-assisted injection molding machines, especially for thin-wall packing applications. One machinery supplier recently developed a molding machine that has a maximum injection velocity of 1,200 mm/s [34]. One source suggests that the old adage of filling a mold as fast as possible consistent with part quality still holds true today [35].

Fill time savings: One seemingly obvious advantage of a fast injection rate is that it takes less time to fill the mold, which will have a positive impact on cycle time. While this may be true, mold fill time typically represents a very small portion of the conventional injection molding cycle. Any time saving does improve the production rate, but this advantage may not be extremely significant. In addition, faster injection rates involve more viscous heat generation, which may cause a small increase in cooling time, which may offset the fast fill rate related time saving.

Residual molecular orientation: Molecular orientation occurs anytime a long chain polymer is exposed to a laminar velocity profile, such as that associated with melt flow into an injection mold. Once the mold is filled and flow stops, the oriented polymer chains can re-randomize in a matter of milliseconds if the melt is hot enough (i. e., fluid enough) to allow this to happen, as discussed earlier in Section 3.3.3. Because the injection molding process is a non-isothermal process, and the hot melt is being injected into a relatively cool mold cavity, some of the molecular orientation associated with the non-isothermal mold filling process will freeze as the mold fills, especially those molecules within the solid skin layers. The frozen-in molecular orientation is typically described as residual molecular orientation. Residual orientation is generally undesirable as it has a negative impact on a plastic's mechanical performance and chemical resistance.

On the one hand, it would seem that injecting a thermoplastic material at a very high fill rate would introduce a greater degree of molecular orientation. While this may be true, a fast fill rate minimizes the time associated with mold filling and therefore limits the solid skin layer thickness that forms during the filling process. The average melt temperature will also be higher with a fast fill rate, which may allow much of this molecular orientation to re-randomize. The net amount of residual orientation is related to the amount of flow-induced orientation minus the amount of orientation that can relax immediately after the mold is filled. This outcome is expected to be more positive when fast fill rates are used.

It should also be noted here that the fibers (e. g., glass fiber) used to provide reinforcement for a fiber-filled thermoplastic, also tend to orient along the flow direction when the injection mold is filled. However, unlike polymeric molecules, fibers will *not* re-randomize after the mold is filled, even if the melt is very hot and fluid. Therefore, the use of fast injection rates may not be advantageous when molding a fiber-reinforced thermoplastic.

Hotter melt temperature: Faster mold filling rates also result in a higher average melt temperature within the cavity. Faster mold fill rates (i) reduce the amount of heat loss to the relatively cool mold as the injection mold fills, and (ii) cause an additional amount of viscous heat generation as the melt flows into the mold cavity. The net effect is typically a hotter and more uniform melt temperature within the cavity, which is beneficial in a variety of respects. Hotter melt temperatures help minimize residual molecular orientation as described above. Hotter melt temperatures also allow for better molded part surface replication and surface gloss. It has also been shown that hotter melts will generally improve knit line appearance and strength [36]. A more even degree of crystallinity can also be obtained when a hotter and more uniform melt temperature can be realized when molding a semi-crystalline plastic [37]. A faster fill rate will tend to be beneficial for all of these outcomes.

The effect of fill speed on melt temperature and melt temperature uniformity has been described above but only in a qualitative way. Injection mold filling simulations can provide additional and very valuable quantitative information regarding the effects of injection velocity and other process variables on melt temperature, both over the flow length and through the thickness of the molded part. Having access to injection mold filling simulation results can be invaluable when setting injection molding parameters, such as injection velocity. While molding simulations are commonly used for both plastic part and injection mold design, the practice of reviewing molding simulation results or conducting molding simulations on the shop floor is unfortunately more limited than it could be.

One downside associated with faster injection rates that result in a hotter melt temperature, is an increase in the potential for flash. In one respect, a hotter, less viscous melt should reduce the clamp tonnage requirement for the process, since the pressure required to fill an injection mold decreases at a higher melt temperature. On the other hand, a hotter, lower viscosity melt will be more likely to flash if the mold parting lines do not shut-off precisely.

Possible shear degradation: Another concern associated with fast injection mold filling rates is the potential of shear induced degradation. The potential for shear degradation is both material dependent and process dependent. Shear degradation is known to be more common with some thermoplastic materials, such as viscous grades of rigid PVC. PVC formulations tend have a narrow processing temperature range and can exhibit shear degradation if the flow paths are not streamlined and if viscous heating is excessive. The potential for shear degradation is greatest when shear stresses are excessively high. For example, forcing a viscous melt through a small diameter runner or gate at high injection rates leads to both locally high shear rates and shear stresses that could result in shear degradation. Shear degradation is far less likely to occur when molding higher MFR (i. e., lower viscosity) thermoplastic materials. Some material suppliers will publish maximum recommended shear rate values for their material grades but, unfortunately, this is not the norm. The presence or absence of shear degradation is rarely recognized by the molding engineers unless there is some visual indicator. Unexpected color changes, splay, or blush, especially in the region of the part just downstream from the gate, can be indicators of shear degradation.

Flow hesitation: A common rule of thumb associated with injection molded part design is to maintain a uniform wall thickness whenever possible. In reality, this goal is rarely achieved, as one of the main advantages of the injection molding process is that it can produce thermoplastic parts that have very complicated geometries. Other common thermoplastic part design rules of thumb provide contradictory guidance. For example, reinforcing rib design guidelines suggest that the rib wall thickness should be less than the thickness of the nominal wall from which the rib projects in order to limit the size of the sink mark opposite the rib. If the rib wall thickness becomes too thin (relative to the nominal wall thickness of the part), mold filling issues, such as flow hesitation can occur. Consider the mold cavity cross section shown in Figure 3.29. The cavity has a relatively thick nominal wall, and a thinner rib extending from the nominal wall. As the mold cavity fills, there will be preferential flow into the thicker wall section due to its lower pressure drop. The rate of flow into the thinner rib section will be slower and it will start to cool and may stall or hesitate as the thicker section of the mold cavity fills. Once the thicker wall section is filled, the cavity pressure will increase. Hopefully, the spike in cavity pressure will cause the stalled flow front in the thin rib to re-start as shown in Figure 3.29 (outcome 1). However, if the stall or hesitation time is too long, the rib may have cooled or solidified to the point where the rib flow front cannot re-start. The chance of re-starting the rib flow is more likely if the mold cavity is filled at a fast injection rate, since this minimizes the time associated with cooling of the stalled rib flow front. The fast injection rate helps prevent a short shot (i. e., outcome 2 in Figure 3.29). The flow hesitation would likely not be an issue if the cavity had a more uniform wall thickness.

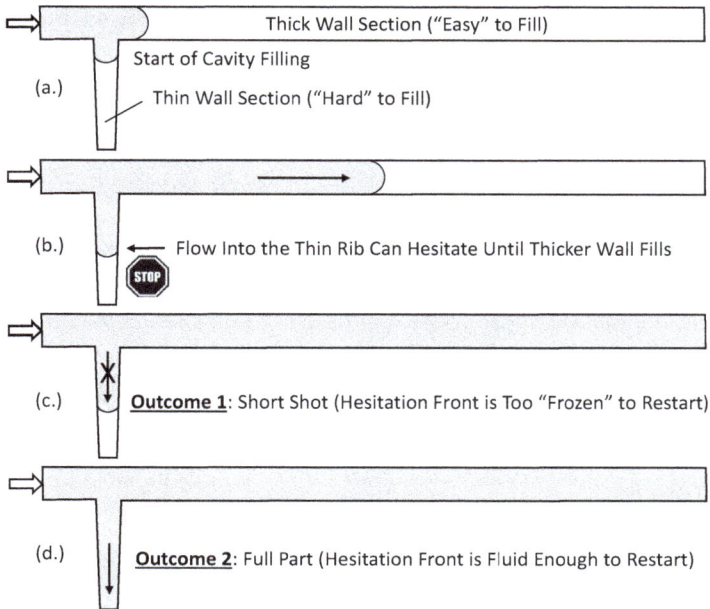

Figure 3.29 Certain thermoplastic part geometries, particularly those with variable wall thickness or thin-wall features (e. g., thin ribs), can exhibit flow hesitation during mold cavity filling. Melt flow into the thin feature may hesitate while the lower-pressure-drop thick nominal wall sections fill. Melt flow into the rib can re-start once the rest of the part has filled, provided that the melt in the thinner feature has not solidified. In such cases, fast injection velocities (i. e., short fill times) are more likely to result in a full part (outcome 2).

3.9.2.3 Mold Venting Considerations

Injection molds must include provisions for venting air and volatiles as the injection mold runners and cavities fill. When an injection mold is filled, the melt flow front forces the air within the runner and mold cavity, and the volatiles, to flow through the vents and into the atmosphere. Injection mold venting guidelines and recommendations are given in Section 2.3.2.2 of this text. Molds that have inadequate venting will require higher-than-normal fill pressures as the flow front progresses during the mold filling process. The higher mold filling pressures associated with inadequate venting, particularly at fast injection rates, can lead to part defects that include short shots, gas trap voids or burn marks (or dieseling). Burn marks are scorched or blackened material defects that occur at locations on the part that are last to fill. They are caused by the sudden, adiabatic compression of air and volatiles that off-gas from the melt front. Therefore, at a minimum, injection molds must incorporate vents that are located opposite the gate or in regions of the mold cavity that are last to fill as shown in Figure 2.73. While this is an age-old rule of thumb, such limited venting is completely inadequate for most applications. If fast injection rates are to be used, injection molds must be very well vented. Burn marks are more likely to occur at fast injection rates, but *only* if the mold has inadequate venting. Figure 2.73 (right) shows a mold cavity that has numerous vents distributed fairly evenly around the perimeter of the cavity. Vents are also located at strategic locations, such as adjacent to a knit line or at a runner cold slug well. It is also good practice to vent volatiles before the melt flow front enters the mold cavity. Cold runner venting will help reduce the appearance of splay on the molded part. Molds that are well vented allow for faster mold fill rates and lower filling pressures. There really is no downside to having more vents, provided the vent depths are appropriate for the material to be molded (i. e., they must be shallow enough to prevent vent flash). It can be difficult for a process engineer to determine if a mold has optimum venting. However, if venting related defects such as burn marks are observed, it is likely that venting is inadequate. Venting efficiency can be improved by adding additional vents or by reducing clamp tonnage if that option is available.

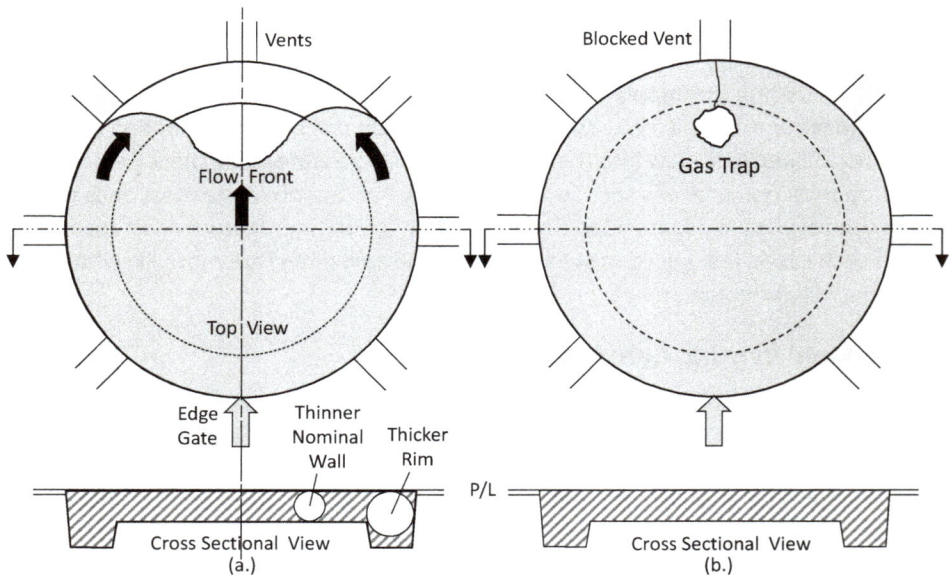

Figure 3.30 Race-tracking is a condition that can occur during injection mold filling if the part has a non-uniform wall thickness. In this case, the racing melt flows around the thicker outer rim of the mold cavity and leads to a gas trap. This is really a part design issue related to the non-uniform wall thickness. It would be good practice to mold parts at a variety of injection velocities to determine if there is an injection velocity where the gas trap does not form. If not, adding flow leaders to alter the flow front progression or adding a venting pin at the gas trap area may be required.

An additional mold filling and venting related phenomenon is known as race-tracking. The edge-gated mold cavity shown in Figure 3.30 has a relatively thin-wall center section and a thicker outer rim section. The mold cavity is well vented but during mold filling, the melt tends to flow around the thicker rim section due to its lower pressure drop. If the raci-tracking melt flow front in the outer rim section reaches the end of flow before the center section flow front, a gas trap (or compressed gas void) will form. In such a case, parts should be run at a series of injection velocities to determine if there is an injection velocity where this condition does not occur. This is really a part design issue that should ideally be corrected by adjusting the part wall thicknesses or by adding flow leaders. Alternatively, the injection mold could be modified with a venting pin, such as that shown in Figure 2.72, in the gas trap area.

3.9.2.4 Experimental Injection Velocity Study

While the use of fast injection rates can offer a number of advantages, as discussed above, it is common to run an experiment where parts are molded at a variety of different injection rates. There are different ways to conduct such an injection mold filling study, one of which has been described in great detail [1]. Basically, a mold filling study involves molding parts at a wide range of injection velocities. Hold and pack pressure times are typically set to zero and the transfer position is set such that the mold is filled to somewhere between 95–99% of the full shot. Cooling and plastication conditions are set such that the machine will produce short shots while operating in a single cycle or automatic cycle mode. The mold fill times and peak mold fill pressures at transfer are recorded at each injection velocity. Parts are also saved for cosmetic inspection. The fill velocity settings are typically varied from the maximum possible injection velocity, to a slow fill time (typically up to 5–10 sec).

Once the study is complete, it is common to plot relative viscosity (melt pressure x fill time) as a function of relative shear rate (1/fill time) as shown in Figure 3.31. The molded short shots are also inspected visually for overall appearance, signs of splay, gate blush, burn marks and other cosmetic defects. Other part defects, such as sink marks, should not be considered as part of this study since the packing and holding pressures phases of the process that influence these defects have not been utilized. If the thermoplastic material is a glass fiber reinforced plastic, burn out tests (done using a crucible and vented muffler furnace) can also be conducted using a small section of the molded part to determine how injection velocity influences the glass fiber length distribution in the molded part.

While injection mold filling experiments of this type or similar variants are common, there is no universal agreement as to how the optimum mold fill time is to be selected based on the experimental mold filling data [38]. Most sources agree that the optimum injection velocity setting is somewhere to the right of the knee region of the curve (i. e., somewhere on the quasi plateau). Because thermoplastic melts exhibit pseudoplastic (shear thinning) flow behavior, the material's viscosity is lower at the higher injection velocities. Natural variations in melt temperature or material lot changes will have less of an effect on the injection mold filling behavior at injection velocities to the right of the knee. Various methods have been used to select the optimum injection velocity from the curve shown in Figure 3.31. These methods include, but are not limited to, (i) selecting a velocity halfway between the knee and the end of the curve, (ii) selecting an injection velocity just to the right of the knee, or (iii) selecting the furthest point to the right where melt viscosity is lowest. Cosmetic issues associated with the molded parts should also be taken into consideration when selecting an appropriate injection velocity.

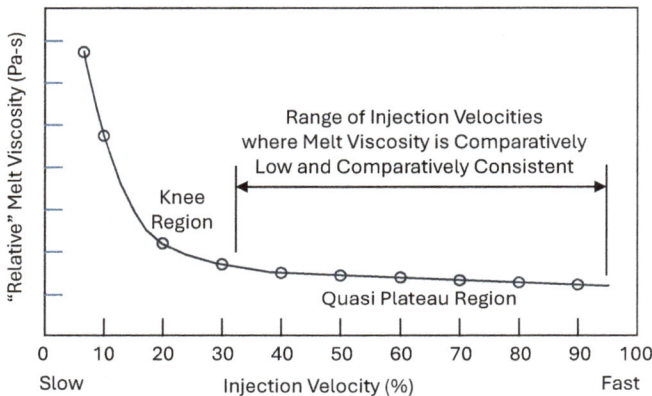

Figure 3.31 Once the injection mold filling study is complete, it is common to plot relative viscosity (melt pressure x fill time) as a function of relative shear rate (1/fill time). Molded parts are also collected and inspected for filling related cosmetic defects [1].

3.9.2.5 Multi-stage Injection Velocity Profiling

Up to this point, all of the discussion regarding injection mold filling and injection velocity has assumed that a single, constant value of injection velocity, whether fast or slow, is used to fill the injection mold. Modern injection molding machines have the capability to inject at multiple injection velocities over the course of the injection stroke. As mentioned earlier, there is no universal agreement when it comes to setting a single constant injection velocity. Likewise, there are no universal guidelines when it comes to setting a multi-stage injection velocity profile. One very common philosophy regarding injection velocity is keep it simple and stick with using one single injection velocity unless there is a clear need for more than one injection rate [39]. However, injection velocity profiling can sometimes be beneficial in terms of part quality; a few examples are provided below.

While profiling injection velocity can offer advantages, it is important to be realistic in terms of the molding machine's injection rate capabilities. All injection units have physical limitations in terms of acceleration rate, deceleration rate, inertia effects and the like. An injection molding machine's process controller may allow for six or more different injection velocities, but that many velocity changes may be unrealistic (and probably unnecessary) if the shot size or injection stroke is relatively short. Drastic changes in the sequential injection velocities can also be unrealistic. Several multi-stage injection velocity profile applications are given in the following sections.

3.9.2.5.1 Jetting Elimination

A classic example of a multi-stage injection velocity mold filling application is that used to eliminate the molding defect known as jetting. Jetting is a phenomenon that can occur when thermoplastic melt is injected into a relatively thick and open mold cavity, especially if an undersized gate is used. An example of the molding defect known as jetting is shown in Figure 3.32 (left). If the melt entering the cavity does not come in contact with the mold cavity walls, or impinges upon an obstruction, it will not develop a flow front. Instead, a stream of molten thermoplastic material will shoot or "jet" across the thick mold cavity and begin to fold upon itself. The folding of the jetting stream of melt will continue until if it causes enough resistance for a melt front to form. The mold cavity will eventually fill normally, but the partially cooled jetted material leaves a poor surface appearance and compromised mechanical properties. Gating schemes or gate types that promote divergent flow help prevent jetting. Gates having short land lengths can also help eliminate jetting by promoting die swell or the memory effect. However, if jetting is observed during the start-up of a new mold, a profiled injection velocity can normally be used to eliminate this unwanted defect without the need for mold modification.

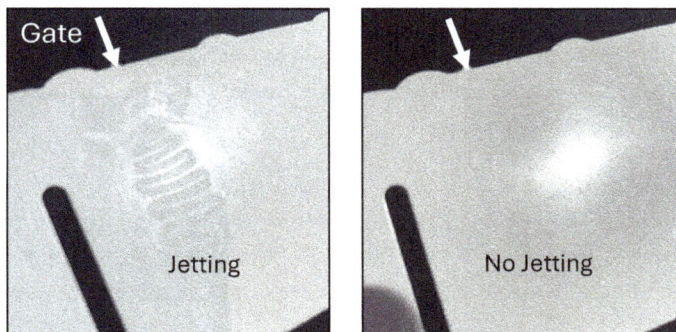

Figure 3.32 Jetting is visible on the ABS part shown above left. This part is molded using a single injection velocity. The gate location is indicated by the arrow. The part on the right is injection molded from the same material but without jetting. Jetting for the part on the right is eliminated using a three stage, "fast – slow – fast" injection velocity profile.

The basic idea in using injection velocity to eliminate jetting is that if a very slow injection rate were to be used during mold filling, a flow front would develop and jetting would not occur. However, a very slow injection rate could lead to a variety of other mold filling problems associated with cooling of the melt during the slow fill. This problem can be solved using a three-step injection velocity profile (in the case of a cold runner mold), as shown in Figure 3.33. The first portion of the shot (i. e., the first injection velocity segment) is injected at a high fill rate to fill the runner system quickly. The injection velocity for the second injection velocity segment is then slowed as the flow front moves through the gate. This second slow injection speed allows a melt front to develop as material flow from the gate into the cavity. Once the flow front develops, the injection velocity is then increased again for the third injection velocity segment. The result of this "fast – slow – fast" injection velocity profile is an overall fast mold fill time, and no jetting.

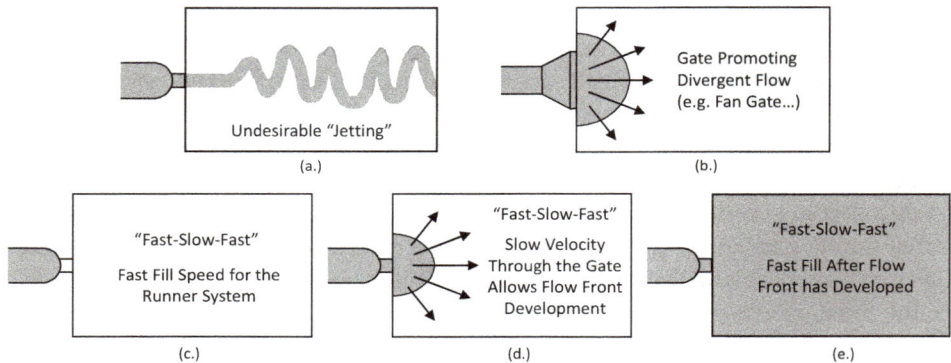

Figure 3.33 Profiled injection velocity can be used to eliminate the condition known as jetting. Jetting can occur when injecting a thermoplastic melt into a relatively thick injection mold cavity from a small gate where the melt does not initially form a flow front. This results in a folded serpentine defect visible on the molded part's surface. A "fast – slow – fast" velocity profile can be used to eliminate the jetting phenomenon and allow for a relatively fast overall mold filling rate.

3.9.2.5.2 Uniform Flow Front Velocity

Another injection mold filling related philosophy that has been embraced by some is to maintain a uniform flow front velocity during the mold filling phase of the injection molding process. It is thought that maintaining a constant flow front velocity can reduce flow lines or flow marks, and reduce part distortion caused by residual molding stress [40,41]. Using a single, constant injection velocity will not result in a uniform melt flow velocity for most part geometries. Gating schemes that involve radial flow (e. g., three-plate mold gating) also result in a non-uniform flow front velocity. In such cases, a profiled injection velocity can be used to bring the mold filling process closer to the goal of achieving a constant flow front velocity. The key with this approach to injection mold filling is to have a complete understanding of the flow front position in relation to the axial screw position.

Consider the mold cavity shown in Figure 3.34 (top). If the mold cavity shown in the figure is filled at a constant injection velocity, the melt front velocity will accelerate as it passes through the narrower center section of the mold cavity. The flow front velocity will then decelerate again as the flow front approaches the end of mold cavity. The regions of the molded part associated with acceleration and deceleration can exhibit non-uniform performance. Figure 3.34 (bottom) shows that a more uniform flow front velocity can be achieved if the injection velocity is profile, using three velocity segments in this particular case. The use of a "fast – slow – fast" injection velocity profile will result in a more uniform flow front velocity, which should improve part quality by helping to eliminate flow lines and residual stress, based on the philosophy that a uniform flow front velocity is beneficial in terms of these molded part attributes.

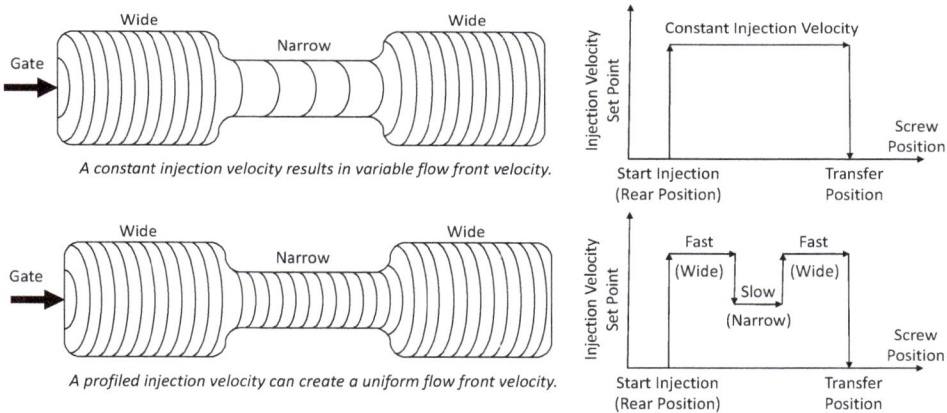

Figure 3.34 One philosophy in setting injection velocity is to *maintain a uniform flow front velocity*. Such a condition can only be achieved by profiling the injection velocity, particularly for parts having a complicated geometry or parts involving radial flow (e. g., filling a circular disc with a center gate).

3.9.3 Low Injection Velocity Constant Pressure Molding

Very slow mold filling speeds have also been shown to be beneficial for the injection molding process in some applications. As an example, slow mold filling speeds have long been common when injection molding very thick-wall parts such as thick optical lenses. Another low injection velocity molding process has been developed that functions by filling and packing the mold at a relatively low *and* constant melt pressure, not at a constant injection velocity as with more convention injection molding processes [42–44]. With this low fill rate molding process, the molded part tends to pack as it fills. The process is said to be less sensitive to normal variations in process variations, material variation or variations in environmental conditions. A comparison of melt pressure vs. time curves for the low fill velocity process and a more conventional injection molding process is shown in Figure 3.35. The slow fill velocity reduces shear heating which can result in a reduced cycle time in comparison with conventional molding. The process is said to eliminate flow hesitation and reduce clamp force requirements due to the lower fill, pack and hold pressures utilized. The low injection rate process is also said to improve molded part consistency, particularly when material viscosity drifts occur. Viscosity drifts can be caused by a variety of reasons including; barrel temperature cycling, material lot variation, or in applications involving recycled materials. A nozzle melt pressure sensor, or a cavity pressure sensor, is used with the proprietary controller to maintain a low and constant fill pressure [42–44].

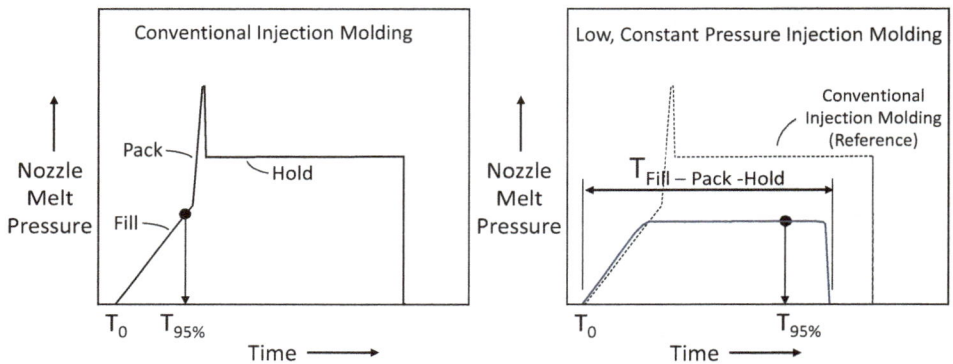

Figure 3.35 A slow fill rate molding process has been developed that involves filling and packing the mold at a relatively low *and* constant melt pressure. The low melt pressure process is said to offer a number of advantages that include a reduction in shear heating and cycle time, clamp tonnage reduction and improved part consistency relative to a more conventional injection molding process [42–44].

3.10 Packing Pressure and Packing Time

The velocity-controlled portion of the mold filling phase for the injection molding process is typically used to fill the mold cavity to somewhere between 95% to 99% of the mold cavity volume. The velocity-controlled portion of the mold filling phase is depicted as Point 1 to Point 2 in Figure 3.36. Once the mold is mostly filled, the mold filling process transfers from velocity control to a pressure-controlled packing pressure phase. The transfer is typically initiated based on screw position. The packing pressure phase is depicted as Point 2 to Point 3 in Figure 3.36. The pack pressure term is a bit of a misnomer as this phase of the molding process has two main functions, only one of which is packing. The first function of the pack pressure phase is to complete filling of the mold cavity and ensure that the mold cavity is 100% full (i. e., the pack pressure phase completes the mold filling process). The second function of the pack pressure phase is to ensure that additional melt is squeezed into the mold cavity. This melt compression will help ensure that the molded part will exhibit good surface replication (e. g., texture replication) and exhibit knit line integrity. The compressed melt will also help to control and compensate for mold shrinkage. This melt compression function is possible because plastic melts are very compressible fluids, especially at the high melt temperatures and melt pressures associated with the injection molding process (up to 250 MPa in some cases).

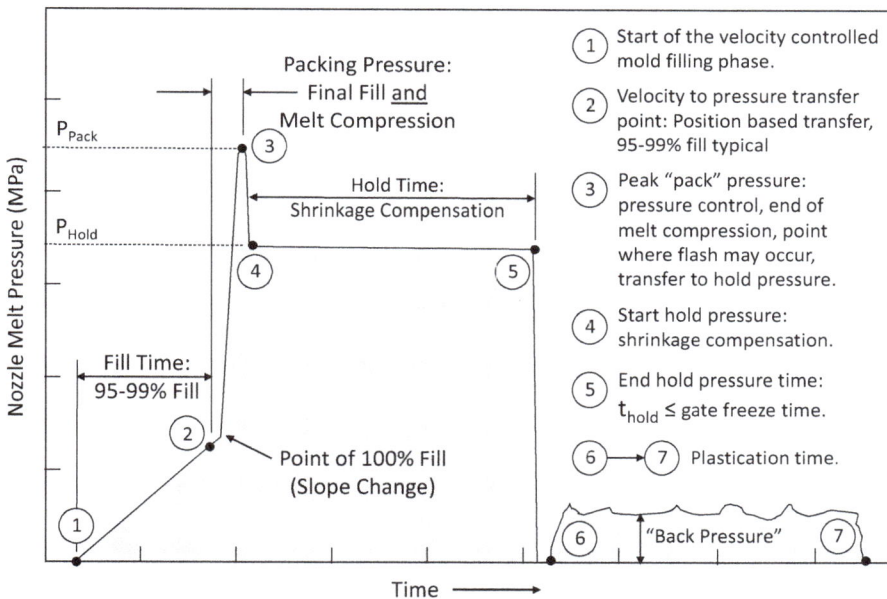

Figure 3.36 The velocity-controlled mold filling phase of the injection molding process (Point 1 to Point 2) is typically used to fill the mold cavity most of the way (e. g., 95–99% full). The packing pressure phase of the molding process (Point 2 to Point 3) is then initiated (i) to complete filling of the mold cavity, and (ii) to squeeze more compressible thermoplastic melt into the mold cavity. This compression phase helps to compensate for mold shrinkage and to ensure that the molded part is properly packed out (e. g., to ensure surface replication and knit line integrity).

The packing phase of the molding process is typically has a very short duration. The packing pressure phase is followed by the holding pressure phase as shown in Figure 3.36 and described in Section 3.11. The holding pressure phase can be initiated based on time, peak hydraulic pressure or peak melt pressure. Transfer based on peak cavity melt pressure (if available) is optimum since the transfer pressure (Point 3 in

Figure 3.36) will be the most consistent, even if there are variations in melt viscosity caused by material variation or temperature variations. Transfer based on nozzle melt pressure (or hydraulic pressure) is generally preferred over a time-based transfer. Small changes in melt viscosity can cause a large change in peak pack pressure (due to the steepness of the pack pressure vs. time curve) if time-based transfer is utilized.

3.10.1 Pack Pressure: Final Mold Filling Considerations

As stated above, the first function of the pack pressure phase is to complete the final portion of the cavity filling process (i. e., the last 1% to 5% of the fill). The peak pack pressure value must be high enough to fill the extremities of the mold cavity, but it must also be low enough so that the mold does not flash (based on the available clamp tonnage). Ideally there will be a relatively wide process window associated with the peak pack pressure value. The pack pressure window is limited by short shots on the low end and evidence of flash at the high end. A wide separation between these two extreme pressure values is desirable.

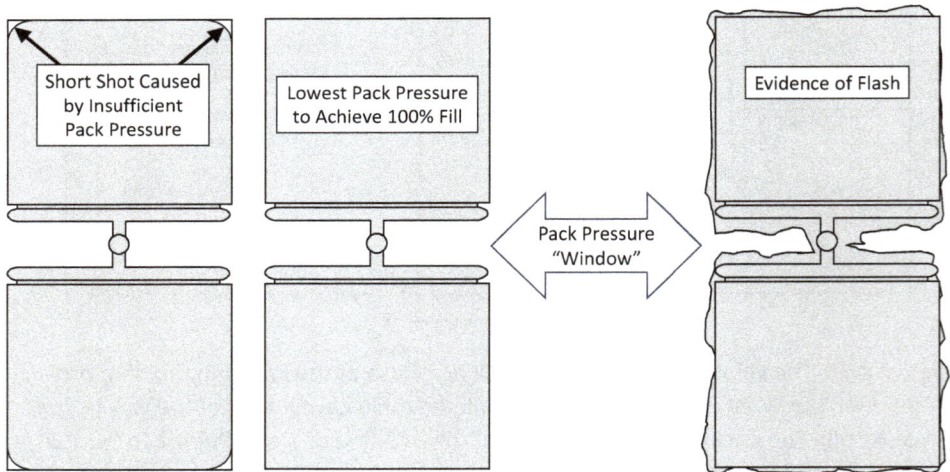

Figure 3.37 The peak pack pressure value utilized here must be high enough to complete the mold cavity filling process (i. e., the final 1% to 5% fill), but it must also be low enough so that the injection mold does not flash (based on the available clamp tonnage).

The range of suitable pack pressures (or pack pressure process window) is typically described (at least theoretically) using a graphical image such as that shown in Figure 3.38. The figure is typically described as a molding area diagram. It can be very time consuming to generate the experimental data required to produce a molding area diagram, but it is helpful to have a theoretical understanding of the behavior depicted in the figure. The upper and lower x-axis melt temperature limits for the pack pressure molding area diagram are typically taken to be the low and high melt temperature limits recommended by the thermoplastic material supplier. Alternatively, these temperature limits may be based on previous experience with the same thermoplastic material. The minimum fill pressure and the maximum fill pressures both decrease at higher melt temperatures since the thermoplastic material will have a lower melt viscosity at the higher melt temperature. A hotter, less viscous melt will have a lower minimum fill pressure requirement and will also tend to flash at a lower pack pressure.

While the lower packing pressure value is relatively easy to determine experimentally. The minimum pack pressure is the lowest pack pressure value that leads to a full part. However, one should use caution when trying to establish the upper packing pressure limit. In such an experiment, a series of parts is molded at gradually increasing packing pressure settings looking for evidence of flash. The upper packing pressure limit is influenced by a number of factors including the amount of clamp tonnage utilized. Increasing the packing pressure to the point where full perimeter flash occurs is *not* recommended. Full perimeter flash occurs when the injection force overcomes the clamp force. Full perimeter flash occurs when an excessively high pack pressure forces some of the cushion material into the parting line gap. The excessive injection force that causes the flash that can cause tool damage, especially when tool construction is complex and softer tooling materials are used. Instead, it is better to look for signs that suggest the onset of flash is near, rather than wait for the actual flash. For example, as packing pressure is added in each molding cycle (while conducting a packing pressure window study) the part may begin to stick in the mold cavity or become distorted due to high ejection forces. There may also be evidence of sharpness or micro-flash at the parting line. The packing pressure value associated with the onset of these events can be considered as the upper packing pressure limit for the process window study (at that melt temperature and clamp force).

Once the overall shape of the pack pressure – melt temperature molding area diagram has been established (or approximated), operating conditions can be selected. Ideally, the packing pressure window, or region of full parts will be large (especially the vertical separation) as this indicates that the process is fairly robust. Selecting a melt temperature and peak pressure value toward the center of the process window is common in many injection molding applications. It is also beneficial to reduce clamp tonnage if possible, using only enough clamp tonnage in order to prevent flash. Reducing clamp tonnage can extend both the life of the injection mold and the injection molding machine clamp unit.

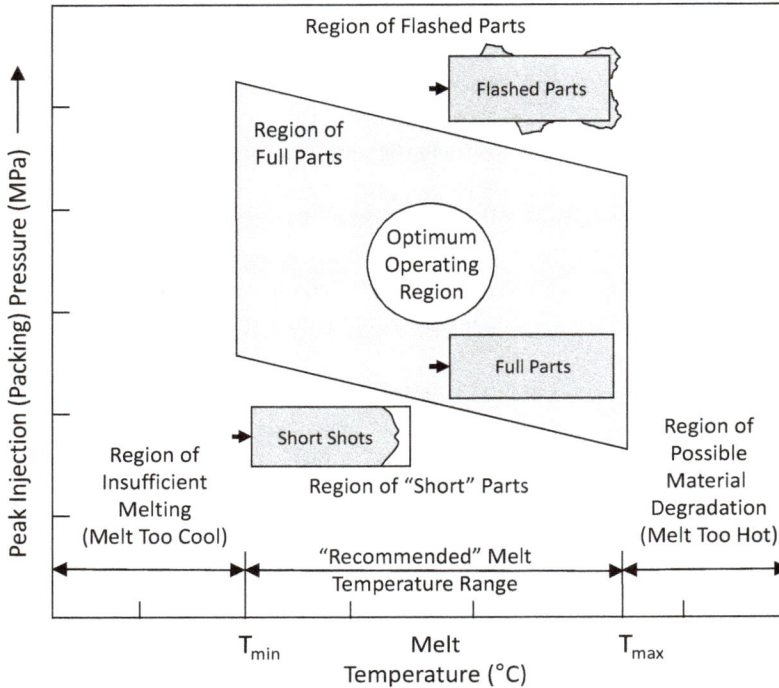

Figure 3.38 The low and high packing pressure limits associated with the final mold filling are both influenced by the thermoplastic material's melt temperature (and melt viscosity). The upper packing pressure limit is dictated by both melt temperature and by the applied clamping force. While the upper packing pressure limit is theoretically the pressure associated with the appearance of flash, purposely flashing an injection mold is not recommended. The upper pressure limit is typically based upon pre-flash conditions such as part ejection difficulties.

It should also be noted here that there are a variety of possible causes of flash for injection molded parts. The molded parts and cold runner system shown in Figure 3.39 all have parting line flash but the causes of the flash vary. The molding shown in Figure 3.39 (c.) has full perimeter flash, which indicates the flash was caused by the injection force overcoming the molding machine's clamp force. In such a case, the mold parting line opens slightly and cushion material squeezes into the parting line gap. This type of flash can be eliminated by increasing clamp tonnage or reducing the peak packing pressure. The flash shown in Figure 3.39 (b.) is non-symmetrical localized flash. Localized flash is typically associated with localized physical mold parting line damage at only that location. It may be possible to eliminate or minimize localized flash by lowering melt temperature, injection velocity or reducing peak packing pressure. However, repairing the injection mold is the most appropriate solution to localized flash. The flash shown in Figure 3.39 (a.) is centrally located flash. Central flash is most likely associated with platen or injection mold plate deflection. For example, a mold design that has an insufficient support plate thickness (or support plate depth) can result flash towards the center of the injection mold. Support plate deflection can be reduced by adding support pillars between the support plate and the B-side clamp plate. The addition of the support pillars should help prevent the appearance of central flash.

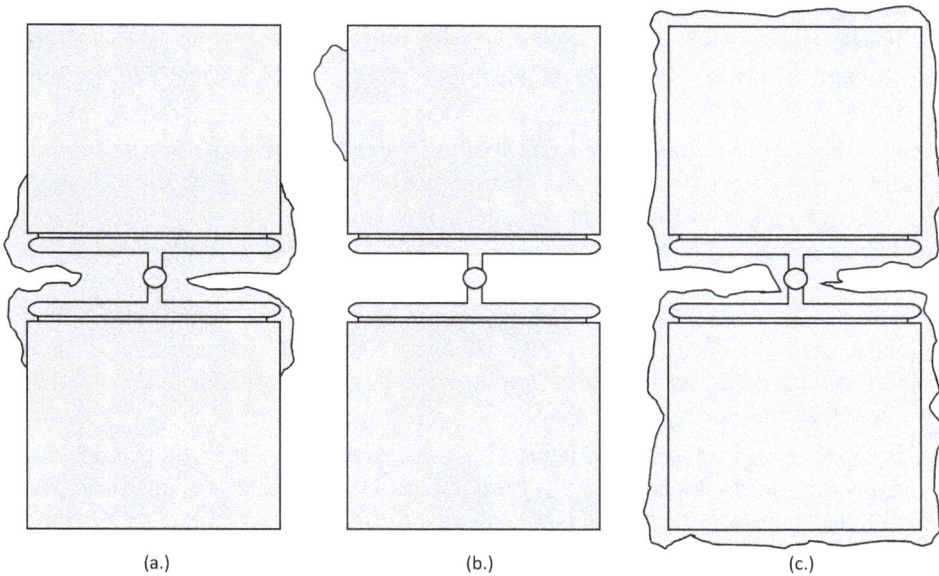

(a.)   (b.)   (c.)

Figure 3.39 Parting line flash can be caused by a variety of different reasons. The centrally located flash shown in (a.) is most likely associated with support plate deflection. The localized flash shown in (b.) is most likely associated with localized injection mold damage at that specific location. The full perimeter flash shown in (c.) is indicative of the applied injection force overcoming the applied clamp force.

3.10.2 Pack Pressure: Compression Phase Considerations

As liquid fluids go, thermoplastic melts are very compressible compared to most other liquid fluids. The significant compressibility stems from the fact that most thermoplastics have relatively high volume coefficients of thermal expansion (CTEV). This means that there is a great deal of free volume when a thermoplastic is heated to its melt processing temperature. Increasing the temperature of the melt for a particular thermoplastic will increase its free volume, which reduces intermolecular bonding and therefore reduces its melt viscosity. Intermolecular forces drop off dramatically as the distance between molecules increases. As an example, the volumetric compressibility of a neat general-purpose polystyrene at a process temperature of 225 °C is about 6% when pressure is increased from atmospheric to 600 bar (61 MPa) [36].

The high CTEV values exhibited by a thermoplastic plastic melt is also responsible for some or all of the mold shrinkage observed when molding thermoplastics. The overall mold shrinkage of an amorphous thermoplastic is due only to the free volume decrease that occurs as the material cools in the injection mold. The mold shrinkage for a semi-crystalline plastic is far more complex since it is due to both (i) the free volume decrease associated with thermal contraction as the molded part cools, and (ii) the free volume decrease that occurs as portions of the melt crystallize when the semi-crystalline material cools.

Because thermoplastics are very compressible, the cavity pressures for the injection molding process can have a very significant effect on the observed mold shrinkage. In fact, the cavity pressure history experienced by the thermoplastic material during the molding process can be used to fine-tune mold shrinkage. A thermoplastic material's cavity pressure history will influence the amount of mold shrinkage compensation. Other molding process variables that impact the cooling rate also have a significant influence on mold shrinkage for semi-crystalline thermoplastic materials. Variables such as mold temperature influence the degree of crystallinity and mold shrinkage for a semi-crystalline thermoplastic.

The compressibility of a plastic melt (and solid) can be determined in a comprehensive and quantitative way using a P-v-T test apparatus. A typical P-v-T curve for a neat amorphous thermoplastic is shown in Figure 3.40. While the expected mold shrinkage for an amorphous thermoplastic material is usually based on linear shrinkage data provided by the thermoplastic material supplier, it is possible to theoretically predict a material's mold shrinkage given a P-v-T curve, and the anticipated molding process condition. At a minimum, P-v-T data can help answer questions related to how process variables (such as peak packing pressure) impact the theoretical mold shrinkage value for the material. Figure 3.40 shows a typical molding injection molding process path superimposed onto a P-v-T curve for a typical neat amorphous thermoplastics.

The injection molding process path depicted in Figure 3.40 is effectively a cavity pressure – time curve superimposed on the P-v-T graph for an amorphous thermoplastic material. The numbered designations represent the sequential steps and transitions for a typical injection molding process.

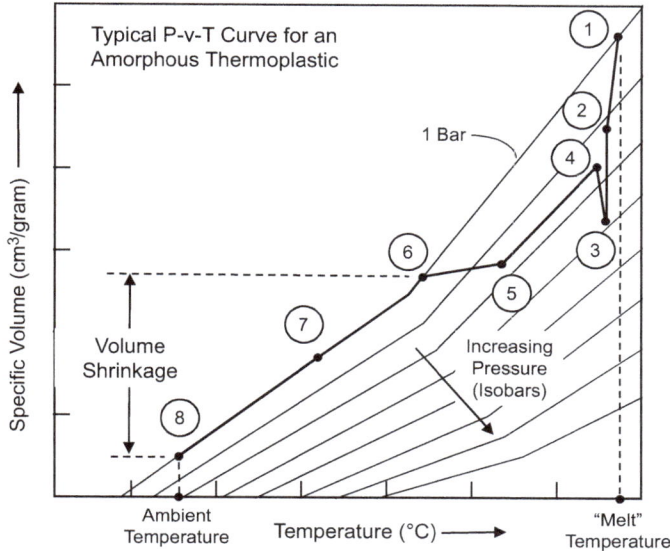

①		Start of the melt injection phase. "Hot" melt into a relatively "cool" mold.
①	②	Mold filing phase (95-98% full).
②		Velocity to pressure transfer point.
②	③	Final fill and packing pressure (or compression) phase.
③		Peak packing pressure and initiation of transfer to holding pressure.
③	④	Transfer from peak pack pressure to holding pressure.
④		Start of the holding pressure phase.
④	⑤	Holding pressure phase. Constant applied pressure as the molded part cools.
⑤		End of holding pressure phase (typically dictated by the gate freeze time).
⑤	⑥	Cavity pressure drops as the part cools and shrinks without compensation.
⑥		Atmospheric pressure. Part volume = cavity volume. Mold shrinkage (as defined) begins
⑥	⑦	Isobaric mold cooling and "in-mold" shrinkage.
⑦		Part ejection from the mold cavity.
⑦	⑧	Post mold isobaric cooling.
⑧		Part reaches room temperature. Any post mold dimensional changes follow

Figure 3.40 A typical injection molding process path is superimposed on a P-v-T curve for an amorphous plastic material. The graph can be used to theoretically predict how injection molding process variables, such as packing pressure or gate freeze time, influence overall volume mold shrinkage for the thermoplastic.

The P-v-T behavior for a semi-crystalline thermoplastic is more complex than that of an amorphous thermoplastic, as shown in Figure 3.41. Like amorphous thermoplastics, semi-crystalline thermoplastics are also very compressible and they have relatively high CTEV values that account for a portion of the material's mold shrinkage. However, semi-crystalline plastics also exhibit a very large step-like volume change at their melting temperature (i. e., the temperature associated with the crystalline melting point). The density of the crystalline regions of a semi-crystalline thermoplastic is significantly greater than that of the amorphous regions. Semi-crystalline thermoplastics can shrink 2–6 time more than amorphous thermoplastics depending on their degree of crystallinity. The mold shrinkage for a semi-crystalline thermoplastic will be influenced by process conditions (such as the packing pressure value utilized), but mold shrinkage is largely determined by the degree of crystallinity achieved by the molded part.

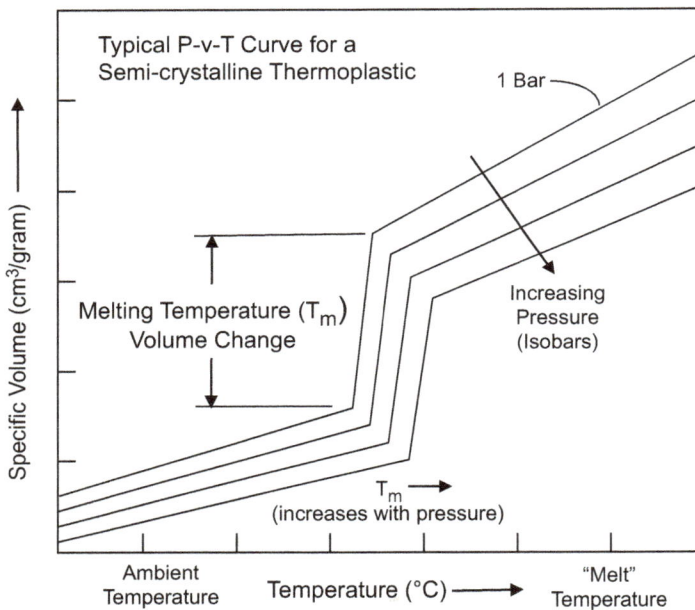

Figure 3.41 The P-v-T behavior for a semi-crystalline thermoplastic is more complex than that of an amorphous thermoplastic as these materials exhibit a large step-like volume change at the material's crystalline melting temperature.

3.11 Holding Pressure and Holding Time

At the time when the packing phase of the injection molding process is complete, the injection mold cavities should be (i) 100% full, and (ii) the melt within the mold should have been packed or compressed into the cavity. The magnitude of the *pack* pressure affects part attributes such as surface replication, knit line integrity, and the like. Because fill and pack times are relatively short compared to the overall cycle time, the amount of cooling or solidification that will have occurred at the end of the packing time is minimal. At the end of the packing time, the moldings (including the cold runner system and gate) will have relatively thin solid skins, however they will also have a molten core. If pressurization of the mold does *not* continue after the packing phase of the process is complete, some of the compressed and fluid melt core material will tend to flow out from the mold cavity, back towards the injection unit's shot chamber. If this occurs, the melt backflow causes the thin solid skins of the molded part to collapse inward, leading to moldings that have excessive sink marks. Shrinkage void formation and poor overall dimensional control are also issues if continued pressurization of the melt contained within the mold does not continue during the holding pressure phase of the molding process. The need for this applied holding pressure becomes greater for thicker wall plastic parts.

As a result, a post-pack follow-on pressure, known as holding pressure, is applied immediately after the packing phase is complete, as shown in Figure 3.36. Fundamentally, the purpose of the holding pressure phase of the process is to hold the molten thermoplastic in the mold as the melt cools and solidifies. During the holding phase, the screw and check valve push on the melt cushion to pressurize the melt in the mold runner system and in the mold cavities. Holding pressure variables are a primary means of control for dimensional issues (e.g., sink marks, shrinkage voids, part dimensions). In the simplest mode of operation, holding pressure is set to a constant value for a fixed period of time (i.e., the hold time) as shown in Figure 3.36 above. In order for holding pressure to be effective, the melt cushion must be maintained for the entire hold time. Loss of the melt cushion during the hold time (e.g., due to check valve leakage) is completely unacceptable since the melt cushion is required in order to pressurize the melt that is within the injection mold.

3.11.1 Effect of Holding Pressure Magnitude

The holding pressure value utilized will have a significant effect on a molded part's dimensions, especially for thicker-wall plastic parts. The holding pressure setting can be similar or even greater than the pack pressure setting, but this would likely lead to an over pack situation where the part may be too highly stressed or may be difficult to eject from the injection mold. In most cases, the magnitude of the applied holding pressure is set to be less than the packing pressure value. As a very rough starting point, the holding pressure is often set to be somewhere between 40% and 80% of the peak packing pressure value as shown in Figure 3.43.

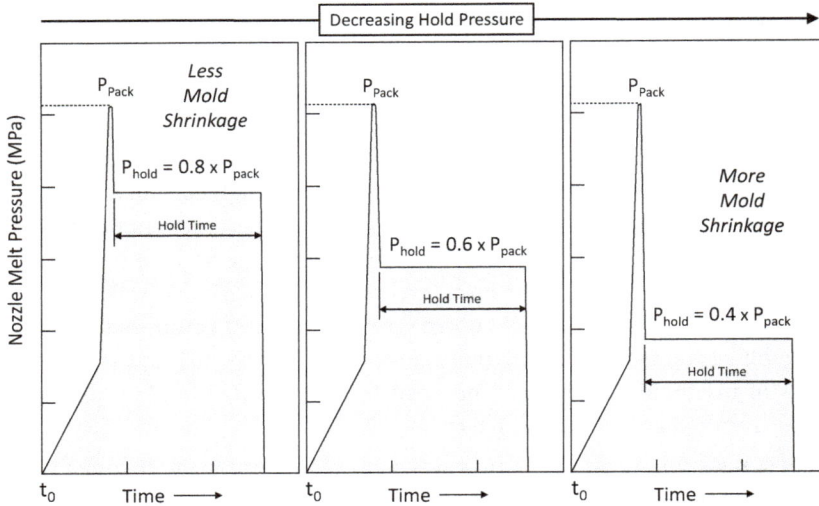

Figure 3.42 The holding phase of the injection molding process normally begins immediately after the packing phase ends. Holding pressure acts to keep the compressed and pressurized thermoplastic melt in the mold cavity as the melt cools (solidifies) and shrinks. Holding pressure helps compensate for mold shrinkage and is used to fine-tune part dimensions. A higher holding pressure history will result in less mold shrinkage.

The holding pressure variables become more important for thicker-wall parts where mold shrinkage issues are more significant. In fact, very thin-wall parts may require little if any holding pressure. Thicker-wall parts take longer to cool and tend to shrink more, especially in the case of semi-crystalline materials. The holding pressure variables (i. e., holding pressure and holding pressure time) are used to fine-tune the final part dimensions for conventional wall thickness and thicker-wall thermoplastic parts. Undersized parts may be brought into specification by using higher holding pressures and/or longer holding times. In fact, molded part dimensions are very much related to the area under the melt pressure – time curves, such as those shown in Figure 3.43. The two most effective ways to control the area under such a curve is by (i) adjusting the magnitude of the holding pressure, and (ii) adjusting the holding pressure time. Peak pack pressure also influences a part's dimensions but to a lesser extent as packing time is typically very short in comparison to the applied holding pressure time, as shown in Figure 3.43.

The window of possible hold pressure times ranges from 0.0 seconds on the low side, to a maximum time where mold cavity pressurization is no longer possible. For example, if any portion of the flow channels (i. e., runner system) connecting the shot chamber to the injection mold cavity freezes or solidifies, pressure transfer from the melt cushion to the mold cavity becomes impossible. In most cases, the upper holding pressure time limit is dictated by the time it takes the gate to freeze or solidify. Gates are normally thinner than both the runner system or the parts, and, as such, they cool first. In some cases, freezing of the nozzle tip may be the limiting factor for holding pressure time.

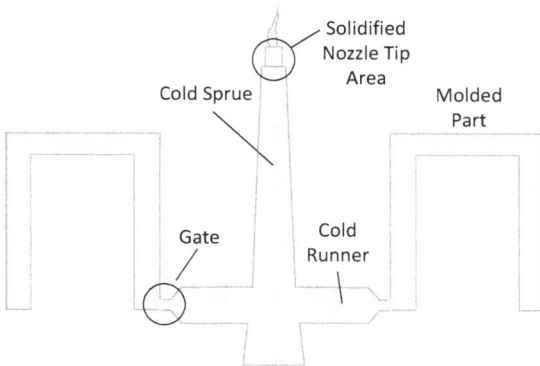

Figure 3.43 Mold cavity pressurization is possible as long as there is a fluid connection between the melt cushion and the mold cavity. In most cases, the gate freeze time limits the time available for mold cavity pressurization as gates are normally thinner than the part or runner and therefore cool more quickly. Small diameter nozzle tips orifices can also freeze and limit the time available for mold cavity pressurization.

3.11.2 Effect of Holding Pressure Time

As stated above, the upper limit for hold time is typically dictated by the time it takes the molding's gate to freeze or solidify. Very small gates (e. g., gates less than 40% of the part's nominal wall) will cool relatively quickly. If the gates are small, they cool too quickly, providing very little control over the possible holding times and the molded part's dimensions. Small gates cool quickly and the resulting area under the melt pressure – time curve is low, as shown in Figure 3.44 (left). The pressure history window (and dimensional control) will be wider, and more forgiving if deeper gates are utilized. Deeper gates take longer to cool and allow the use of longer holding pressure time. However, deeper gates are more difficult to de-gate and may be a cosmetic concern. These conflicting factors must be balanced when selecting a molded part's gate geometry.

It is worth noting here that injection molds are often built with smaller than optimum sized gates for a number of reasons. The concept of designing a mold with a smaller than optimum gate size is common since it is safe practice with steel. Small gates offer improved aesthetics and easier de-gating in the case of a cold runner mold. On the other hand, small gates can lead to both mold filling and pre-mature gate solidification, which decreases the holding pressure phase molding process window. Figure 3.44 shows that increasing gate depth, extends the time a process engineer has to control cavity pressure history and therefore the molded part's dimensions. In general, longer hold times are beneficial for molded part's dimensional control. One source indicates that cavity pressure history can alter the size of a plastic part molded from an amorphous material by up to 2% [45]. The impact of cavity pressure history for part dimensions for a semi-crystalline material can be up to 0.5%. Gate size should be optimized during initial mold qualification trials so that the holding phase process window is sufficiently wide. On the other hand, holding pressure times must also be short enough to ensure that there is sufficient time available for the shot plastication (i. e., shot generation for the next molding cycle) before the injection mold opens as shown in Figure 3.23.

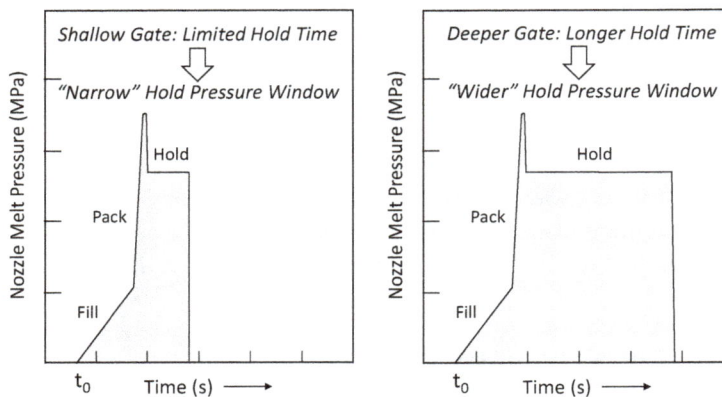

Figure 3.44 The holding pressure time that can be used is limited by the time it takes the gate to solidify. Deeper gates allow for a longer holding pressure time, allowing for greater control over cavity pressure history, and therefore the molded part's dimensions.

It is also worth noting here that the use of a valve-gated hot runner can also be very advantageous is terms of a molded part's dimensional control. For example, valve gates can be relatively large, providing for good mold filling characteristics and a very wide cavity pressure history process window. Valve gates can also be closed on-demand at any time during the injection molding cycle. With valve gates, the holding pressure time utilized can then be short, long or anywhere in between. This gives a process set-up engineer increased control over the molded part's mold shrinkage and part dimensions when a valve-gated hot runner mold is utilized.

3.11.3 Experimental Gate Seal Study

The upper limit for a holding pressure phase time setting is typically determined by the time required for the relatively shallow gate(s) to freeze or solidify. Once the gate has solidified, no material can enter or exit the injection mold cavity (i. e., the gate has closed). Applying additional holding pressure time after the gate solidifies will allow for additional runner system pressurization, but will have no impact on the molded part's dimensions. This could cause runner ejection difficulties, wastes energy, and reduces the time available for shot plastication. As a result, the gate freeze time is usually considered to be either the upper holding pressure time limit or the optimum holding pressure time. Shorter holding pressure times can be used if the part is deemed to be within dimensional specification.

The gate freeze time can be estimated theoretically using conductive heat transfer equations or molding simulations, but it is typically determined experimentally. A gate freeze study is used to determine the gate freeze or gate seal (solidification) time. This experimental study is conducted once the pack phase variables have been determined and set. To begin a gate freeze study, the molding machine is typically set to run in the semi-automatic or automatic mode using a sufficient cooling time and appropriate plastication conditions. It is very important to keep the machine on cycle during the gate freeze study as any sort of cycle delay (in the semi-automatic mode) may confound the study results.

A gate freeze study begins by setting the holding pressure to an appropriate value. At this point in the start-up process, the optimum holding pressure setting has yet to be determined. A holding pressure setting of about 50–60% of the peak packing pressure is typical for a gate freeze study. The holding pressure magnitude does affect part quality, but has a very limited impact on the time it takes for the gate to freeze or solidify. Once the holding pressure value is set, the first shots are molded with a holding pressure time setting of zero seconds. These molded parts should appear full in terms of flow length, but will exhibit sinks or surface depressions since some of the melt that was squeezed into the mold cavity by the packing pressure will flow back towards to the shot chamber (known as reverse flow). The gate and runner cross sections remain molten at this early point in the molding cycle. As the gate seal study progresses, hold time is increased by 0.5 or 1.0 seconds for the next molding cycle and parts are saved for inspection. The holding pressure time continues to be increased for subsequent shots until the hold time reaches about 10 to 12 seconds. All of the parts molded during the gate freeze study are collected, inspected and weighed.

As the applied holding pressure time is increased, the molded part weight will increase for two reasons. First, as holding pressure time increases, the solid skins continue to grow, at the expense of the molten core as depicted in Figure 3.45. This reduction in molten core cross section causes an increase in the reverse flow pressure drop. Increasing the holding pressure time also provides more time for melt to be pushed into the mold cavity as the thermoplastic material shrinks. It is important to independently confirm that the melt cushion is being maintained over the entire course of the gate freeze study because a loss of cushion will prevent pressure transmission from the melt cushion to the mold cavity.

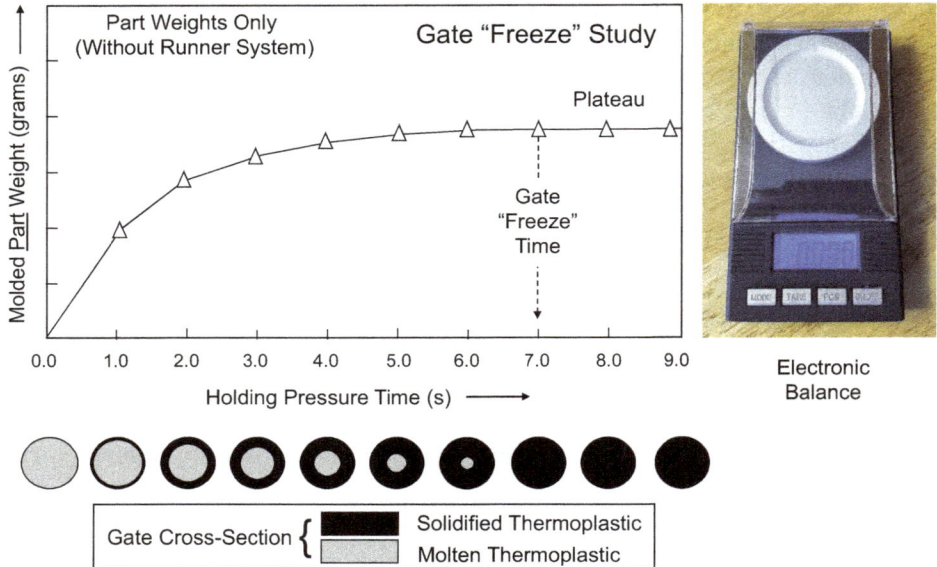

Figure 3.45 The upper limit for the holding pressure time can be determined by performing a gate freeze study where parts are molded at various holding pressure times and weighed. Once the gate freezes or solidifies, the part weight will remain constant since the solidified gate prevents material from entering or exiting the injection mold cavity. The gate freeze time is generally considered to be the maximum and often optimum holding pressure time.

Once all of the gate freeze study parts have been molded, the parts are carefully degated (in the case of a cold runner mold) and weighed using an electronic balance or scale. The molded part weight data is then plotted as a function of hold time as shown in Figure 3.45. The part weight graph should plateau at the time associated with gate freeze since no material can enter or exit the mold cavity after gate solidification. The time associated with the onset of the plateau region is the gate freeze time, which is also the upper limit of the holding pressure time for the process (at the particular mold and melt temperatures that were set previously). Holding pressure times less than the gate seal time can be used, but this may lead to dimensional issues (undersize part, sinks etc.). Many sources consider gate seal time to be equivalent to the optimum hold time. Another school of thought is to add an additional second or so to allow for temperature variations. In any case, it is important that only the parts (and not the runner system) are weighed during the gate freeze study since runner weight can continue to increase even if the gate has solidified.

If the injection mold is equipped with an in-mold cavity pressure sensor, the pressure signal can also be used to study gate freeze time as shown in Figure 2.78. In such a case, the gate seal study is run as described above, molding parts at increasing holding pressure times. The cavity pressure-time curve will generally show a gradual decay if the hold pressure time is anytime after the gate freeze. If the holding pressure timer times out before the gate is fully solidified, there will be an observable discontinuity or drop in the cavity pressure – time trace. In any case, the results of a gate freeze study can be effectively used to determine the optimum holding pressure time. However, at this point in the process start-up, the holding pressure magnitude setting is just a placeholder or "ball park" value that has been set to allow the gate seal study to be conducted. The final holding pressure value will most likely need to be dialed in based on the part's visual appearance and dimensional requirements. Holding pressure adjustments can be done at this time but may need to be re-evaluated once the remaining molding process variables have been determined.

3.11.4 Limitations Associated with Constant Holding Pressure

Ordinarily, holding pressure is set at a constant value for the entirety of the hold pressure time. This mode of operation has an inherent shortcoming that can be important, particularly in tight-tolerance molding applications. A melt pressure vs. time trace for a typical constant holding pressure molding process is shown in Figure 3.46. The thermoplastic melt contained within the cavity cools and shrinks over the course of the holding pressure time. In addition, melt within the runner and the gate remains at least partially fluid over the course of the holding pressure time. As the melt in the cavity cools, the pressure in the mold cavity decreases (due to material shrinkage) and the applied hold pressure forces additional melt through the runner and into the mold cavity. While the amount of material being pushed into the cavity during the holding pressure phase is relatively small, there is some flow of melt into the mold cavity during the holding phase. If there is flow into the cavity, there must also be a pressure drop over the length of the cavity, particularly since the melt viscosity increases as the melt cools. The non-uniform cavity pressure associated with a constant applied holding pressure can lead to dimensional variation for the molded part as shown in Figure 3.46 and 3.47 [37].

The existence of pressure drop during the holding pressure phase has also been discussed in studies where the mold cavity is instrumented with multiple pressure sensors [46]. The area under a cavity pressure sensor – time curve for the sensor located near to the gate will be greater than the area under a cavity pressure – time curve for a sensor placed away from the gate as depicted in Figure 3.46. Since mold shrinkage is influenced by the melt's pressure history, parts molded with constant hold pressure tend to exhibit non-uniform shrinkage. While there are several reasons that mold shrinkage can be anisotropic (e. g., molecular orientation, fiber orientation, crystallinity variation) pressure history variation is also a significant factor that results in anisotropic mold shrinkage. This anisotropy shrinkage phenomenon is typically described as area shrinkage variation. The observed mold shrinkage tends to be greater away from the gate than near the gate when the applied hold pressure is constant. Increasing the constant hold pressure value will reduce overall mold shrinkage, but will not eliminate the area shrinkage variation. While this type of area shrinkage variation is known to exist, it is ignored in most applications as the magnitude of the mold shrinkage variation may be very small, and the molded parts may be within dimensional specification. However, in tight-tolerance applications, the area shrinkage variation may be significant enough to cause the molded part's dimensions to be out of specification.

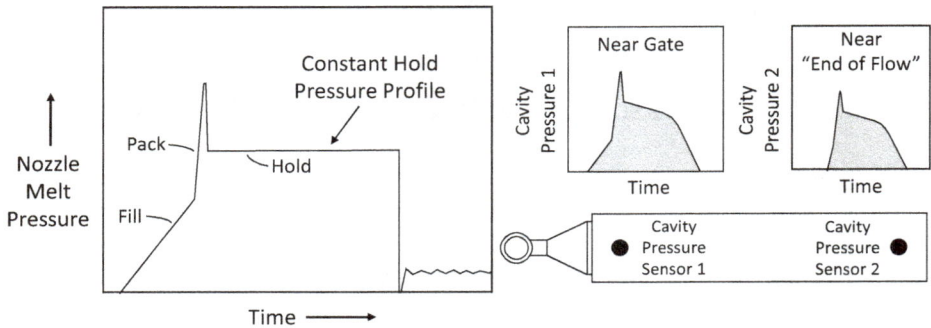

Figure 3.46 One very common mode of control for injection molding is to use a constant hold pressure for the entire holding pressure time. As the melt within the cavity cools, it also shrinks. As the melt shrinks, the mold cavity pressure begins to decay. As the cavity pressure decays, the applied hold pressure pushes more melt into the cavity, especially in the cavity areas near the gate. This results in an anisotropic mold shrinkage phenomenon known as area shrinkage variation.

Further support for the concept of area shrinkage variation is demonstrated if one considers the possible gating schemes used for an injection molded plastic gear. Gear-like parts are ideally molded with a 360° disc gate located at the inside diameter of the gear. This is the most symmetrical gating option and typically results in a gear that exhibits very uniform shrinkage and no knit lines. However, disc gates can require a secondary operation to de-gate smoothly. In other cases, the inside diameter of the gear may be too small for a disc gating scheme.

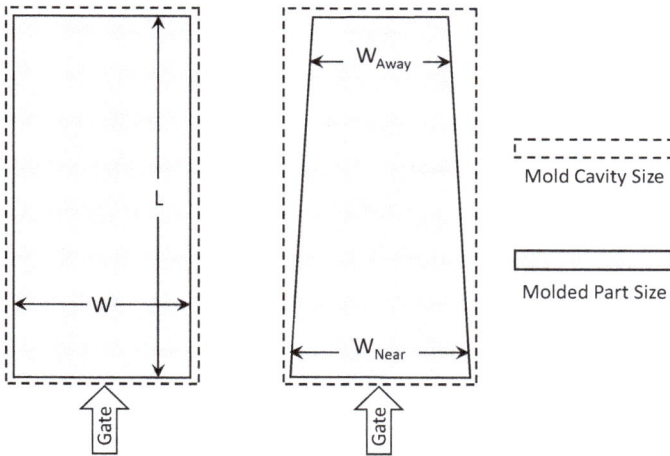

Figure 3.47 Isotropic mold shrinkage, such as that shown above (left) is a goal for most injection molding operations. However, when constant hold pressure is utilized, thermoplastic parts tend to shrink more away from the gate than near the gate. This is the result of a non-uniform cavity pressure history over the length of the molded part. Holding pressure provides for more shrinkage compensation in sections of the mold cavity near the gate. Note that the area mold shrinkage variation associated with the cavity pressure history variation for the figure on the right is exaggerated.

Face gating using a three-plate mold or a hot-runner mold is also common for plastic gears since face gating eliminates the need for a secondary operation to remove the gate (since de-gating is automatic when the injection mold opens). While a small gear with a short flow length could easily be molded with a single gate (from a mold filling perspective), face gated gears usually have at least three gates. A gear molded with a single face gate will tend to be out of round (i. e., egg shaped) as a result of area shrinkage variability as shown in Figure 3.48. Regions of the gear near the gate tend to shrink less than regions of the gear that are away from the gate. Molding a gear with three face gates does add additional knit lines (and runner scrap for a cold runner mold) but the uniformity of mold shrinkage increases as the number of gates increase. A gear gated with three face gates will not be perfectly round (it will be tri-lobal in shape), but it will be rounder than a gear molded with one or two face gates. The mold cavity pressure history, and therefore the part's mold shrinkage will be more uniform if three gates are used compared with one gate. The philosophy of using multiple gates to minimize flow lengths and cavity pressure history variation is common in tight tolerance molding. The use of multiple gates also has the added benefit of allowing for a faster mold filling velocity.

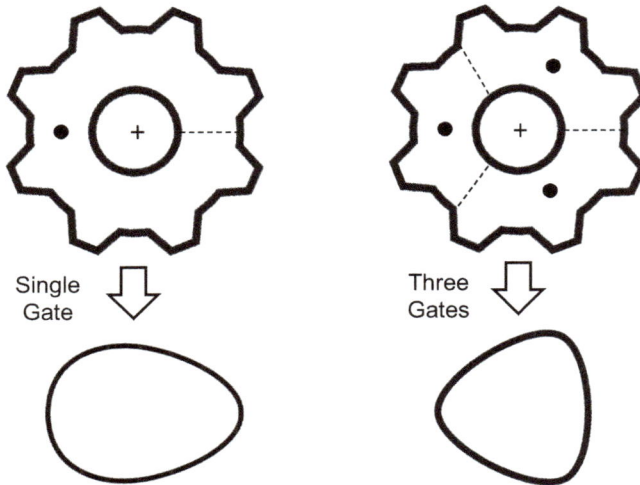

Figure 3.48 A gear molded with three gates will tend to be rounder than a gear molded with one gate. While a small gear can easily be filled using one gate, a gear molded with one gate will tend to be egg shaped due to a greater cavity pressure history for regions of the part near the gate. The use of multiple gates for gears is more common for injection molded gears. This practice minimizes cavity pressure history variation and therefore area shrinkage variability, which results in a rounder gear shape.

3.11.5 Multi-stage Holding Pressure Profiling

Almost all current generation injection molding machines have the capability to profile holding pressure, either as time steps or as a continuous ramp. While the use of profiled holding pressure has been criticized by some, it has also been shown to be advantageous, especially in applications where a more uniform mold shrinkage is desirable [46]. Like a barrel temperature profile, or an injection velocity profile, there are many different ways that holding pressure can be profiled (e. g., an increasing profile, a decreasing profile, a pyramid profile). The possible advantages, practices and rules associated with holding pressure profiling are not very well established compared to these other profiles, but one of the more common holding pressure profile applications is discussed below.

The use of constant hold pressure, as shown in Figure 3.46, is known to produce molded parts that exhibit some degree of non-uniform shrinkage. The mold shrinkage tends to be lower near the gate and greater away from the gate due to the greater cavity pressure history (and therefore shrinkage compensation) near the gate region of the molded part. The cavity pressure history, and therefore mold shrinkage will be more uniform if hold pressure is decreased over the course of the holding time as shown in Figure 3.49. The basic concept of setting a decreasing hold pressure profile is that as the applied hold pressure is decreased, melt will tend to bleed from the cavity as the hold pressure is decreased while the holding time progresses. However, more melt will bleed from the regions of the part that are closer to the gate than away from the gate. The use of the decreasing holding pressure profile minimizes the area shrinkage variability which is positive, but also tends to increase the overall magnitude of mold shrinkage, due to a reduction in the area under the melt pressure – time curve. As a result, when a decreasing hold pressure profile is used, the initial hold pressure is set higher than one would normally choose, if the holding pressure were set to a constant pressure value. A higher pressure at the start of the holding pressure phase will compensate for the lower holding pressure at the end of holding phase. The decreasing hold pressure profile can be implemented by dividing the hold time into multiple time segments as shown in Figure 3.49 or it may be ramped downward continuously if the machine control has that capability. While there is some agreement that the use of a decreasing holding pressure profile is beneficial (at least with regard to mold shrinkage uniformity), there are few guidelines regarding the optimum reduction rate or the shape of the decreasing holding pressure profile curve. Advanced molding simulations can provide guidance with respect to establishing the optimum holding pressure profile.

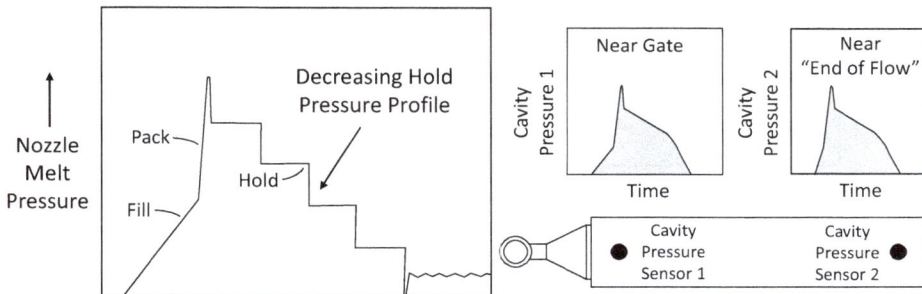

Figure 3.49 Most current generation injection molding machines have the ability to profile holding pressure. The use of a decreasing hold pressure profile has been shown to help minimize area shrinkage variations. A decreasing hold pressure profile results in somewhat greater average mold shrinkage, but a more uniform cavity pressure history over the length of the cavity, and therefore more uniform area shrinkage. The greater overall shrinkage issue can be remedied if a higher than normal starting hold pressure is used to compensate for the loss in applied cavity pressure history that occurs towards the end of the hold phase (compared with a conventional constant holding pressure process).

3.12 Plastication Delay Time

Once the holding pressure time has expired, two concurrent sets of process variables will need to be determined. Referring to Figure 3.23, these variables include (i) the additional cooling timer that will need to be set such that the molded parts can be ejected from the injection mold without deformation, and (ii) a number of plastication variables that occur simultaneously within the additional cooling timeframe. At this point in the process start-up procedure, the additional cooling timer can be set to a time that is somewhat greater than the required cooling time, just so that an automatic injection molding cycle can be utilized. The additional cooling time can be optimized after other concurrent plastication variables are determined.

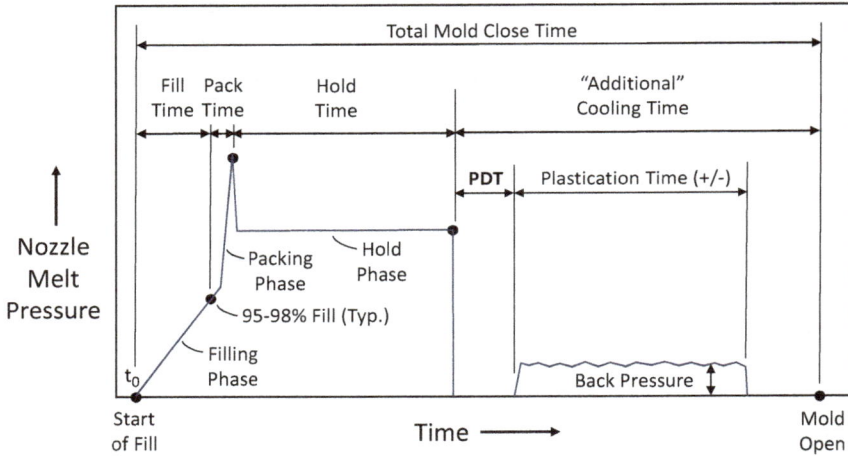

Figure 3.50 Plastication delay time (labeled PDT) is an optional time that can be added between the end of the hold pressure time and the start of plastication. A short plastication time is often added to minimize check ring wear caused by residual holding pressure. A plastication delay time is also commonly used in applications where the additional cooling time is significantly longer than the time required for plastication (i. e. screw recovery).

The additional cooling time (i. e., cooling time after the holding pressure phase is complete) and the plastication variables are very much interrelated. For example, the time available for plastication is normally dictated by the additional cooling time as shown in Figure 3.50 below. This interrelation means that tradeoffs or compromises will need to be made in terms of individual variable optimization. For example, a hypothetical screw rotational speed of 60 RPM may be optimum from a melt quality point of view (for a particular application), but 100 RPM may be required based on the additional cooling time setting (i. e., the time available for plastication). The variable interactions make optimization of all process variable difficult or impossible. There will typically need to be tradeoffs.

One optional plastication variable is plastication delay time. Plastication delay time is the time between the end of the holding pressure phase, and the start of decompress forward, or the start of screw rotation (in applications where decompress forward is not used), as shown in Figure 3.50. In many injection molding applications, the optional plastication delay time is not utilized. However, some sources suggest the use of a plastication delay time is beneficial for most injection molding operations [47, 48]. One reason to use a plastication delay time is to help reduce the potential for wear on the screw check ring and ring seat at the start of plastication. The melt cushion is typically pressurized to hundreds of MPas during the holding pressure phase of the injection molding process. Because thermoplastic melts are highly compressible, the melt cushion pressure does *not* instantly drop to zero MPa when the holding pressure is released. If screw rotation is initiated immediately after the holding phase timer expires, the residual holding pressure will push the check ring against the ring seat as the screw starts to rotate. Adding a short plastication delay time (e. g., 0.1 to 0.5 seconds) will allow some or all of the residual holding pressure to decay. This residual melt pressure reduction should help reduce check ring wear while reducing the initial screw drive motor torque. This is an alternative to the use of forward decompression for the purpose of check ring wear reduction.

Plastication delay time is also commonly used in applications where the additional cooling time required to cool the part sufficiently is significantly longer than the time required for plastication of the next shot. This situation may occur if the molded parts are very thick, but the parts are very small. In such a case, a long cooling time is required, but the time required to generate the small shot is relatively short. Slowing the screw rotational speed would also help in such a case but that alone may not solve the issue. As an example, let's say the additional time required for mold cooling is 40 seconds and the time required for plastication at a very slow screw rotational speed is 25 seconds. If no plastication delay time is used and plastication begins immediately after the holding pressure time has expired, the full shot would be generated 15 seconds before the injection mold is set to open. Such a situation is depicted in Figure 3.51 (top).

One school of thought suggests the plasticated shot will develop thermal gradients if it sits idle in the injection molding barrel for an extended period of time. Many sources suggest that the melt temperature will be most uniform immediately after plastication is complete. As a result, one common practice in setting plastication variables is to set the variables such that plastication is complete shortly before the mold opens (e. g., 1.0–2.0 seconds before the mold opens) as shown in Figure 3.51 (bottom). The short, 1.0–2.0-second dwell time after plastication is used to account for variability in plastication time, particularly in applications where regrind is used. Note that plastication time does not have a set point, it is an outcome that is related to many other variables including; screw rotational speed, back pressure and pellet geometry. The resulting plastication time can vary from shot to shot, although the variability should be minimal if the process is in control. The plastication delay timer (in conjunction with screw rotational speed and back pressure) can be used to ensure that plastication will be completed at the desired time (i. e., shortly before the mold opens), as shown in Figure 3.51 (bottom).

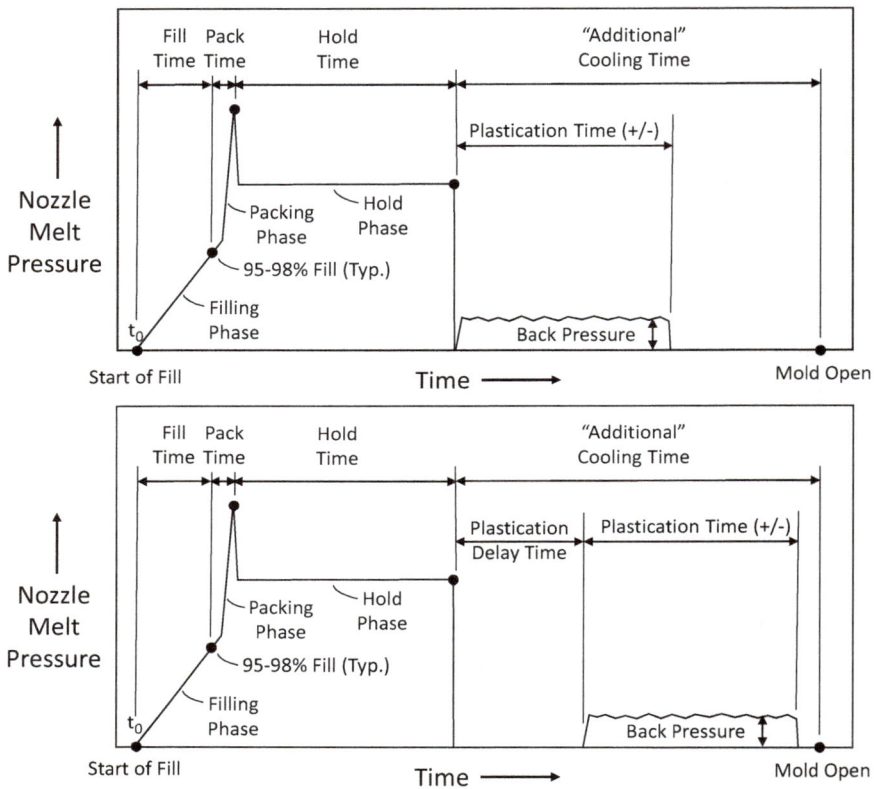

Figure 3.51 If the plastication phase of the molding process is completed long before the injection mold is ready to be opened (top figure), the plasticated shot in the barrel can develop thermal gradients. A plastication delay time can be added so that plastication will be completed shortly before the injection mold opens (bottom figure). The latter can reduce the potential for thermal gradient formation and result in a more uniform melt temperature distribution for the plasticated shot.

3.13 Decompress Before Screw Rotation

Another optional plastication variable is decompression or suck-back before screw rotation (also known as decompression forward). Decompression before screw rotation is a rearward axial movement of the screw that occurs before plastication, as shown in Figure 3.52. The primary decompression variable is the amount of suck-back or the distance of the rearward axial movement. It is relatively common to use a small amount of forward decompression (e. g., several millimeters) in order eliminate any residual holding pressure. Eliminating the residual holding pressure helps minimize or eliminate check ring and ring seat wear that can occur at the start of screw rotation [48]. Forward decompression is a very effective alternative to using a plastication delay time for the same purpose (see Section 3.12). One downside of using forward decompression is the possibility of sucking air into the nozzle or barrel. Decompressing the screw by pulling it rearwards creates a vacuum in front of the screw tip. It is then possible for air to be sucked into the shot chamber through an unsealed area such as the nozzle tip and sprue bushing interface. The trapped air could cause splay, air traps, or void-like defects in the molded part. This may or may not be a significant issue depending upon the design and effectiveness of the mold venting scheme.

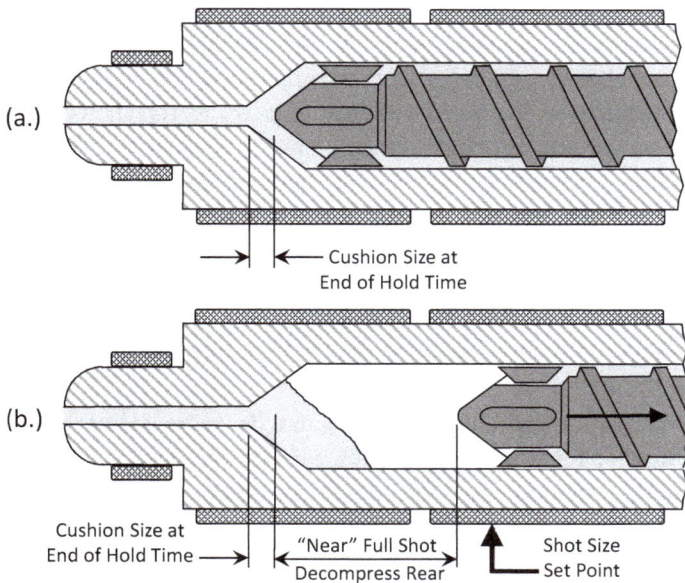

Figure 3.52 The use of a small amount of forward decompression before screw rotation is common to eliminate any residual holding pressure that can cause check ring wear at the start of plastication. Larger forward decompression distances (e. g., near full shot decompression) are sometimes used to improve plasticating screw recovery rates for applications where there is a screw pumping efficiency issue.

Decompression before screw rotation can also be used to assist with plastication. Using a near full shot amount of decompression is a way to essentially eliminate back pressure. Sucking the screw back to a position just short of the shot size set point, as shown in Figure 3.52 (bottom), creates an open chamber within the barrel in which the melt will be extruded. Once the empty shot chamber is filled with melt, the screw will then pump itself back the remaining distance until the screw hits the shot size limit and stops rotating. This method of plastication can be used to limit the plastication energy input into materials where minimum energy input is desired. Using a large forward decompression can also assist with plastication for applications where screw recovery rate is an issue. As an example, decompression forward can help reduce screw recovery times for badly worn plasticating screws. Screw recovery rates can also be low when trying to pump a very low-viscosity (very high MFR) melt using a screw with a deep metering zone. A large decompression stroke can help in both of these cases, although rebuilding the screw or using a more appropriate screw design would be the better solutions.

3.14 Screw Rotational Speed and Back Pressure

Two important plastication process variables are (i) screw rotational speed (RPM) and (ii) back pressure (MPa or lbs/in^2). There are many interactions between these and other plasticating variables, which complicates the determination of the optimum set point for each variable. As stated earlier, there will need to be tradeoffs when establishing optimum plastication variable set points based on outcome priorities. For example, the melt temperature uniformity (for the shot) tends to be better if lower screw rotational speeds are used. However, lower screw rotational speed also increases the time required for screw recovery or shot size generation. There is general agreement that the injection molding cycle time should not be dictated by the plastication rate, but rather cycle time should be determined by the total cooling time required for part solidification. In such a case, a higher-than-optimum screw speed would be a reasonable compromise.

As with many process variables, it is good manufacturing practice to look towards the thermoplastic material supplier for process guidelines, specifically plastication process guidelines. Unfortunately, material supplier information regarding plastication variables is limited, since factors related to screw geometry (e. g., screw diameter, L/D ratio, compression ratio, etc.) all impact on the plastication variable settings. Material suppliers can make recommendations as to the most appropriate screw geometries for their materials, and offer some very rough guidelines on plastication process variable settings. A classic example would be plastication guidelines for a glass fiber reinforced thermoplastic. Almost all manufacturers of glass fiber materials suggest the use of relatively low back pressures. High back pressures increase screw recovery time and therefore the amount of mechanical shear experienced by the material during plastication. Using a low back pressure reduces the shear history associated with plastication which helps maintain glass fiber length by reducing fiber damage during plastication. The mechanical performance of a glass fiber reinforced thermoplastic will be influenced in a negative way if the length of the relatively brittle glass fiber is reduced during the injection molding process.

3.14.1 Screw Rotational Speed

The maximum time available for the plastication phase of the injection molding process is dictated by the amount of additional cooling time, as shown in Figure 3.51. At the earliest, plastication can begin as soon as the holding pressure time ends. The start of plastication can be delayed if either a plastication delay time or decompress forward are utilized. Normally, the plastication process should be completed before the additional cooling timer has elapsed. As a result of this interaction, the time available for plastication will help a process engineer establish the most appropriate screw rotational speed for the process. The most appropriate speed rotational speed (based on time available) may or may not be optimum in terms of melt quality. Other plastication variables, such as back pressure, can also be used to fine tune melt quality of the molten thermoplastic.

In most cases, the goal of the plastication process is to produce a consistently high-quality melt, having a uniform melt temperature or uniform composition (i. e., a properly mixed melt) within the allotted time. This is relatively easy when cooling times are long, but can be more challenging when the time available for plastication is relatively short. As an example, a large thin-wall molding may require a large shot size. However, thin-wall parts require relatively short cooling times, which limits the time available for plastication. In such a case, a faster-than-optimum screw rotational speed may be required.

The plastication model for reciprocating screw injection units is complicated. The barrel heaters provide some energy input, but it is the mechanical shearing action during screw rotation (an injection) that provides the vast majority of energy input for shot plastication. While it may be somewhat counter intuitive, studies have shown that use of lower screw rotation speeds tends to produce the best melt quality, especially in terms of melt temperature uniformity [29,30]. These studies have shown that melt temperature variation within the plasticated shot can be as high as 20 °C at higher screw rotational speeds. This is in large part due to the fact that plastics are relatively good thermal insulators, so a slower rate of plastication allows more time for heat transfer and thermal homogenization to occur within the melt. While such ΔT_m results are material and machinery dependent, it is generally accepted that lower screw rotational speeds are preferred over higher screw speeds in terms of melt quality. It is also important to note that screw rotation speed recommendations vary significantly with the screw diameter. Smaller diameter screws are typically set to higher RPM values than larger diameter screws, since it is the screw's circumferential velocity that correlates with the shear rates imparted to the melt.

One common recommendation for setting the screw rotational speed is to set the speed such that the screw reaches the shot size set point shortly (e. g., 1.0 to 2.0 seconds) before the additional cooling time expires [1,48]. This short time after recovery minimizes formation of shot thermal gradients, but does allow time for small variations in plastication rate. Remember, plastication time is not set. Plastication time is an outcome determined by setting the screw rotational speed, back pressure, and shot size limit. The actual plastication time required (from shot to shot) will vary for a variety of reasons, especially issues related to irregular feeding (e. g., use of regrind etc.). If the molded part is relatively thick and the additional cooling time is very long, the time available for plastication may allow for a screw rotational speed that is too slow to be effective. In such a case, the use of some plastication delay time is recommended, as shown in Figure 3.51 (lower figure).

3.14.2 Back Pressure

Back pressure is the amount of pressure that the screw must generate in order to push itself rearward during plastication (i. e., to reciprocate). Adding resistance to reciprocation increases the back pressure required for plastication as described earlier in Section 2.2.2.3.2. It is important to note here that injection molding screws are *not* positive displacement pumps. The net effects of adding back pressure are two-fold. Adding back pressure (at a constant screw RPM) means that the screw will turn more times to reciprocate the same distance. Adding more back pressure also means that the melt is exposed to a greater compressive force during plastication.

The theoretical volumetric output for a general-purpose reciprocating screw can be approximated using Equation 3.2. The maximum volumetric output rate for the plasticating screw will occur when the pressure differential for the screw metering section is zero (i. e., no added back pressure). In such a case, there is only positive drag flow which is directly proportional to the screw rotational speed. Adding back pressure causes the ΔP terms in Equation 3.2 to increase, which results in a loss in output per screw revolution due to both the increase in negative pressure flow (reverse channel flow) and leakage flow (reverse flow over the screw's flights). On the other extreme, excessive back pressure will cause the negative flows to balance out the drag flow and the screw will rotate without reciprocation. The back pressure used for an injection molding process must lie somewhere between these two extremes.

Screw Output = Drag Flow − Pressure Flow − Leakage Flow (3.2)

$$Q = [AN] - [B\Delta P/\mu] - [C\Delta P/\mu]$$

Where:
Q = Screw volumetric output (mm^3/rev)
A, B, C = Screw geometric constants
ΔP = Metering zone pressure differential (MPa)
μ = Melt shear viscosity (Pa-sec)

Typical back pressures used during plastication range from a few hundred to a few thousand lbs/in^2 melt pressure [48]. At a minimum, a low back pressure is typically recommended to improve plastication and shot size consistency, and to help minimize air entrapment. Adding more back pressure improves melt mixing, but increases the time required for screw recovery (at a constant screw rotational speed), as depicted in Figure 3.53.

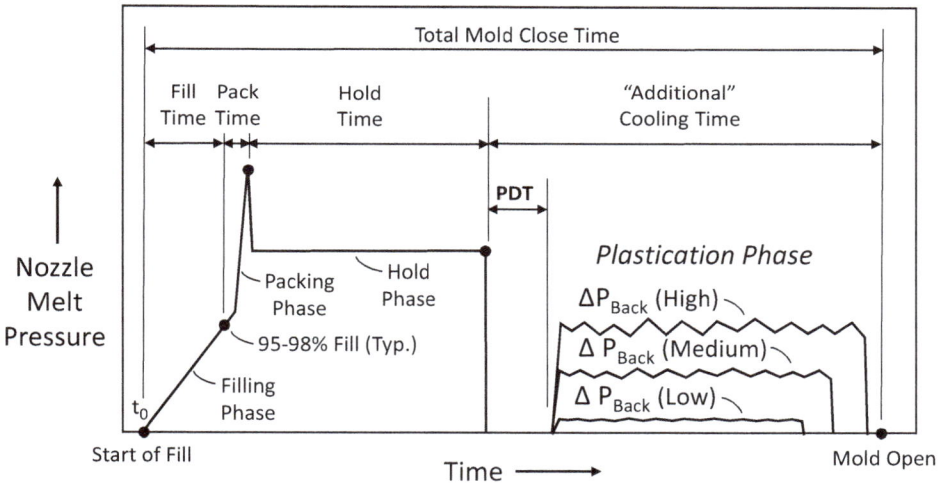

Figure 3.53 Back pressure during plastication can be used to improve shot volume consistency, improve the degree of mixing and provide for additional control over melt temperature. Increasing back pressure will also increase the time required for screw recovery (at constant rotational screw speed).

The increase in plastication time associated with added back pressure will increase the total amount of viscous dissipation or energy input for the shot. The added back pressure increases the total number of screw rotations required to generate the shot. It also causes an increase in melt temperature due to the increased viscous dissipation and an increase in the amount of melt mixing. From a mixing perspective, the use of a moderate or high back pressure would be more important in an application where a color masterbatch additive is used than in an application involving a pre-colored thermoplastic material. On the other hand, the use of a low back pressure would be appropriate when plasticating a glass fiber reinforced plastic (in order to minimize fiber length degradation).

While adding more back pressure will result in a higher melt temperature at the end of screw rotation, some cooling of the melt may occur during the time delay between the end of melt plastication and initiation of the melt injection phase of the molding process. If the addition of back pressure causes the screw recovery time to exceed the time available for plastication, an increase in the screw rotation speed would be required in order to bring the screw recovery time back to the allotted plastication time window. The back pressure setting can also be profiled on some current generation injection molding machines, but there seems to be no universal agreement as to the best practices and procedures for setting the back pressure profile.

3.15 Decompress After Screw Rotation

Another optional plastication variable is suck back or decompress after screw rotation (a.k.a. decompress rear). As with decompress forward, decompress rear does not involve any screw rotation. Decompress rear is a rearward axial suck-back of the screw that occurs immediately after plastication is complete. The primary purpose of decompress rear is to eliminate residual back pressure that can cause melt drool. Pulling the screw back a short distance increases the volume of the shot chamber, which relieves any residual back pressure (or thermal expansion in some cases), as shown in Figure 3.54. Melt drool can also occur if a hygroscopic plastic is not properly dried as the residual moisture causes a foaming action, and a reduction in melt viscosity, for condensation polymers (e. g., nylon 66). While the use of decompression rear is helpful to prevent drool caused by residual moisture, correcting the drying issue would be a more appropriate solution if a hygroscopic thermoplastic is not properly dried. The set distance for decompression rear is commonly a distance value that is close to the throw distance of the plasticating screw's check ring [48].

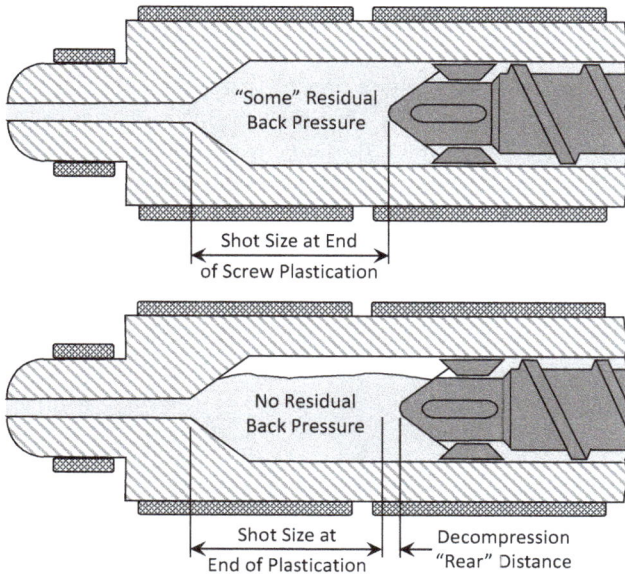

Figure 3.54 Decompress or suck back after screw rotation is an optional molding process parameter that eliminates any residual back pressure. Residual back pressure can cause nozzle tip melt drool when a shut-off nozzle is not used, or gate drool for non valve-gated hot runner molds. Decompress rear eliminates residual back pressure by increasing the shot chamber volume by pulling the screw back a short distance after screw recovery is complete.

Melt drool can occur under a variety of different circumstances if decompress rear is *not* used. Figure 3.55 shows two situations where nozzle tip melt drool can occur. If the injection mold being used has a cold sprue, and sprue break (i. e., carriage retract) is *not* used, melt can drool from the nozzle tip when the solidified sprue is pulled from the sprue bushing as the mold opens; this is shown in Figure 3.55 (left). In this case, the melt drool is not visible and can cause a variety of "head scratching" molding defects. For example, any melt drool will at least partially solidify before the next shot is injected. The partially solidified melt drool could cause a variety of mold filling issues including a mold filling imbalance for a multi-cavity mold or gate clogging for small pin gates. The partially solidified melt drool can also result in a variety of molded part surface defects.

Nozzle tip melt drool can also occur in applications where sprue break (carriage retract) after plastication is used, and decompress rear is not used, as shown in Figure 3.55 (right). This would lead to a situation where the nozzle tip/sprue seat seal becomes compromised. Residual back pressure can also cause gate drool for hot runner molds with thermal gates when the hot runner mold opens. Decompress rear can eliminate that issue. Fundamentally, decompress rear is not required if the barrel is equipped with a shut-off nozzle, or if a hot runner mold has valve-gated hot nozzles. However, the use of decompression after plastication is said to be beneficial for valve-gated hot runner molds, not to eliminate drool, but to depressurize the hot runner in order to minimize valve gate closing forces, thereby extending the hot valve gate nozzle assembly life. Reported downsides to the use of decompress rear include possible air entrapment in the shot due to vacuum effects or an increase in splay on the molding surface [49,50].

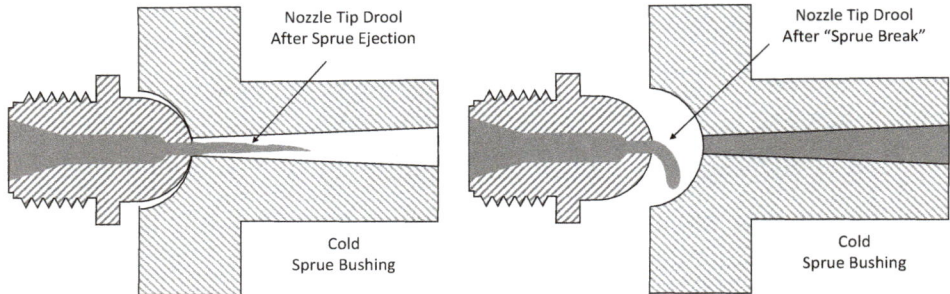

Figure 3.55 Nozzle tip melt drool is likely to occur if decompression after plastication is *not* used, particularly when back pressure is used during plastication. The left figure shows a cold runner mold where both decompress rear and sprue break (carriage retract) are not used. In this case, nozzle tip drool could occur when the cold sprue from the previous cycle is pulled from the sprue bushing as the injection mold opens. The drool would not be visible in such a case and it could lead to some "head scratching" mold filling issues. The right-hand figure depicts an application where sprue break after plastication is used without decompress rear (shown just before the mold opens). In such a case, the melt drool could interfere with the nozzle tip / sprue bushing seat and seal.

3.16 Additional Cooling Time

The overall mold cooling phase of the injection molding process actually begins as soon as the thermoplastic melt is injected into the injection mold. While there is some viscous heating that occurs during the mold filling phase, most of the cooling time is associated with conductive cooling of the hot, molten plasticated melt by the cooler injection mold steel. Ideally, the injection mold's cooling system is able to remove heat from the mold steel as quickly as the melt can deliver heat to the mold steel. In that way, mold temperature would remain constant throughout the cycle, In reality, mold temperature will tend to vary to some degree over the course of the cycle due to factors that include coolant layout restrictions and coolant fluid flow rate limitations.

Much of the mold cooling phase occurs during the time that the molded part is being pressurized during the fill, pack and holding pressure times, as shown in Figure 3.56. Cooling efficiency during these pressurized phases of the molding process is typically quite good since the driving force (ΔT) is large and the internal cavity pressure helps create intimate contact between the thermoplastic part's solid skins and mold's cavity and core steel surfaces.

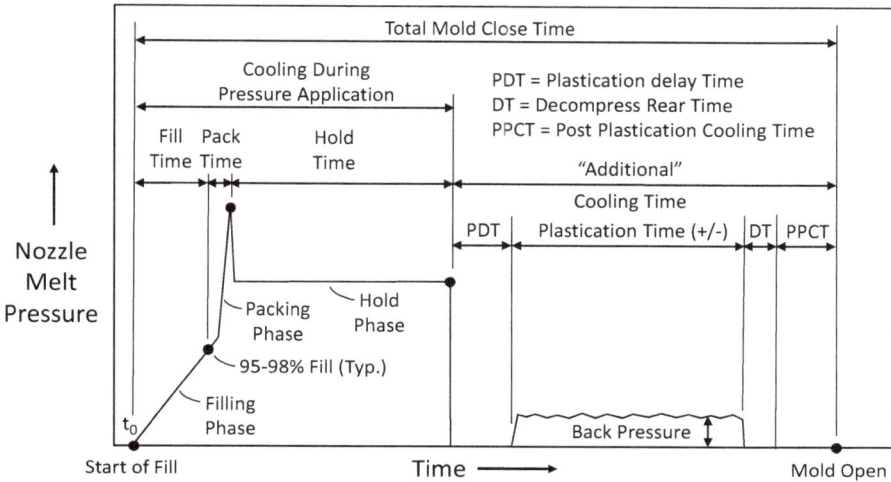

Figure 3.56 The cooling phase of the injection molding process begins as soon as the hot thermoplastic melt is injected into the relatively cold injection mold. While the molded part's gate is typically solidified by the end of the holding pressure time, an additional cooling time is required to ensure the normally thicker molded part has cooled to the point where it will not distort or deform after ejection from the injection mold. The additional cooling time must also be long enough to allow for plastication of the shot for the next molding cycle.

In a typical injection molding process, the holding pressure set time is determined by performing a gate freeze study, as described in Section 3.11.1. At the point in the cycle time where the holding pressure is released, the gate will have solidified, but the normally thicker plastic parts and the runner system (in the case of a cold runner mold) have only partially solidified. An additional cooling time must be added to ensure that the molded part is cool and rigid enough to be ejected from the injection mold without any permanent deformation or distortion. This required mold cooling time could be determined theoretically using a molding simulation or cooling time equations, as discussed in Section 1.9. However, cooling time is most commonly determined experimentally as part of the mold start-up process. Theoretical cooling time calculations are common when designing plastic parts or molds, but experimentation is typically required when starting up a new process. Experimental determination of the optimum additional cooling time typically begins by molding of parts having a cooling time that is longer than required. The additional cooling time is then gradually decreased in each molding cycle, until part deformation issues begin to arise, as shown in Figure 3.57. If the part is too hot when the mold opens, permanent deformation can occur due to:

- The forces associated with pulling the part from the cavity when the mold opens.

- The forces associated with the mold's ejector system pushing the part from the core.

- Residual molding stresses caused by differential shrinkage or molecular orientation.

- The static or dynamic forces associated with inertia or gravity.

Permanent deformation of the injection molded parts due to any of these forces is unacceptable in most injection molding applications. The additional mold cooling time must be long enough such that none of these forces causes permanent deformation or warpage. While permanent deformation of the moldings themselves is normally unacceptable, some cold runner deformation *is* generally acceptable, provided the runner can be fully ejected from the injection mold. For example, full round cold runners are advantageous for efficient mold filling, but they are in fact very difficult to cool since their surface area to volume ratio is very low. In addition, cold runner systems often have diameters that are significantly greater than the molded part's nominal wall thickness (e. g., to minimize mold filling pressure drop). Full round runners are difficult to cool, but they are naturally drafted, and relatively easy to eject, even if they have not fully solidified. Pulling the sprue from the sprue bushing is perhaps the most difficult aspect of cold runner system ejection, especially for brittle, high-modulus thermoplastic materials with a low shrink rate.

The primary consideration when determining the minimum acceptable value for the additional cooling time is ensuring that there is no permanent deformation of the molded parts caused by the factors listed above. There are also other secondary" process considerations that must be taken into account when setting the additional cooling time, but these other considerations should not be the controlling factor for this time setting. Many other injection molding process phases must occur within the additional cooling timeframe, as shown in Figure 3.56. Some of these concurrent process phases are optional, but the conventional molding process shown in Figure 3.56 has (i.) a plastication delay time (ii.) a plastication time (iii.) a decompress rear time and (iv.) a post plastication cooling time. These variables are all typically optimized *after* the minimum additional cooling time has been established where no permanent part deformation occurs. Many sources agree that a variable, such as screw recovery time, should not be the primary factor in determining the additional cooling time. Ideally, the minimum additional cooling time is determined based upon the thermoplastic part's dimensional stability upon ejection. The other secondary parameters (e. g., screw rotational speed, back pressure) should be set within the constraints of the additional cooling time limit.

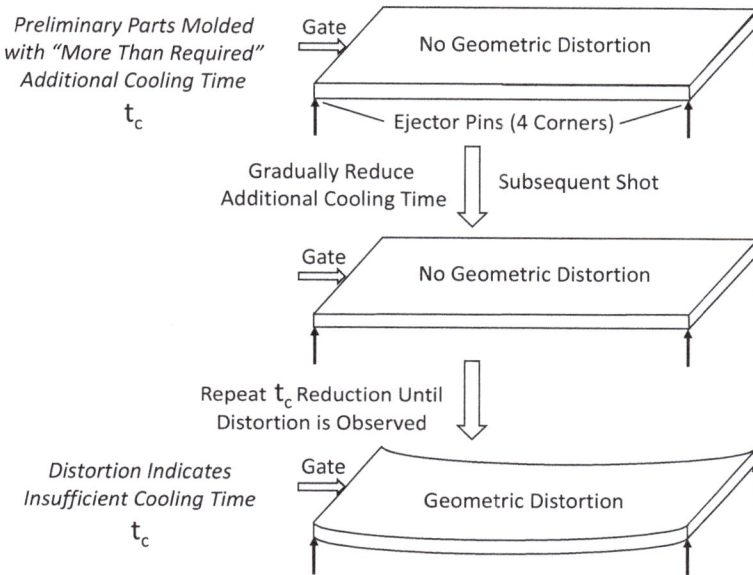

Figure 3.57 The optimum setting for the additional cooling time is typically determined experimentally by molding parts at a variety of additional cooling times, and then inspecting the parts for signs of distortion or deformation. The lowest additional cooling time where permanent deformation does not occur can be considered as the optimum additional cooling time.

For example, a screw rotational speed of 60 RPM may have been found to be optimum in terms of melt quality for a particular injection process. If the screw recovery time at the 60 RPM screw speed is longer than the optimum (or minimum) additional cooling time (based on the no deformation criteria), then a screw rotation speed of, say, 80 RPM may be required in order to complete plastication within the allotted additional cooling time. Additional time must also be available for the other plastication parameters such as decompression, if utilized. It is also important to include a post-plastication time to account for variations in the screw recovery time. As mentioned earlier, the post-plastication time may be a second or two (or less), depending upon the consistency of the plastication process. An alternative to the procedure described above would be to stay with the 60 RPM screw speed setting, and add additional cooling time so that plastication can be completed before the injection mold opens for part ejection. While this is an option for setting the additional cooling timer, it would be difficult to justify based upon economic factors.

In some cases, it is also possible to eject the molded parts while they are still warm enough to deform, but this deformation is somehow restricted. This could involve fixturing the molded parts after ejection or using some sort of post-ejection cooling process. This is not common for most injection molding applications but may be appropriate for some applications, such as very thick-wall parts, particularly if production volumes are limited.

3.17 Part Ejection Parameters

Once the mold cooling time has expired, the molding machine clamp will open and the molded parts and runner system (in the case of a cold runner mold) will need to be ejected. In most cases, the molding is pulled from the A-side cavity when the mold opens as the first step in the part ejection process. Mold opening parameters include positions, pressures and velocities associated with platen movement. The opening stroke is typically set to be just greater than the open space (i. e., daylight) required for part ejection in order to minimize the platen traverse time. The open stroke must include space for end-of-arm tooling in applications where robotic part removal is utilized. The platen traverse velocity is typically slow for the initial breakaway, and then set high to minimize the platen traverse time. The platen velocity is normally reduced towards the end of the platen traverse stroke to allow for more gradual platen deceleration. Once the mold is open, the molding machine ejector plate is then used to activate the mold's ejector plate and ejection devices to push the molded parts from the B-side mold cores.

The ejection sequence is simplest for open and shut molds that do not have other special mold actions such as splits, side actions or core pulls. With an open and shut injection mold, the ejection stroke will be limited by the mold's maximum ejector plate travel distance. In most cases, a single knock out or ejector stroke is used to push the molded parts from the mold's cores. Multiple ejector strokes are sometimes used to help prevent the molded parts from hanging up on the ejector hardware for molds that utilize ejector plate return springs, or if the mold and molding machine ejector plates are tied together. Once the molded parts have been ejected and the ejector plate has been retracted, the mold close positions, pressures and velocities can be set. This includes setting of both low-pressure mold protection variables and clamp tonnage. The ejection sequence is typically more complex for molds with special actions. The mold opening and closing parameters must be set such that the ejector system is reset before the mold closes in order to prevent mold damage. Special mold actions are sometimes instrumented with a position sensor to ensure that the mold action is in the home position prior to activating the mold closing sequence.

3.18 Part Dimensional Qualification

The various sections of this chapter describe some of the main considerations associated with setting the primary process variables associated with the injection molding process. Following these guidelines, an initial molding process can be established. Parts can be molded at these process conditions, but at this point, there is no confirmation that the molded parts are within specification. If the part specifications are very loose, there may be no need to tweak the process parameters any further. However, if the molded parts have cosmetic or dimensional tolerance requirements, then some of the molding process parameters may need to be adjusted to bring the part dimensional or other attributes into specification. Making changes to variables like melt temperature or mold temperature at this point is not optimum since (i) these variables have interaction with most other molding process variables, and (ii) these temperatures take time to stabilize. Even so, such changes may be required for certain part attributes, such as the molded part's surface finish.

Perhaps the most common reason to tweak the process at this point is to ensure that the parts meet or exceed the required dimensional specifications. As discussed earlier, mold shrinkage is relatively difficult to theoretically predict with certainty, and mold shrinkage is often non-uniform. Most injection molded parts do have dimensional tolerance requirements and these tolerances are becoming tighter as the plastics industry continues to evolve. The most common molding process variable used to control part dimensions is the holding pressure value (or holding pressure profile). Holding pressure can be dialed up if the parts are undersized, while holding pressure can be reduced if the parts are oversized. Constant holding pressure values influence all of a part's dimensions in a global way, while a profiled hold pressure may be used in applications where some part dimensions are in specification, and others are out of specification. In terms of dimensional control, holding pressure changes are effective, instantaneous, and have a more limited number of interactions with many other injection molding process variables.

3.18.1 Measurement Errors

Establishing whether a newly developed injection molding process is producing plastic parts that are within dimensional specification can be difficult for several reasons. First, thermoplastic materials are significantly more flexible (or softer) than metals. For example, a typical ABS plastic has a modulus that is about one hundred times less than that of carbon steel. Traditional mechanical measuring tools, such as general-purpose micrometers or calipers, impart jaw forces that can deform an injection molded thermoplastic part, particularly for more flexible thermoplastics. For example, it is very difficult to obtain repeatable and reliable measurements for elastomeric parts, even when the molded part dimensions themselves may be consistent. Low and constant force measuring tools, such as the adjustable force digital micrometer and caliper shown in Figure 3.58 provide both more accurate and more consistent results when measuring thermoplastic parts, especially those molded from more flexible materials. Optical measuring instruments are also widely used for measuring injection molded thermoplastic parts since optical measuring techniques are non-contact.

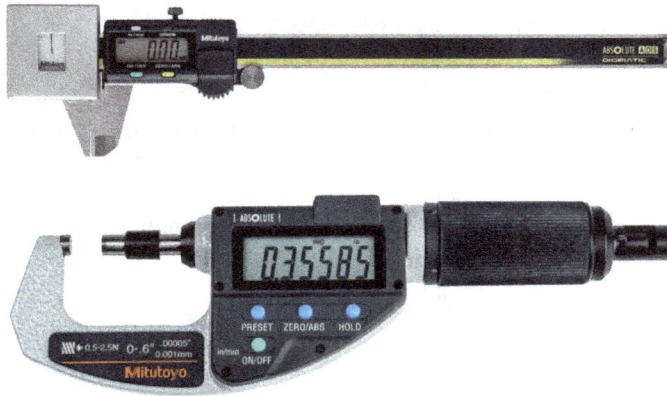

Figure 3.58 Low force measuring instruments, such as the precision adjustable low force micrometer and low force caliper shown, provide both more accurate and more consistent results when measuring injection molded parts that are produced from low modulus thermoplastic or elastomeric materials. (Images provided courtesy of Mitutoyo America Corporation, Aurora, IL 60502, https://www.mitutoyo.com/).

3.18.2 Dimensional Stabilization Time

A second (sometimes major) issue associated with dimensional measurements is the amount of time required for dimensional stabilization after molded part ejection. The total or overall shrinkage for a thermoplastic part is the sum of three separate shrinkage components. Overall shrinkage is the sum of:

- The amount of shrinkage at the time of demolding or part ejection.

- The amount of additional shrinkage that occurs as the part cools to room temperature.

- The amount of additional shrinkage (or expansion) that occurs after reaching room temperature.

The importance of accurate dimensional measurements varies with the application and the thermoplastic material type. However, in applications where dimensional requirements are important, there is a waiting period associated with dimensional stabilization. This waiting period can be hours or even days for some thermoplastics. This waiting time issue is most certainly material dependent as outlined in Section 1.4.3.3.

Plastic parts are typically ejected from the injection mold when they are at a temperature close to the material's DTUL or Vicat softening temperature (typically well above room temperature). It will generally take some number of minutes for the warm thermoplastic parts to cool down to room temperature after the parts are ejected from the injection mold. The ejected parts continue to shrink as they cool (outside of the injection mold) so it is common to wait 20–30 minutes or an hour before measuring the ejected parts. Note this recommended waiting time holds true *only* for materials that do not exhibit post molding shrinkage (i. e., an additional amount of shrinkage that occurs after the ejected part reaches room temperature). For example, general-purpose polystyrene is an amorphous thermoplastic that is not hygroscopic. As such, general purpose polystyrene exhibits little or no post mold shrinkage. A polystyrene part's dimensions are generally considered to be stable once the part temperature reaches room temperature as shown in Figure 3.59.

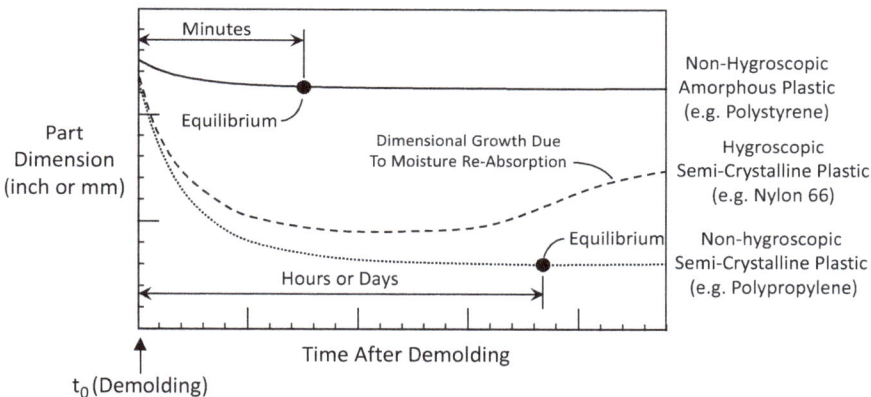

Figure 3.59 Thermoplastic materials shrink as they cool in the injection mold cavity and after the partially cooled molded parts are ejected from the injection mold. All thermoplastics exhibit both in -mold shrinkage and additional out of mold shrinkage as they cool from the ejection temperature to room temperature. Some thermoplastics, especially semi-crystalline thermoplastics with a low glass transition temperature, will also exhibit post molding shrinkage for hours, or even days, after cooling due to increase in their percent crystallinity. Hygroscopic plastics can exhibit dimensional growth due to moisture re-absorption after ejection from the mold.

Other thermoplastics, particularly semi-crystalline thermoplastics having a low glass transition temperature, will continue to shrink after they reach room temperature. Plastic parts molded from a material such as a semi-crystalline polypropylene can continue to shrink for hours or even days after reaching room temperature due to post mold crystallization. Post mold crystallization is an asymptotic function meaning the rate of change is most significant in the first few minutes or hours after ejection from the mold. The rate of dimensional change decreases dramatically over time eventually reaching zero. With some materials, it can take days in order to obtain an equilibrium value for post mold shrinkage. In such a case, a good approach is to estimate the amount of post mold shrinkage using shorter term shrinkage data. An experiment can be conducted where part dimensions are taken over a relatively short time frame (say over an hour or two after the part temperature reaches room temperature). The part dimensions (Y-axis) can then be plotted as a function of time (X-axis) and with an exponential curve fit that is then extrapolated to longer times. The curve should be asymptotic as depicted in Figure 3.59, where the extrapolated plateau dimensional value is a good estimate of the final or equilibrium part dimension. This type of experimental shrinkage data can also be used to help design hot gauges that can be used as go – no go gauges. Gauges of this type can be used as a production monitoring tool to determine dimensional compliance before the molded production parts have reached their final dimensions [51].

Some thermoplastics can also exhibit dimensional growth after molding. This is most common for polar, hygroscopic thermoplastics that re-absorb moisture from the environment after the parts are ejected from the mold. Hygroscopic plastics are normally dry as molded, and will re-absorb moisture over time at a rate that is dependent upon many factors including; the material type, the surface area to volume ratio of the molded part, and the environmental conditions to which the part is exposed. A very limited number of thermoplastics exhibit significant growth due to moisture re-absorption since most thermoplastics do not absorb a high concentration of water at equilibrium. For example, a material like ABS will re-absorb a low concentration of water (e. g., 0.01–0.04%), which has only a minimal effect on the molded part's dimensions. However, a thermoplastic part molded from a polar semi-crystalline thermoplastic material like nylon 66 can absorb one or more percent water. This level of moisture absorption will result in measurable dimensional growth. Precise prediction of the dimensions of a nylon 66 part can be challenging because nylon 66 can exhibit *both* post mold crystallization (i. e., post mold shrinkage) and post mold dimensional growth due to moisture re-absorption. Fortunately, most thermoplastics do not exhibit such complicated dimensional variations over time.

3.19 Molding Process Shut-Down Procedures

The procedure and practices used when shutting down an injection molding process vary depending upon variables such as the thermoplastic material being molded, machine type, mold type (e. g., hot runner mold or cold runner mold) and the duration of the injection process shut-down. While different molders have different shut-down procedures, there are a number of common practices used when shutting-down an injection molding process.

3.19.1 Short Term Shut-Down Procedures

There are times where a molding process will be temporarily shut down for a matter of hours or a few days (e. g., overnight or for a weekend). If the same cold runner injection mold and thermoplastic material will be used when production resumes, the shut-down procedure is relatively simple since it does not involve an injection mold change. It is common to close the hopper slide gate a few shots before the end of production to avoid excess purge material waste. If the thermoplastic material being molded is (i) non-corrosive and (ii) thermally stable, the barrel can be emptied as much as possible by purging the material. If the material is corrosive or thermally unstable, the barrel should be purged with a non-corrosive, stable thermoplastic or purge compound that has a similar melt temperature range to the material to be molded for production. The purge compound itself should then be purged from the barrel. Once the barrel is as empty as possible, a small amount of decompression can be used to eliminate contact between the check ring and its seat so that these surfaces do not gall when restarting the injection molding process. The barrel temperatures and feed throat cooling water can then be shut off. The shut-down and purge procedures that are used with a hot runner mold vary widely depending on the hot runner design. Some hot runner suppliers recommend closed mold purging while others allow for open mold purging. The hot runner supplier should be consulted for the best shut-down and purging practices for the hot runner mold.

The injection mold itself also deserves attention to detail during a short-term shut down. After barrel purging, the injection mold can be opened for visual inspection and maintenance of the open injection mold faces. The exact level of mold care varies depending on many factors, but this is a good opportunity for at least some preventative maintenance. At this point, the injection mold coolant system (and heating system in the case of a hot runner mold) can be shut down. For example, the injection mold should be examined for any coolant leaks or wiring issues. While purging the mold cooling circuit is generally considered to be a good manufacturing practice, it is also common not to purge a mold's cooling circuit for a short-term process shut-down. In such a case, the coolant flow is shut off, but the coolant is not expunged from the injection mold for the short term shut-down.

A non water-based injection mold cleaner (or dry ice blasting) can then be used to clean the cavity and core surfaces along with any exposed parting line vents. The use of cotton gloves during cleaning will help avoid finger prints and skin oils that are acidic and can cause corrosion unless they are neutralized. Only soft copper or brass tools and brushes should be used for injection mold cleaning or repair. It is also good practice to spray the mold with a rust corrosion preventative (especially for the cavities and cores), even when the down time is only hours or a day. Spraying the mold with a corrosion inhibitor adds a cleaning step to remove the corrosion inhibitor during the next mold start-up, but it is inexpensive insurance against what could be a major and costly issue (i. e., rust or corrosion). Leader pin cleaning and re-lubrication is also recommended as an additional step in the mold shut-down process.

After ensuring that the mold ejection system has been properly reset (e. g., the slides are in the correct position) the injection mold can then be closed, but NOT fully clamped. Leaving a paper-thin gap between the mold faces is typical. This near closing of the mold will help the mold halves expand more uniformly during the next mold start-up procedure, and it protects the mold faces from any sort of mechanical damage.

3.19.2 Longer-Term Mold Storage Considerations

Injection molds that will be removed from the injection molding machine and stored for extended periods of time (i. e., many days, weeks, or years) require additional considerations and care. If an injection mold is to be removed from the molding press, it can either be maintained while it is in the molding machine clamp (before removal) or it can be brought to the tool shop after it has been removed from the molding clamp. In-press mold maintenance, such as that described above in Section 3.19.1, is less common for longer-term mold storage. Molds that are sent to the tool shop before long-term storage (along with their safety straps) can undergo far more in-depth mold inspection and maintenance procedures. Tools that are disassembled for cleaning and maintenance have many more cleaning options. For example, smaller mold components can be ultrasonically cleaned which is not possible when the mold is in the press. Only acid neutralizers and rust preventatives that are approved for longer-term storage should be used. Some industrial rust preventatives are rated based on the number of months or years that they provide corrosion protection. Rust preventatives with tracing dyes are also available to help ensure that the coating coverage is complete and uniform [52].

It is assumed here that the injection mold storage will be long-term, but the injection molding machine shut-down short, since the molding machine will be used for production with other molds. It is assumed that it is only the injection mold that is to be stored for the long term. One major difference between an overnight shut-down and long-term storage is coolant water purging. As cooling fluids can be corrosive, the coolant should be removed before or soon after the mold is removed from the injection mold clamp. This can be done using vacuum or dry compressed gas (typically compressed air). If dry compressed air is used to purge the mold's cooling circuits, it should be done neatly and safely as depicted in Figure 3.60. The atomized coolant shown in Figure 3.60 (left) could cause worker injury or corrosion of other metal components. Coolant purging should be done in a more controlled fashion as depicted in Figure 3.60 (right).

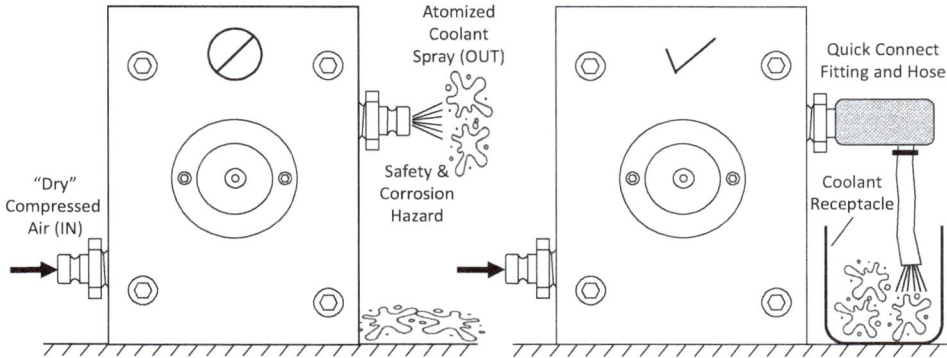

Figure 3.60 Injection mold cooling lines should be evacuated or purged prior to long-term mold storage. This should be done with care and caution. The purged coolant should be collected in a controlled manner (right) to avoid the potential for corrosion and worker injury from the atomized coolant spray if compressed air is used to purge the lines.

One supplier offers a unique corrosion resistant cooling circuit valve that provides for a very convenient and efficient method of clearing the coolant from the injection mold cooling circuits [53]. The valve is installed upstream of the cooling fluid supply manifold as shown in Figure 3.61. During normal injection molding operations, the valve allows the mold coolant to flow freely through the injection mold. Once the production run has been completed, the valve stem is moved to a position that allows compressed shop air to purge the coolant from the injection mold before the mold is removed from the molding machine clamp. The valve also has a vent port so that any residual mold coolant can be drained into a container. A visual flow meter can also be installed downstream of the return manifold for visual conformation that all of the injection mold coolant has been removed by the compressed air purge. An air separator can also be added to the circuit.

The SWAP® valve simplifies and reduces mold changeover times and is especially useful for applications where injection mold changes are frequent. This technology helps eliminate injection molding machine platen and injection mold corrosion caused by accidental coolant discharge. This technology also helps to improve worker safety by helping to eliminate the potential for slip accidents by keeping the decks and floors around the injection molding machine dry.

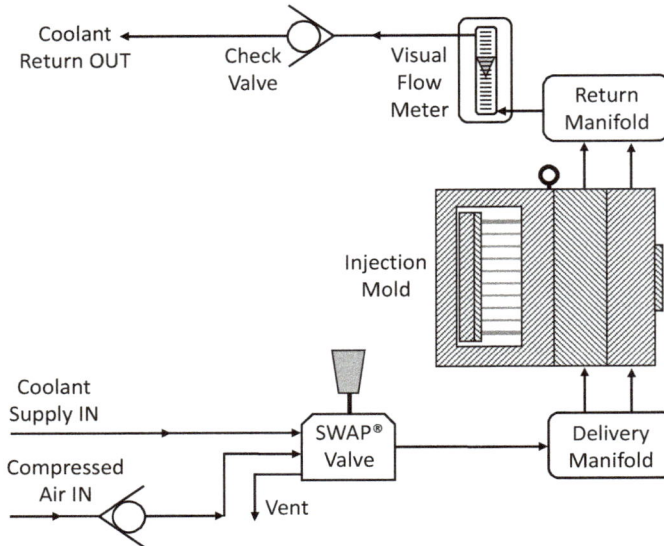

Figure 3.61
One supplier has developed a unique corrosion resistant valve technology that facilitates the clearing of mold coolant from the injection mold cooling circuit before a mold change [53].

Once the injection mold has undergone all of its maintenance procedures (coolant evacuation, cleaning, coating with corrosion prevention), it can be closed, safety-strapped and stored. The injection mold should be numbered and its exact storage location recorded. Ideally, the storage location will have a relatively dry and climate-controlled environment as even atmospheric humidity can lead to mold corrosion over the long term. It is also good practice to have an injection mold accessory case that can be stored in the same proximity as the mold. The numbered case can be used to store injection mold specific items such as the correct nozzle tip, ejector rods, spare inserts, custom tools, end-of-arm robot tooling and the like [54]. As a last note, it is also a good practice to save and label a few molded shots (i. e., retain shots) before the mold is shut down and stored for a long time period. These retained shots can serve as indicators of the condition of the injection mold cavities and cores (e. g., surface finish, presence of localized flash) before the injection mold was stored, as a reference for comparison with parts produced in future production runs [1].

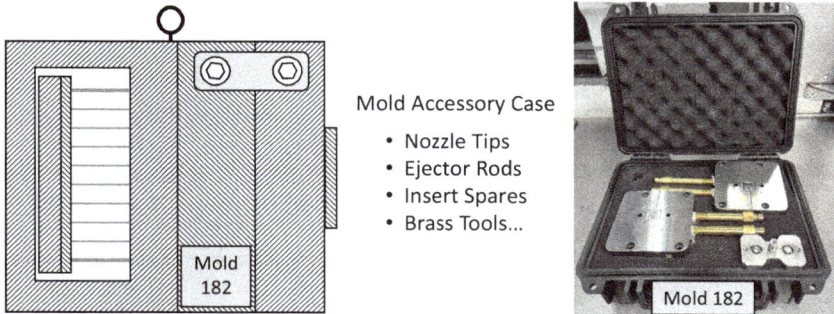

Mold Accessory Case

* Nozzle Tips
* Ejector Rods
* Insert Spares
* Brass Tools...

Figure 3.62 It is a good practice to create a mold specific accessory case that can be placed in the same proximity as the injection mold for during mold storage. The case can contain a variety of accessory items specific to that particular injection mold (e. g., knock out rods, insert spares, end-of-arm robot tooling).

References

[1] Kulkarni, S., "Robust Process Development and Scientific Molding", Hanser Publishing, Munich (2017).

[2] Schiller, G., "A Practical Approach to Scientific Molding, 2nd Ed.", Hanser Publishing, Munich (2024).

[3] "Cosmetic Specifications of Injection Molded Parts", SPI (now PLASTICS), Washington, D.C. 20006.

[4] "Pros and Cons of Handling Plastic Injection Molding In-House", RevPart, Fayetteville, GA 30214.

[5] *https://www.sabic.com/en/products/polymers/polycarbonate-pc/lexan-resin*.

[6] Technical Bulletin, "Copolyester Processing Guide", Eastman Chemical Company, Kingsport, TN 37662 (2014).

[7] Blog, "Optimal Mold Temperatures for Injection Molding: A Comprehensive Guide", Boyan Manufacturing, Yongkang, China, January 5 (2014).

[8] Technical Bulletin, "Tempro Mold Temperature Controllers", WITTMANN USA, Inc. 1 Technology Park Drive, Torrington, CT 06790.

[9] IMS Company, 10373 Stafford Road, Chagrin Falls, Ohio 44023.

[10] Anonymous, "High-Temperature Hoses", *Plastics Technology*, January 24 (2022).

[11] Longley, R., "Manifolds for Mold Cooling Water", Burger and Brown Inc., Grandview, MO 64030.

[12] *https://www.smartflow-usa.com/turbulent-flow-rate-calculator/*.

[13] Bales, S., "Reducing or Maintaining Cycle Times with Mold Coating", *Mold Making Technology*, April 1 (2007).

[14] Welsh, M., "How to Keep Injection Molds Clean", Chardon Labs, Reynoldsburg, OH 43068.

[15] Yamanaka, N. and Braig, S., "Hot/Cold Injection Molding: Which Approach is Best Fit for You?", *Plastics Technology*, April 25 (2016).

[16] Macedo, C. et. al., "The Potential for RHCM Technology in Injection Molding Using a Simple Conventional Heating and Cooling System", *Results in Engineering*, **19** (2023).

[17] Greener, J. and Wimberger, R., "Precision Injection Molding", Hanser Publishing, Munich (2006).

[18] Technical Bulletin, "Annealing of Molded *Makrolon*® Polycarbonate Parts", Covestro LLC, Pittsburg, PA 15205 (2005).

[19] Allison, K., "What is the Annealing Process for Plastic Molded Parts?", Crescent Industries, New Freedom, PA 17349 (2023).

[20] https://imscompany.blog/2022/11/02/how-to-set-barrel-zone-temps-in-injection-molding/.

[21] https://www.toolsid.com.

[22] Sentry Air Systems, Inc., Cypress, TX 77433.

[23] Bozzelli, J., "Shutting Down and Starting Up an Injection Molding Machine", *Plastics Technology*, February 25 (2014).

[24] Denzel, D., "Follow These Tips to Maximize Purging Efficiency", *Plastics Technology*, April 1 (2022).

[25] *Asaclean* Blog, "Purging Tips for Cleaning Your Injection Molding Hot Runner Systems", Asahi Kasei Plastics North America (2024).

[26] Pizzo, J., "Tips and Techniques: Break the Taboo on Purging Hot Runners", *Plastics Technology*, January 4 (2011).

[27] Pizzo, J., "Tips of the Trade: Purging Troubleshooting for Injection Molding", *Plastics Today*, August 23 (2010).

[28] Bozzelli, J., "Screw Decompress Before Screw Rotate", *Plastics Technology*, March 25 (2013).

[29] Amano, A. and Atsugi, S., SPE Annual Technical Conference, **33** 250 (1987).

[30] Jeon, J. et. al., Melt Temperature Variation in the Barrel of Injection Molding Machine, SPE Annual Technical Conference, **62** (2016).

[31] MTMS, LLC, www.PlasticMeltTemp.com, Janesville, WI 53546..

[32] Johannaber, F., "Injection Molding Machines – A User's Guide", Hanser Publishing, Munich, p.141 (2008).

[33] Malloy, R., Chen, S., and Orroth, S., "A Study of Injection to Hold Pressure Switch-over Based on Time, Position, and Pressure", SPE Annual Technical Conference, 33, May (1987).

[34] https://www.engelglobal.com/en/us/products/injection-molding-machines/thin-wall-injection-molding-machine.

[35] Williams, J., "Injection Molding Cooling Time: A Breakdown", RJG Inc., Traverse City, MI 49686.

[36] Malloy, R., "Plastics Part Design for Injection Molding, 2nd Ed.", Hanser Publishing, Munich, Germany (2011).

[37] Osswald, T., Turng, T., and Gramann, P., "Injection Molding Handbook", Hanser Publications, Munich (2022).

[38] Hoffman, D. and Beaumont, J., "A New Look at Evaluating Fill Times for Injection Molding", *Plastics Technology*, July 29 (2013).

[39] Bozzelli, J., "Should You Profile Injection Velocity?", *Plastics Technology*, June 26 (2012).

[40] Chen, X., "Automatic Velocity Profile Determination for Uniform Filling in Injection Molding", *Polymer Engineering and Science*, 50 7 (2010).

[41] Mensler H., et. al., "A Method for Determining the Flow Front Velocity of a Plastic Melt in an Injection Molding Process", Polymer Engineering and Science, April 9 (2019).

[42] Altonen, G., "A New Way to Mold Better Parts Faster and Easier", *Plastics Technology*, March 2 (2018).

[43] Altonen, G., "Advancing Autonomous Molding with a Uniform-Low-Pressure Process", *Plastics Technology*, September (2019).

[44] Travitz, J., "An Introduction to iMFLUX Technology: Welcome to the Green Curve", Beaumont Technologies, Erie, PA 16506, October 19 (2020).

[45] "The Fundamentals of Shrinkage in Thermoplastics", www.plastics.covestro.com (2016).

[46] Malloy, R., Nunn, R., and Bandreddi, M., "An Experimental Evaluation of Profiled Holding Pressures", SPE Annual Technical Conference, **40**, May (1994).

[47] Schiller, G. "A Practical Approach to Scientific Molding", Hanser, Munich Germany (2018).

[48] Bozzelli, J., "Screw Speed Versus Recovery Time", *Plastics Technology*, March 21 (2014).

[49] Gattshall, R., "Using Decompression to Your Advantage—Your Questions Answered", *Plastics Technology*, August 23 (2019).

[50] Gattshall, R., "Understanding—and Using—Decompression to Your Advantage", *Plastics Technology*, August 1 (2019).

[51] Sepe, M., "Dimensional Stability After Molding: Part 1", *Plastics Technology*, December 28 (2012).

[52] Technical Bulletin, slideproducts.com, Slide Products, Inc., Wheeling, IL 60090.

[53] Technical Bulletin, SWAP® Valve, smartflow-usa.com, Grandview, MO 64030.

[54] Uline, uline.com, Pleasant Prairie, WI 53158.

4

Other Injection Molding Processes

4.1 Introduction

The conventional thermoplastic injection molding process described in the previous sections of this text is versatile and appropriate for the production of most plastic part geometries from a very wide variety of thermoplastic materials. However, the conventional molding process does have limitations with respect to certain part geometries and certain thermoplastic materials. For example, the conventional molding process is not very well suited for the production of exceptionally small or micro-size plastic parts. In such a case, it is advantageous to utilize injection molding machinery that has been modified or designed specifically for the production of small or micro-size part molding.

The conventional injection molding process is also most appropriate for plastic parts having a conventional nominal wall thickness (typically up to several millimeters). While very thick parts can be molded using the conventional injection molding process, long cooling times and excessive mold shrinkage for thick parts lead to less-than-optimum part quality and poor economics. If thicker parts are required for a new application, for reasons such as greater bending stiffness, other modified injection molding processes, such as structural foam molding or internal gas assist molding, may be more appropriate and should be given consideration. The sections that follow in this chapter describe the basic concepts associated with a number of these unconventional or modified injection molding processes. The discussion in this chapter is limited to modified thermoplastic injection molding processes. Modified molding processes used for the production of thermoset plastic parts, such as reaction injection molding (RIM), or liquid silicone rubber (LSR) injection molding, are therefore not included.

4.2 Small Part and Micro Injection Molding

In recent years, there has been an increase in the demand for very small plastic parts, many having a complex geometry. Industry trends towards miniaturization; thin-walling, lightweighting and the like have pushed the limits of the conventional injection molding process. Molding very small, thin-wall plastic parts with micro-features can be difficult or impossible using conventional injection molding machinery. As a result, injection molding machinery specifically designed for producing micro part geometries has become available in recent years.

The increasing demand for micro plastic parts has led to the development of injection molding machinery specifically designed for the production of these sometimes-microscopic plastic parts. However, these micro molding machinery developments are relatively recent. Prior to the development of this specialized micro molding machinery, molders of very small or micro plastic part geometries were limited to using conventional injection molding machinery for these applications. The micro parts shown in Figure 4.1 were produced way back in 1968, long before the development of purpose-built micro molding machinery.

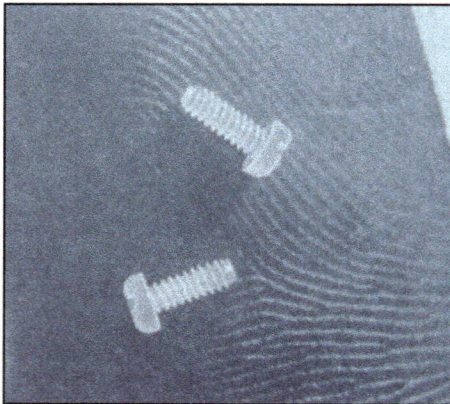

Figure 4.1
The thermoplastic micro machine screws shown have a thread pitch of 90 threads per inch. These parts were produced by Gries Molding of New Rochelle NY in 1968, long before the development of micro molding machines.

Molding high-quality, tight-tolerance micro or small plastic parts using conventional injection molding machinery can be difficult for a number of reasons. Concerns associated with micro molding range from tool production, to molding process variables, to parts handling. Two of the more important injection molding variables associated with small or micro-part production are:

- Low shot size utilization factors (that can lead to excessive material residence time).

- Shot-to-shot variation due to the limited injection stroke associated with a small shot volume.

Most conventional injection molding machines utilize a reciprocating screw injection unit fitted with a non-return valve (typically a check ring). Many sources suggest that the shot utilization for conventional injection molding should be somewhere between 20–80% of the injection unit's shot capacity. The residence time experienced by the plastic melt in the injection unit barrel is influenced by both the shot utilization percentage and the overall cycle time for the process. When molding very small or micro parts using a conventional injection molding machine, a shot utilization factor well below the recommended 20% lower limit is likely. In such a case, the material can experience an excessively long barrel residence time that can lead to a variety of issues, most notably thermal or hydrolytic degradation (in the case of condensation polymers). The use of a smaller shot capacity barrel can help minimize the degradation issue when producing micro-parts with conventional molding machinery, but the residence time may still be excessive. Starve feeding can also be beneficial to reduce residence time when molding very small parts.

While smaller shot capacity injection units and starve feeding can be beneficial for micro-part molding, adding runner volume is one of the simplest and most common ways to reduce material residence time when molding micro-parts using conventional injection molding machinery, as depicted in Figure 4.2. While the large cold runner volume (relative to the micro-part volume) does result in a residence time reduction, the added material cost associated with the runner scrap is a very significant disadvantage of this approach. In addition, the use of cold runner regrind in such an application is not practical based on relative part and runner volumes (i.e., the runner volume is far greater than that of the micro-parts). Adding runner volume has the additional benefit of increasing the magnitude of the shot stroke when molding micro-parts using conventional molding machinery. The observed shot-to-shot injection stroke variation tends to be greater when the injection stroke is short. Ideally, injection strokes should be significantly greater than the axial distance associated with check ring closure in order to reduce shot-to-shot variation.

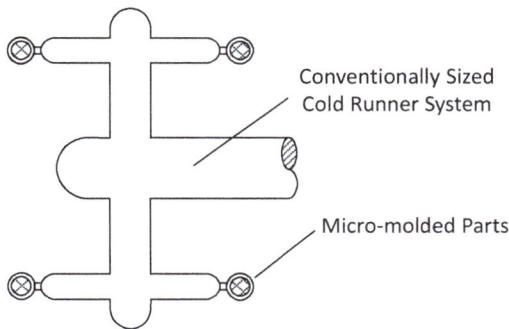

Conventionally Sized Cold Runner System

Micro-molded Parts

Figure 4.2
Molding small or micro-size thermoplastic parts using conventional size injection molding machinery is not optimum due to both excessive residence time and shot-to-shot repeatability issues. In such cases, adding runner volume is beneficial in terms of micro-molded part quality, but material scrap becomes an economic issue.

While molding small or micro-size parts using conventional injection molding ma-chinery is possible, it is certainly not optimum in terms of molded part quality and molding tolerances. The use of purpose built micro-molding machinery is a better solution for small-part applications. The importance of using purpose built micro-molding machinery becomes greater as the micro-part size decreases. Some micro-size parts are so small that dozens of micro-parts can be produced from a single plas-tic pellet.

Micro-molding machines are available from a number of machinery suppliers. The machine designs vary from supplier to supplier, but most micro-molding machines have several key and common features. In the simplest case, micro-molding machines are simply scaled down versions of conventional molding machines (with the excep-tion of the injection unit). Clamp forces are typically several tons. The micro-molding machinery platens and other machine components are physically smaller.

Aside from the smaller physical size, micro-molding machines are also designed to:

- Minimize material residence time.

- Provide repeatable injection strokes.

- Provide very high injection pressure capabilities.

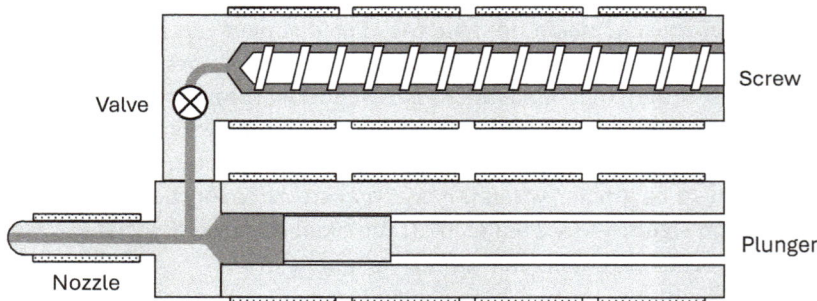

Figure 4.3 Most micro-injection molding machines utilize some variant of a two-stage injection unit, where the melt plastication and the mold filling machine components operate independently.

Unlike most conventional molding machines, most micro-molding machines have in-jection units that *do not* use a reciprocating screw fitted with a check ring. While small capacity injection units equipped with a locking check ring are sometime used for micro-molding, purpose built micro-molding machines are most commonly used for micro-molding applications. Their relatively small shot sizes result in short resi-dence times and greater precision. It is more common for micro-molding machines to utilize a two-stage injection unit, where plastication is accomplished using a small-di-ameter screw. The rotating screw pumps the melt into a small diameter (e. g., pencil size) plunger pot as depicted in Figure 4.3. The two-stage injection unit also includes a provision (e. g., a valve) that prevents back flow along the screw channels during the melt injection and packing stages of the process. The two-stage screw and plunger injection unit design eliminates the variability associated with a sliding check ring and allows for very high injection pressure capability (typically up to 300 MPa melt pressure). While flow lengths associated with micro-molded parts are relatively low, high injection pressures are often required for thin wall sections and micro-feature replication. Micro-molding machines are also available as multi-shot machines. The reader is directed to a micro-molding text for more details regarding micro-molding machine design and micro-tooling [1].

4.3 Structural Foam Injection Molding

Conventional thermoplastic injection molding processes produce parts that have a solid cross section. The structural foam molding process is a modified injection molding process that is used to produce thermoplastic parts that have solid skins and a foamed core, as depicted in Figure 4.4 and Figure 4.5. While there are several different structural foam process variants, the low pressure structural foam process is discussed here. This foam molding process offers a number of advantages over conventional injection molding, one of which is the ability to produce plastic parts that have a very high bending stiffness to weight ratio. Much like an I-beam, structural foam parts (that have solid skins and a foam core) are very efficient in terms of bending stiffness. Hence the use of the term structural to describe this foam molding process. Structural foam parts offer other advantages over their fully solid injection molded counterparts. These advantages include; reduced part warpage, sink mark elimination, design flexibility and low cavity pressure. However, there are also many disadvantages that limit the use of this modified molding process that are discussed below.

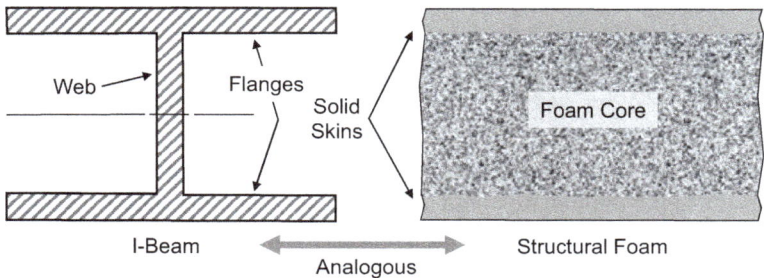

Figure 4.4 The structural foam molding process is used to produce thermoplastic parts that have solid skins and a foam core. Like an I-beam, structural foam parts offer very good stiffness-to-weight ratios, particularly in bending applications.

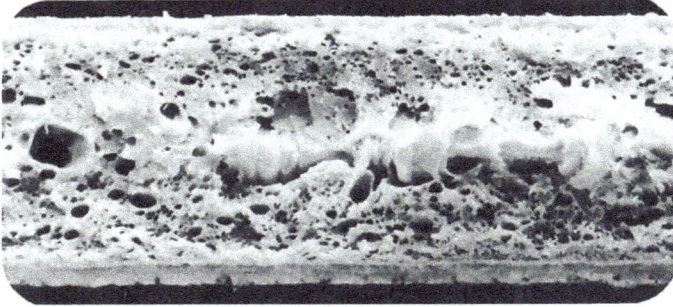

Figure 4.5 Fracture surface (cross section) of a typical low-pressure structural foam plastic part showing the integral solid skin layer and the foam core [2].

4.3.1 Structural Foam Material Considerations

The structural foam molding process can be accomplished using either (i) conventional injection molding machinery with minor modifications, or (ii) using purpose-built structural foam injection molding machines (which have been optimized for foam molding in terms of machine design). In either case, the first consideration is the material formulation to be molded.

Almost any thermoplastic material that can be injection molded can also be used in a structural foam molding application. Some type of foaming agent (or foaming additive) is also required for structural foam molding. The foaming agent can be a physical blowing agent, such as compressed CO_2 or N_2, or a powdered chemical blowing agent (CBA), such as azodicarbonamide. CBAs are powdered compounds that decompose at a specific temperature yielding large quantities of gas as the product of decomposition (decomposition to N_2 or CO_2 gas is typical). The CBA's decomposition reaction can be either endothermic or more commonly exothermic. The chemical foaming agent can be added to the base thermoplastic either directly as a powdered additive, or blended with the base thermoplastic as a masterbatch additive. The use of a masterbatch CBA is convenient, since the CBA is a fine powder used at a very low concentration (typically about 1%). Feeding or blending a low concentration of fine powder with plastic pellets may be more cost effective, but it is technically more difficult to achieve a consistently uniform powder CBA concentration. If a masterbatch CBA is used, it is important to ensure that the masterbatch carrier resin is compatible with the base thermoplastic being molded (to avoid defects such as delamination). It is also important that the active CBA has both an adequate gas yield and a decomposition temperature that is just below the normal processing temperature of the thermoplastic to be molded [3]. A nucleating additive, such as TiO_2, can also be incorporated into the material formulation in order to achieve a more uniform cell structure for the structural foam core.

4.3.2 Low-Pressure Structural Foam Injection Molding Process

As stated, there are several variants of the structural foam molding process, but the low-pressure structural foam molding process is most commonly utilized to produce structural foam plastic parts. As an example, the structural part shown in Figure 4.6 could be produced using either a conventional injection molding process or a structural foam molding process. The parts produced by the two processes would be similar, but details associated with the part design, tool design and part performance would differ.

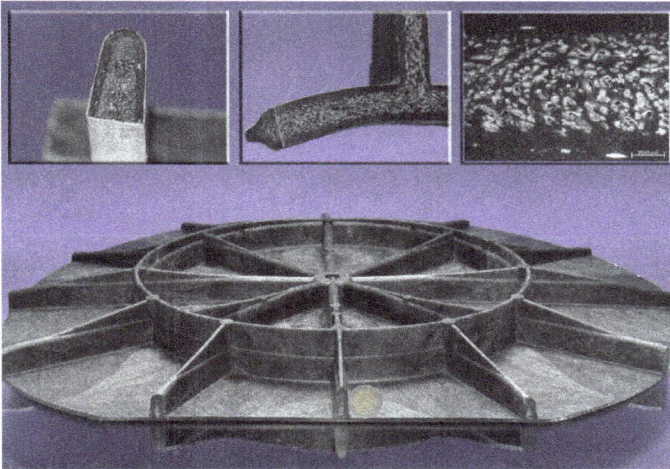

Figure 4.6
Low pressure structural foam part made from a long fiber reinforced polypropylene (source: Moller Tech).

In terms of molding machinery, the low-pressure structural foam molding can be achieved using slightly modified conventional injection molding machinery. Most importantly, a shut-off nozzle (see Chapter 2) is required for structural foam molding in order to limit gas pressure loss during portions of the molding cycle when the nozzle is not blocked off (e. g., when the mold opens to eject the part).

The structural foam molding process begins by preparing the material formulation using either batch mixing or multiple feeders. Hygroscopic materials would require pre-drying in the same way as they would for conventional injection molding. The barrel temperature profile and plasticating conditions are very critical process variables for the structural foam molding process. An increasing barrel temperature profile is typical for structural foam molding in order to limit pre-mature CBA decomposition and gas loss through the barrel feed throat. A moderate screw rotational speed and a relatively high back pressure are typically used to help keep the decomposed CBA gases in solution (under pressure) as the shot is being generated during plastication. A shut-off nozzle (or a valve gate in the case of a hot runner mold) is used to prevent the gas pressure from causing nozzle drool when the mold is open. The gas must be kept in solution (under pressure) while the shot is waiting for the injection signal. Also note that the mold temperatures used for structural foam molding are typically similar to those that would be used with conventional injection molding (for the same base plastic material).

Once the shot has been plasticated, the plastic melt (containing compressed gas) can be injected into the closed mold for the next cycle. The mold filling and pressurization phases of the structural foam molding process are significantly different than those used for conventional injection molding, as depicted in Figure 4.5. Conventional injection molding involves a number of distinct pressurization phases, namely:

- Mold volumetric filling under velocity control.

- Packing or melt compression immediately after mold filling.

- Holding pressure to help compensate for mold shrinkage (requires a cushion).

The conventional injection molding process is a relatively high cavity pressure process. Clamp tonnage requirements for conventional injection molding are often described as being at least 2–5 tons per square inch of projected area in order to prevent flash (a typical rule of thumb). Unlike conventional injection molding, the structural foam molding is a short-shot process where less than a full shot of melt is plasticated and injected. Shot sizes for structural foam molding are typically reduced by 10% to 30% compared to solid injection molding for the same part. During the injection phase of the process, the screw (and check ring) bottom out after the mold cavity is partially filled with the short shot. Once the melt is in the mold cavity, the gas pressure within the melt forces the melt into the unfilled portions of the mold cavity. This expansion also leads to the formation of the foam core.

There is no packing pressure phase or holding pressure phase in the traditional sense with low pressure structural foam molding. Packing and holding pressures are associated with the internal gas pressure (not the screw) since there is no cushion with this short-shot process. This internal gas pressure is relatively low compared to the packing and holding pressures used for conventional injection molding, as shown in Figure 4.7. As a result, clamp tonnage requirements for low pressure structural foam molding are typically less than 1 ton per square inch, far lower than conventional injection molding. The low clamp force requirement for structural foam molding is a significant advantage in terms of tooling cost and tool maintenance. Aluminum tooling is commonly used for structural foam due to its good machinability and good heat transfer characteristics. The low clamp force requirement makes structural foam molding an attractive process for very large plastic parts. It is for this reason that purpose built structural foam molding machines tend to have a larger-than-normal platen sizes, but lower-than-normal clamp capacities. Purpose-built structural foam molding machines usually have a relatively large shot capacity compared to conventional injection molding machines, since they are commonly used to produce larger-size parts.

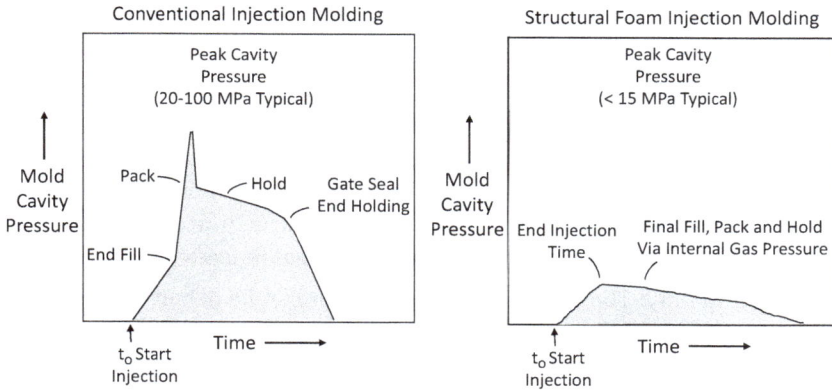

Figure 4.7 The conventional injection molding process is a relatively high cavity pressure process. The low-pressure structural foam molding process is a short-shot process where cavity pressures are significantly lower than those encountered in the conventional injection molding process.

An additional benefit of the structural foam molding process is that mold shrinkage issues, most notably (i) part warpage and (ii) sink marks are all but eliminated. In conventional injection molding, sinks and warpage issues are caused by non-uniform material shrinkage than can be caused by non-uniform cavity pressure and/or non-uniform material or mold temperatures. With structural foam molding, gas pressure, which is relatively uniform throughout the cavity, helps to minimize shrinkage variation, warpage and sink marks. Structural foam parts are typically sink free and have very predictable and uniform mold shrinkage. There is little or no tendency for sinks to form since the internal gas pressure helps to force the solid skin layers against the mold surface throughout the cycle. This is true even when the part wall thickness is non-uniform. Warpage is also minimized since the foam serves as a buffer between the two solid skins.

While the structural foam process does offer a number of advantages over conventional injection molding (most notably low cavity pressure and uniform mold shrinkage behavior), the process has a number of drawbacks that limit the use of this process in many applications.

Nominal wall thickness: The structural foam process is most suitable for plastic parts that have a nominal wall thickness of at least 4.0 mm, which is significantly greater than the wall thickness for most injection molded parts. The thick wall allows for the proper development of the structural foam's solid skins and foam core zones. The relatively thick nominal wall also reduces the fill pressure requirement, which is particularly important for the final portions of the mold filling (due only to the low internal gas pressure). It should also be noted that molds used for structural foam molding should be very well vented to help minimize the resistance to the final mold filling, since internal gas pressures are relatively low.

It is noted here that structure foam parts may or may not offer any significant weight savings compared to solid (but thinner) injection molded parts. The structural foam process does involve a material density reduction of around 10–30%, but the nominal wall thickness of a structural foam part is significantly greater than that of a conventional injection molded part. Mold cooling times for structural foam parts also tend to be far longer than those associated with conventional injection molding due to the greater nominal wall thickness for structural foam parts. Structural foam parts that are ejected from the mold too soon will tend to puff up or balloon after part ejection.

Molded part surface quality: The part surface quality that can be achieved with low-pressure structural foam molding is inferior to that achievable with conventional injection molding. This is due to several factors. First, low-pressure structural foam parts tend to exhibit surface defects such as excessive splay, swirling, or silver streaking. The splay is the result of gas bubbles that break through the melt front during the filling or injection phase of the molding process. Using a fast injection velocity can help reduce the severity of the surface defects.

A second issue that impacts surface quality for structural foam parts is the very low-pressure nature of the process as shown in Figure 4.7. The relatively high packing pressure associated with the conventional injection molding process provides for outstanding surface replication, particularly for lower viscosity, high melt flow rate (MFR) materials. Mold surface features, such as a fine surface texture, can be easily replicated with the conventional injection molding process, due to the high packing pressure used with conventional injection molding. However, the relatively low gas pressure associated with the structural foam molding process is insufficient for fine feature surface replication. That said, surface textures are sometimes used for structural foam parts since surface textures can help mask surface defects. While most structural foam parts are pigmented or colored using masterbatch pigments, structural foam parts used in applications where surface appearance is critical, are sometimes painted.

Mechanical performance: Unlike their injection molded counterparts, structural foam parts are not uniformly solid throughout their cross section. Instead, they have solid skins and a foamed core as shown in Figure 4.5. As a result, the properties of the foamed part will be influenced by both the bulk plastic material properties, and the degree of density reduction. Unfortunately, it is difficult or impossible to predict the solid skin thickness and uniformity of cell structure with any certainty before a structural foam part is molded. As stated earlier, the solid skins on the foam core of a molded structural foam part will result in a good stiffness to weight ratio, but other mechanical properties, such as strength or toughness, can be impacted in a negative way (in comparison with conventional injection molded parts). Both the foam cell size and uniformity of cell structure are also important factors related to the mechanical performance of structural foam parts.

4.3.3 Gas Counter Pressure Structural Foam Injection Molding Process

One of the limitations of the low-pressure structural foam process described above in Section 4.3.2 is the limited molded part surface quality that can be achieved. One of the common surface defects associated with structural foam molding is surface splay, often excessive surface splay. The surface splay forms as the foaming gas bubbles break through the melt flow front as the mold fills, as depicted in Figure 4.8 (top). While the degree of splay or swirl-like appearance associated with the low-pressure structural foam molding process varies from process to process, it can be a significant issue in many applications. Low pressure structural foam parts are sometimes painted in order to achieve a higher quality surface appearance, but this can be a very expensive secondary operation.

Conventional "Low Pressure" Structural Foam Injection Molding

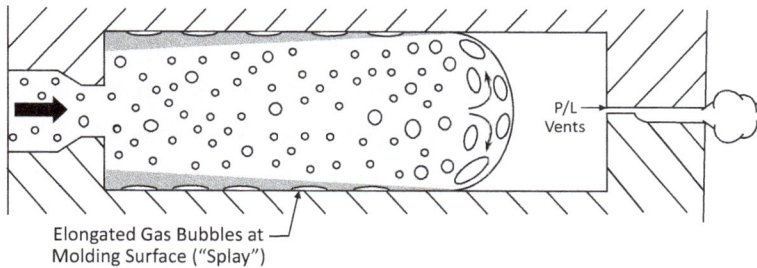

Inert Gas "Counter Pressure" Structural Foam Injection Molding

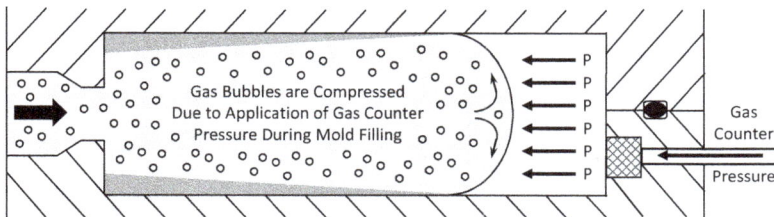

Figure 4.8 The surface defect known as splay is a common defect associated with the low pressure structural foam molding process. Splay forms as internal gas bubbles break through the melt flow front during mold filling. The appearance of splay can be minimized or eliminated if the mold cavity is pressurized with an inert gas prior to mold filling. The magnitude of the gas counter pressure utilized needs to be high enough to prevent splay formation, but low enough not to impede mold filling.

Over the years, a number of other structural foam molding techniques have been developed in an effort to produce structural foam parts that exhibit an improved surface appearance. Some of these structural foam molding techniques are described as gas counter pressure structural foam processes. Gas counter pressure foam molding processes involve pressurizing the injection mold with an inert gas, such as nitrogen, prior to injecting the molten plastic foam formulation. The gas counter pressure within the injection mold cavity is high enough to prevent pre-mature gas expansion during the mold filling process, as shown in Figure 4.8 (bottom), yet it is low enough that it does not impede the mold filling process to any appreciable degree. Keeping the gas bubbles in solution during the mold filling phase minimizes splay formation, and will therefore improve the molded part's overall surface appearance. Like conventional structural foam molding, gas counter pressure structural foam molding is a short-shot process. The counter pressure gas is then vented towards the end of the mold filling process, thereby allowing the melt's internal gas pressure to fill out the extremities of the mold cavity [4,5].

Gas counter pressure structural foam processes are somewhat more complex than the conventional low pressure structural foam molding technique described in Section 4.3.2 above. Auxiliary gas counter pressure control equipment and a pressurized nitrogen source need to be integrated with the structural foam molding process. The mold cavity must also be capable of being pressurized with the inert counter pressure gas as shown in Figure 4.7 and Figure 4.8. This can be difficult to accomplish in practice, particularly for molds of complex construction (e. g., molds with many inserts, side actions). Mold parting lines, ejection devices (e. g., ejector pins), core pins, and the like must all be sealed with gaskets or o-rings to allow for mold cavity pressurization. This gas counter pressure process also requires a controller to initiate mold cavity pressurization, as well as gas venting at the end of the mold filling phase, as shown in Figure 4.9.

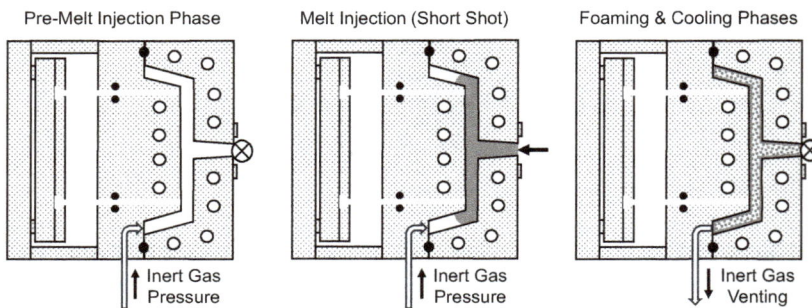

Figure 4.9 The gas counter pressure structural foam injection molding process involves pressurizing the mold cavity with an inert gas prior to melt injection in order to limit splay formation. A short shot is then injected into the mold. After injection, the inert gas is vented allowing the melt's internal gas pressure to fill out the extremities of the mold cavity.

The foamed plastic parts produced with the gas counter pressure structural foam process tend to exhibit a somewhat higher density, improved surface finish, and a more uniform skin-core structure. These factors all lead to improvements in molded part performance compared to parts produced using the traditional low pressure structural foam process. The gas counter pressure structural foam molding process retains the fundamental benefits of the traditional low pressure structural foam process (i. e., high bending stiffness to weight ratios, sink mark elimination, reduced warpage, low cavity pressure), but offers very significant improvements in both the surface quality and the mechanical performance of the molded part.

4.4 Gas Assist Injection Molding Process

Another group of modified injection molding processes are commonly described as gas assist injection molding techniques (GAIM). There are a number of gas assist process variants, but they all involve the use of a compressed, inert gas to somehow improve of the capabilities of the conventional injection molding process. While there are different gas assist molding process variants, the most common processes described as either (i) contained channel GAIM, (ii) open channel GAIM, and (iii) external GAIM. The basic concepts associated with each of these gas assist processes are described in the sections that follow. Gas assist molding processes offer advantages that are similar to those of the structural foam processes described above, but without the disadvantages of surface splay and a long cycle time. Like the structural foam process, the gas assist molding processes can be utilized with most injection moldable thermoplastic materials [6–8].

4.4.1 Contained Channel Gas Assist Injection Molding Process

The process known as contained channel gas assist molding is commonly used to produce plastic parts that have very thick wall sections or features. Consider the shovel handles shown in Figure 4.10. The hand grip region of the shovel handle has a relatively large diameter for ergonomic reasons. The grip region of the part is excessively thick for the conventional injection molding process. Very thick sections require more material, a long cooling time and exhibit excessive mold shrinkage. Parts of this type are commonly produced by the conventional injection molding process, but the thick sections are normally cored out in some way to reduce material consumption and decrease the cooling time requirement, while still achieving the desired ergonomic function. The molded part shown in Figure 4.10 (left) is molded conventionally using two side actions to core out the center section of the hand grip, but the mold is relatively complex. The molded part shown in Figure 4.10 (center) is also molded conventionally without the need for special mold actions. The grip region of the center part has a series of ribs that reduce material consumption, average nominal wall thickness and cooling time (compared to a solid part). However, the ergonomics functions associated with this design are compromised compared to the other parts shown in Figure 4.10.

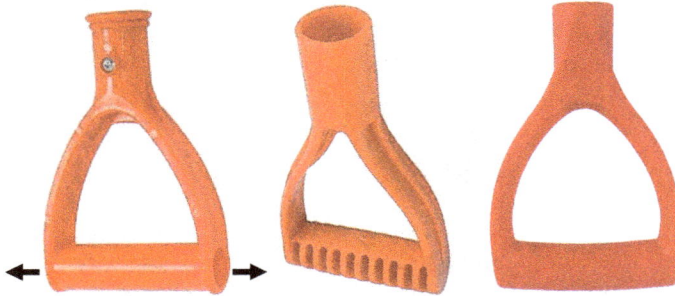

Figure 4.10 The grip region of a typical shovel handle has a relatively large overall thickness or diameter. Such a part could be molded with the conventional injection molding process, but the handle would typically be cored out in some way to reduce the nominal wall thickness, material consumption and mold cooling time (left and center). The part on the right has a more solid appearance and could be a candidate for either the structural foam molding process, or the contained channel gas assist molding process. The gas assist process allows for the production of a hollow grip section.

The grip region of the part shown in Figure 4.10 (right) does not appear to be cored out in any way and could be molded using a variety of injection molding techniques. It could be a solid injection molded part, but material consumption and cycle time would be excessive. It could be molded as a structural foam part as described in Section 4.3. The structural foam part would offer advantages in terms of material consumption, but the mold cooling time would still be very long and the part would tend to exhibit surface splay. The part could also be produced using the contained channel gas assist molding process. This modified injection molding process allows for the production of a handle grip region that is hollow (without the need for coring). A bulky, handle-like part of this type that is molded with the contained channel gas assist molding process offers a number of advantages. The use of this gas assist process can help minimize material consumption and mold cooling time, while allowing for design flexibility and good ergonomics.

The basic contained channel gas assist molding process is depicted schematically in Figure 4.11. Like the structural foam molding process, the gas assist molding process is a short-shot process. As a result, the molding machine clamp force requirements for the gas assist molding process are significantly lower than those of the conventional molding process. The contained channel gas assist process begins with the plastication of the melt in the usual manner, except that a less-than-full shot is plasticated. The short shot is then injected into the mold runner and cavity. After injection, a compressed inert gas (typically dry nitrogen) is then injected into the center of the melt stream. The compressed gas forces material into the unfilled sections of the mold cavity as it expands. The internal gas pressure also serves as the packing pressure and holding pressures for the process (since there is no cushion with a short-shot process). The gas can be injected at various locations depending on the specific process variant. Gas is sometimes injected into the center of the molding machine nozzle, or it can be injected directly into the mold runner system or into the mold cavity itself. The valve pin through which the compressed gas is injected has both open and closed positions. A controller is used to regulate the gas injection timing, the gas injection pressure profile, and the gas venting function prior that must occur prior to molded part ejection. Note that the molded part will end up having a through-hole if the gas valve pin is located within the mold cavity.

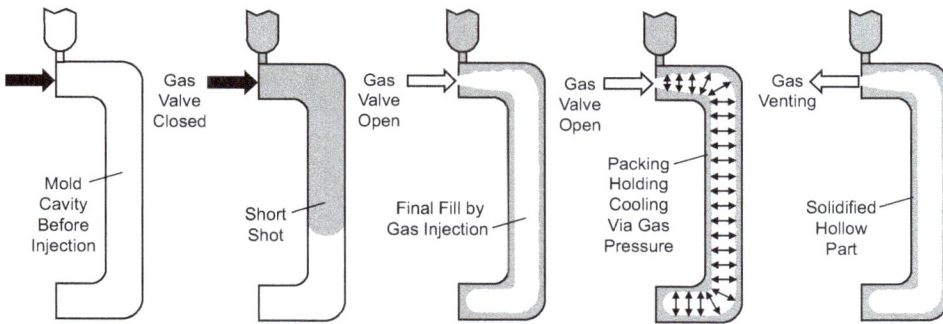

Figure 4.11 The contained channel gas assist molding process is primarily used for globally thick cylindrical or other high aspect ratio parts (e. g., suitcase handles, coat hangers). The process begins with a short-shot injection phase, followed by injecting an inert compressed gas. The compressed gas expands and displaces the hot core of the short-shot melt, causing the flow front to advance, thereby filling out the unfilled sections of the mold cavity. The gas is then vented before mold opening and part ejection. The net result is a hollow plastic part having a conventional molded part wall thickness, reasonable cycle time and good surface finish.

In this type of contained channel application, the average (final) nominal wall thickness of the plastic part will be dictated by the shortness of the injected shot (or percentage of underfill). A larger shot volume will produce a molded part with a thicker (average) nominal wall thickness. It is very important to note here that one important goal associated with contained channel gas assist molding is to achieve a uniform wall thickness. The compressed gas will always take the path of least resistance. If we assume the part shown in Figure 4.11 has a cylindrical cross section, the gas should flow symmetrically down the center of the part where the melt temperature is highest, and the material viscosity the lowest. However, if the melt temperature is not uniform, and/or the mold cavity surface temperatures are not uniform, the resulting nominal wall thickness values will be non-uniform due to the different cooling rates. In the end, the uniformity of the part wall thickness is dictated by the uniformity of the melt and mold temperatures, as well as the gas injection pressure profile utilized.

As stated earlier, the contained channel gas assist molding process offers a number of advantages that are similar to those associated with the structural foam process. Both molding processes are low-pressure processes suitable for globally thick plastic parts. However, with contained channel gas assist molding, the compressed gas should never break through the melt flow front. As a result, gas assist parts should not exhibit the splay issues associated with low-pressure structural foam molding. Both processes are capable of producing sink-free parts with little or no warpage as the packing and holding pressures are relatively uniform throughout the mold cavity. The uniform mold cavity pressure history helps minimize area mold shrinkage variations.

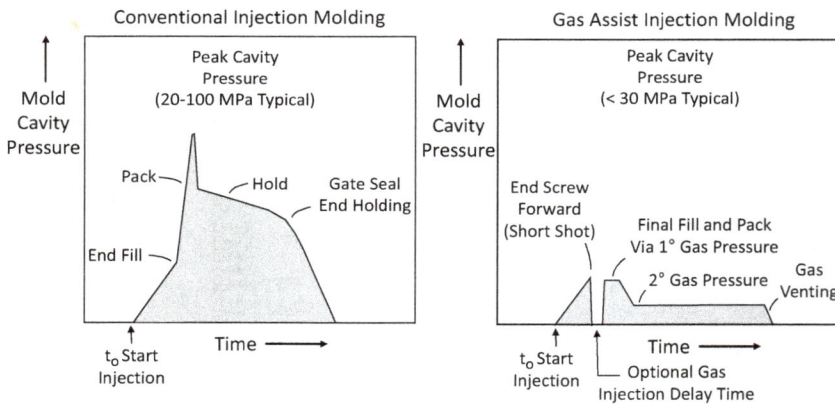

Figure 4.12 The contained channel gas assist molding process is a short-shot process. As such, the cavity pressures associated with the contained channel gas assist molding process are significantly lower than those encountered with conventional or traditional injection molding processes.

Figure 4.12 shows a typical cavity pressure – time curve for both a conventional injection molding process (left) and a contained channel gas assist molding process (right). The short-shot nature of the gas assist process results in significantly lower cavity pressures than the conventional injection molding process. The overall shape of the cavity pressure – time curves for gas assist molding can vary depending on the application. For example, the point in time where gas injection begins is an important process variable. Gas injection is sometimes initiated while the short shot is being injected, in order to avoid hesitation marks on the part surface. Alternatively, gas injection can sometimes begin just after short-shot injection or after an optional gas injection delay time. The delay time can help prevent the gas from blowing through the melt front in applications where the melt's viscosity is very low. The gas pressure can also be profiled in much the same way that packing and holding pressures can be profiled with a conventional injection molding process. The gas pressure used for the packing phase of the process is commonly described as primary (1°) gas penetration, while the gas pressure used for the holding phase of the process is generally described as secondary (2°) gas penetration. The internal gas pressure is typically vented towards the very end of the molding cycle, prior to part ejection.

4.4.2 Open Channel Gas Assist Injection Molding Process

While the contained channel gas assist injection molding process described in Section 4.4.1 is most commonly used for the production of globally thick, long cylindrical or high aspect ratio plastic parts, the open channel gas assist molding process is more appropriate for the production of larger plate-like or three dimensional parts having a more conventional nominal wall thickness. Like the contained channel gas assist process, the open channel gas assist molding process is a short-shot molding process; as such, it is a lower cavity pressure molding process compared to conventional injection molding. Plastic parts that are commonly produced with the open channel gas assist molding process include patio furniture, business machine housings, tote bins and the like.

The general concept of open channel gas assist injection molding is depicted in Figure 4.13. The single cavity injection mold shown produces a large, centrally-gated, plate-like plastic part that has a conventional nominal wall thickness (e. g., 2.0–3.0 mm). The part geometry also contains a number of strategically placed semi-cylindrical gas channels that have a significantly greater thickness than the part's nominal wall thickness. The open channel gas assist molding process begins with the injection of a short shot into the mold cavity. The resulting short-shot flow pattern would be similar to that shown on the left in Figure 4.13. The melt flow pattern associated with the short-shot injection is dictated by the mold filling pressure drop in all directions. The pressure drop within the thicker gas channels is lower than it is for other regions of the part due (i) to their greater depth and (ii) warmer average melt temperature.

After the short shot has been injected, compressed inert gas is then injected into the runner or mold cavity, as shown in Figure 4.13 (center). The gas injector pin is located at the center of the part in this case. The injected gas will follow the path of least resistance. The gas will tend to flow along the gas channels due to their greater depth and warmer average melt temperature. The melt that is displaced by the gas flows into unfilled sections of the mold cavity. When the gas filling phase is complete, the entire network of gas channels within the part should have a hollow cross section surrounded by a solid skin as shown in Figure 4.13 (right). Once the mold filling process is complete, the compressed gas pressure is then used to apply the packing and holding pressures required to control sink marks and part dimensions. The gas is then vented prior to part ejection. A typical cavity pressure – time curve for the open channel gas assist process would be similar to that shown in Figure 4.12.

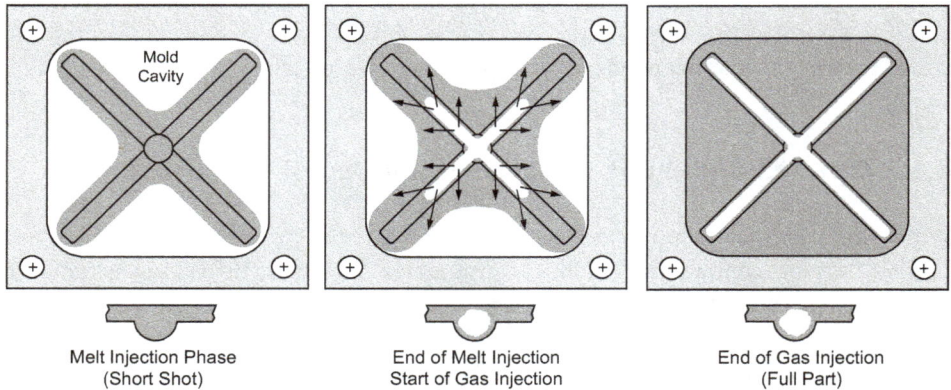

Melt Injection Phase   End of Melt Injection   End of Gas Injection
(Short Shot)   Start of Gas Injection   (Full Part)

Figure 4.13 The mold filling phase of the open channel gas assist molding process is a multi-step process. A short-shot injection phase is followed by a compressed gas injection phase for the final portions of mold filling.

This open channel gas assist molding process is very different than the conventional injection molding process and requires a great deal of engineering work when designing the part and the injection mold. The strategic layout and dimensions of the gas channels are critical and are generally determined using a gas assist molding process simulation program before the mold is built.

The open channel gas assist molding process competes with structural foam molding for the manufacturing of large plastic parts. Both the structural foam and open channel gas assist processes are low cavity pressure processes. This can be advantageous in terms of machine clamp force requirements and may allow for the use of aluminum molds. However, unlike structural parts that generally have a wall thickness greater than 4.0 mm (therefore a long cooling time), open channel gas assist parts have a conventional wall thickness and therefore a conventional mold cooling time. Structural foam parts also exhibit surface defects such as splay or swirling, while the gas assist parts have no such issue. Both open channel gas assist parts and structural foam parts exhibit limited warpage, largely due to the relatively uniform mold cavity pressures associated with each process. In many ways, the open channel gas assist molding process provides the benefits of both the conventional molding process and the low-pressure structural foam molding process, without the disadvantages associated with each process (the best of both worlds).

The gas channels used for the open channel gas assist molding process can have a variety of sizes and channel geometries. The primary purpose of the gas channels is to promote and control the melt flow pattern during mold filling, but the channels can also provide the secondary benefit of improving the stiffness or rigidity of the part. The gas channels in many ways function like reinforcing ribs, and may themselves incorporate a rib as shown in Figure 4.14 (right). Like all ribs, a rib integrated with the gas channel will increases bending stiffness of the part. Another advantage of such a rib placement is that the integrated rib does not cause the sink issue normally associated with ribs, since the base of the integrated rib is isolated from the visible part surface opposite the rib.

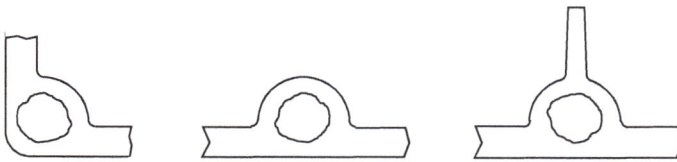

Figure 4.14 A wide variety of gas channel geometries are common with open channel gas assist injection molding. The gas channels are sized and strategically located to achieve the desired mold filling pattern. Gas channels also provide structural rigidity and sometimes incorporate an integrated rib to further enhance structural performance.

There can be a number of molding defects associated with the open channel gas assist injection molding process. Unlike the contained channel gas assist molding process described in Section 4.4.1, where gas penetration *must* follow a defined path, the gas flow pattern associated with open channel gas assist molding is *not* as well defined. With open channel gas assist molding, the gas should preferentially flow along the gas channels, and not through the core regions of the thinner, cooler part wall. However, there is no guarantee that this will be the case with the open channel process. If the compressed gas does penetrate into regions of the part other than the gas channels themselves, the part's structural performance would be compromised, as depicted in Figure 4.15. The unwanted gas penetration forms finger shaped branches adjacent to the gas channels. Note that this gas penetration defect is internal and could go unnoticed in pigmented gas assist parts without careful inspection.

This defect, commonly referred to as gas fingering or gas spill over, can occur for a number of reasons. For example, localized fingering can occur if the mold cavity surface temperature is non-uniform. Gas fingering is more likely to occur in areas of the cavity where the mold cooling is limited. Process variables, such as gas injection delay time, can be used to minimize the possibility of gas fingering. The melt within the thinner nominal wall will become cooler and more viscous as a result of the added gas injection delay time, reducing the tendency towards gas fingering [9].

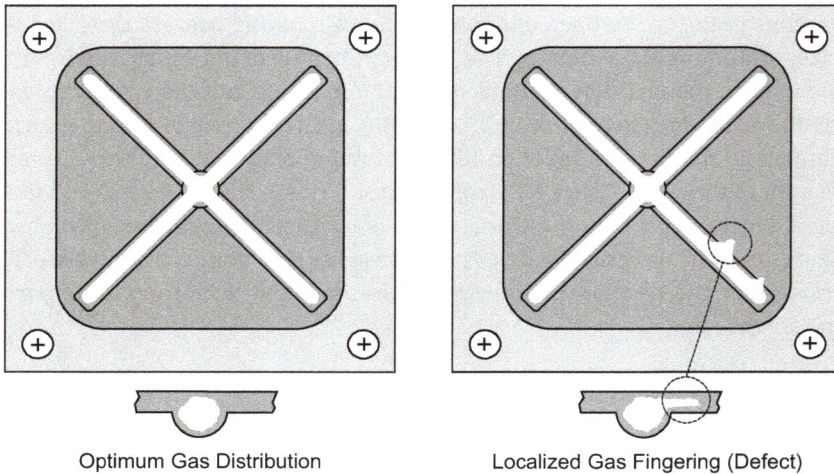

Optimum Gas Distribution   Localized Gas Fingering (Defect)

Figure 4.15 One of the more common defects associated with open channel gas assist molding is that of unwanted and localized gas penetration into the thinner part wall adjacent to the gas channel. This defect, known as gas fingering, is a structural defect that can be caused by a number of different factors, including a non-uniform mold surface temperature.

4.4.3 External Gas Assist Injection Molding Process

External Gas Assist Injection Molding is a very different process than the gas assist processes described above. The external gas assist process is most easily described with reference to Figure 4.16. This process involves injecting compressed inert gas (typically compressed nitrogen) into the mold cavity, but not into the core of the plastic part. The gas pressure is injected into the microscopic gap between one of the part's surfaces and the mold cavity (or core) surface. This process is really a modified conventional injection molding technique [7, 10, 11].

The gas is normally injected into the microscopic gap after the mold filling and packing (or compression) phases of the conventional injection molding are complete. This blanket of gas is used as a supplement to the normal holding pressure that is applied to control part dimensions, enhance surface replication and reduce sink mark formation. Unlike the gas assist processes described above, the compressed gas used with the external gas assist process does not penetrate into the molding material.

The gas enters the mold from only one side of the mold, usually the non-appearance (core) side. The gas pressure helps force the molten polymer against the opposite surface of the mold (appearance surface). This supplemental gas pressure helps compensate for mold shrinkage and eliminate mold shrinkage-related defects. The process is particularly well suited for applications where surface quality and surface replication are critically important. The process also helps reduce the formation of sink marks and voids that are associated with locally thick wall sections (e. g., wide ribs).

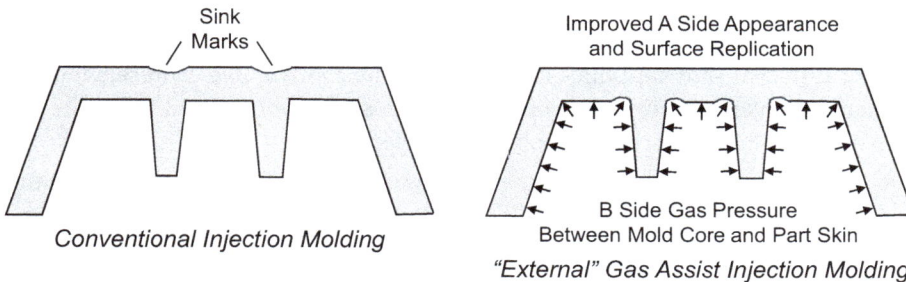

Figure 4.16 External gas assist injection molding involves the injection of a microscopically thin layer of high pressure inert compressed gas between the core and the underside of the plastic molding after the mold cavity is filled with melt. The gas pressure helps force the molten plastic against the cavity (or appearance) side of the tool resulting in an improved surface finish (e. g., less sink, improved surface replication).

The gas pressure is typically applied for the duration of the holding and cooling phases of the molding process. With this process, the external gas pressure is typically vented before the mold opens for part ejection, although a reduced external gas pressure may be used as an ejection assist. The most significant issue or downside associated with this external gas molding process is the need to pressure-seal the mold along the parting lines and other areas of the mold that are normally not pressure tight (e. g., ejection pins, slides).

4.5 Co-injection Molding Process

Co-injection molding processes represent one sub-group of multi-material injection molding processes. While there are several variants of the co-injection molding technology, they all produce multi-layer plastic parts comprised of at least two different plastic material grades. A co-injection molding process is loosely analogous to a gas assist injection molding process, where the gas is replaced by a second plastic material. Co-injection molded plastic parts end up having a layered structure, that will have overall performance characteristics that are influenced by the different plastic material layers.

The basic concepts associated with a typical two material, co-injection molding process are depicted in Figure 4.17 [2, 7]. A standard co-injection molding machine is a purpose-built molding machine with two separate injection units, each having its own actuators and controller. Co-injection molding processes have also been created by adding a second retrofit barrel onto a conventional injection molding machine, equipped with a machine controller that allows for such as modification. In either case, one of the two injection units (the primary injection unit) will plasticate and inject the outer skin material, while the second injection unit (the secondary injection unit) will plasticate and inject the core material. The two injection units can have similar or different shot capacities depending on the shot volume requirements for each material layer. The two injection units are in some way isolated from one another using a multi-port valve assembly or valve gates.

The co-injection molding process begins by injecting a partial shot (short shot) of the primary (or skin) material into the fully closed mold cavity. The shortness of the primary short shot that is injected will influence the skin's thickness and determine the amount of core material that will be subsequently injected into the mold cavity. Immediately after the skin material has been injected, the valve is activated such that the secondary core material is then injected into the center of the runner system or mold cavity. Like the compressed gas used in gas assist molding, the core material will flow wherever it is easiest to flow, typically through the central regions of the runner and part where the material is the hottest and has the lowest viscosity or resistance to flow. There is no mixing of the two materials as the melt flow into an injection mold is laminar or layered flow, not turbulent flow. The primary material also ends up as the skin material throughout the part due to the fountain flow behavior of plastic melts during mold filling. The secondary core material injection continues until the mold cavity is full. After the secondary core material has been injected, the valve may return to the starting position to clear the core material in preparation for the next molding cycle. In the end, the multi-material molded part that is produced should consist of a primary skin material that fully encapsulates the secondary core material.

The co-injection molding process offers a number of advantages relative to the conventional injection molding process. While there are many possible advantages, the co-injection molding process is more complex, and requires the use of more capital-intensive machinery and tooling. It should also be noted that the relative distribution of the two materials within the mold cavity is not easily predicted. The core material will flow wherever it is easiest when injected into the mold. This could result in non-uniform core material distribution if the melt or mold temperature was not uniform. It can also be difficult to achieve uniform core material distribution in applications where the part geometry is complex. For example, dead-end geometric features, such as bosses or ribs, may tend to fill fully with skin material, thereby preventing core material penetration into the dead-end feature. In the end, the total volume of both the skin and core materials within the part are known, but the exact distribution of core material cannot be determined without sectioning the part. Co-injection process simulation programs are commercially available as an engineering tool to help optimize the design and process conditions for co-injection molded plastic parts.

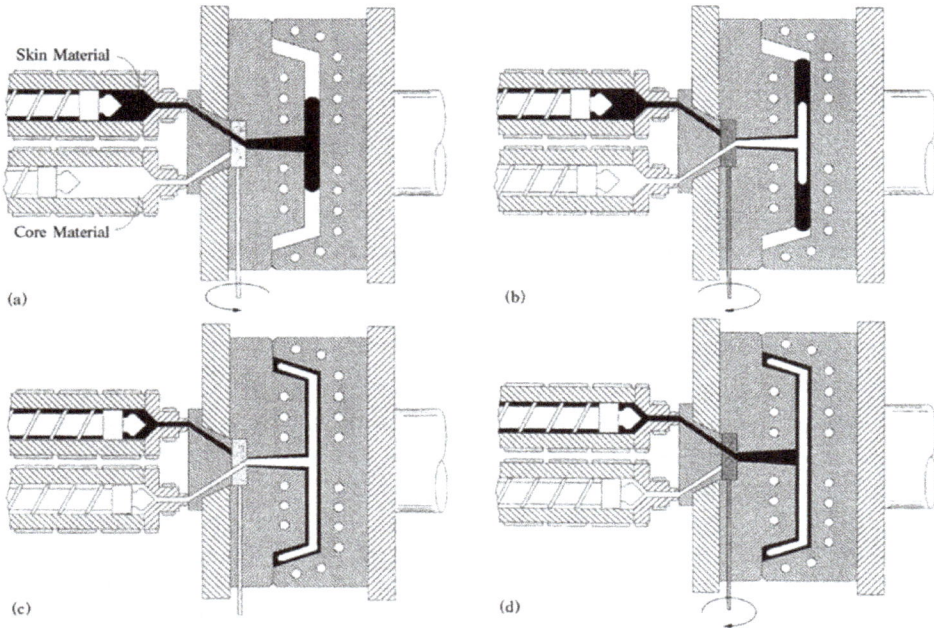

Figure 4.17 The co-injection molding process is used to produce multi-material plastic parts that have a layered structure. A typical co-injection molding process uses two separate, independently controlled injection units. The primary injection unit is used to plasticate and inject the skin material, while a secondary injection unit is used to plasticate and inject the core material. In the end, the molded part that is produced consists of a skin material that fully encapsulates the inner core material [2].

In theory, almost any combination of plastic materials could be co-injection molded. However, in practice there are many issues that limit the choice of materials. For example, it is typically easiest to work with materials having similar processing temperatures and melt viscosities. The chemical compatibility of the materials is also important, particularly when interfacial adhesion is desirable. It is also ideal if the two materials exhibit similar mold shrinkage values in order to limit shear stress at the interface where the two materials meet.

While there are a wide variety reasons why one might consider using the co-injection molding process for the production of plastic parts, some of the more common applications associated with the co-injection molding process include:

Material cost savings: Many co-injection molding processes utilize a core material that is less expensive than the primary skin material. The core material could be a lower cost regrind or recyclate, or perhaps an unpigmented version of the skin material. The recent demand for plastic parts that have recycled content has been one of the driving factors in expanding the use of the co-injection molding process.

Improved barrier properties: Many plastic parts, particularly molded preforms and containers used in the food and drink packaging sector, require good barrier properties. While most plastics are good moisture barriers, most have limited resistance with respect to oxygen or carbon dioxide permeation when compared to metals or glass. High-barrier plastic packaging can be produced by incorporating a thin, co-injected layer of a high-barrier core material, such as ethylene vinyl alcohol. Plastic packages with improved barrier properties can be created by co-injecting a thin high-barrier material layer sandwiched between the package skin material. The increasing demand for high-barrier plastic packaging is another driving factor for the co-injection molding process development. While multi-layer plastic packages can extend product shelf life, the recycling of multi-layer or multi-material plastic parts is generally more difficult than that of monolayer plastic packaging products [2].

Improved surface appearance: Some plastic material formulations develop a less than optimum surface finish when injection molded. For example, the surface finish that can be achieved with a glass fiber reinforced thermoplastics is generally inferior to that of an unreinforced material grade. Likewise, plastic parts produced using the low-pressure structural foam process will have a surface that contains splay or swirls. These surface appearance issues become moot when the material having surface defects is used as the core material in a co-injection molding process. A high-quality surface finish and appearance can be achieved in such applications if a neat thermoplastic is used as the co-injected skin material.

Electromagnetic interference (EMI) or radio frequency (RF) shielding applications: Many electronic enclosures or business machine housing applications require some level of EMI or RF shielding. Since most plastics are not inherently electrically conductive, this EMI or RF shielding performance is typically accomplished using conductive paints or coatings sprayed onto conventionally molded plastic parts. Plastics can be made conductive using additives such as carbon black, stainless steel fibers, nickel coated glass fiber etc. These additives enhance the material's electrical conductivity, but tend to introduce appearance and mechanical performance issues. Co-injection molding is sometimes used in such applications where the conductive material is used as the co-injected core material, and a compatible neat thermoplastic, with its improved surface appearance and mechanical performance, is used as the skin material. As with all co-injection molding processes, it is important to confirm that the distribution of the core material is uniform throughout the finished molding.

4.6 Multi-shot Injection Molding Process

Like co-injection molding, multi-shot injection molding is a multi-material injection molding process that requires molding machinery with multiple injection units. It differs from the co-injection molding process that is discussed in Section 4.5 of this text, in that the multi-material plastic parts it produces do not have a layered or sand-wich-like structure or material distribution. Rather, multi-shot molding processes, of which there are many variants, are more like an insert molding processes, where a second material (and a third material in some cases) is molded onto or adjacent to a solidified part that was molded from a first (primary) material.

The terminology associated with multi-shot molding can be a bit confusing as the process is sometimes described as (i) multi-component molding (ii) multi-shot molding, (iii) over-molding, or (iv) two shot molding [2, 7]. The two-shot molding process, which involves only two molding materials, is by far the most popular multi-shot molding process. The multi-material molding process discussion here is limited to the more common two-shot injection molding process.

Figure 4.18 One of the first applications for two-shot injection molding was the production of computer keys. Computer keys having printed-on characters (letters, numbers, symbols) would tend to wear over time, while those manufactured via two-shot molding will not show signs of wear since the light color plastic lettering has depth.

The two-shot molding process offers number of advantages in terms of both production economics and end product performance compared to conventional injection molding. The process does require more expensive, specialized molding machinery and more expensive tooling, but its popularity is growing rapidly due to the economic and performance advantages that can be realized with this process (especially in higher volume applications). One of the earliest applications for the two-shot molding process is that of multi-color computer key manufacturing. All computer keys (or pushbuttons) have one or more characters (e. g., letters or symbols) on their top surface as shown in Figure 4.18. Keys of this type could be manufactured using a variety of different manufacturing technologies, each offering its own advantages and disadvantages. For example, a pigmented thermoplastic key could be molded using a conventional injection molding process, and then have the lettering printed (e. g., pad printed) onto the top surface. The letter could also be applied as an adhesively bonded label or sticker. None of these early computer key marking methods are ideal in terms of longevity as the lettering quality could deteriorate over time due to the repeated abrasive action associated with typing or data entry. The two-shot injection molding process offers a great performance advantage for such an application since both the dark color computer key body, and the light color lettering, such as that shown in Figure 4.18 are molded part sections that are fused together as a single-piece part. With two-shot molding, the surface appearance of the light-colored lettering will not deteriorate over time since the lettering itself is a plastic part (having depth) that is intimately joined with the dark plastic key component surrounding it.

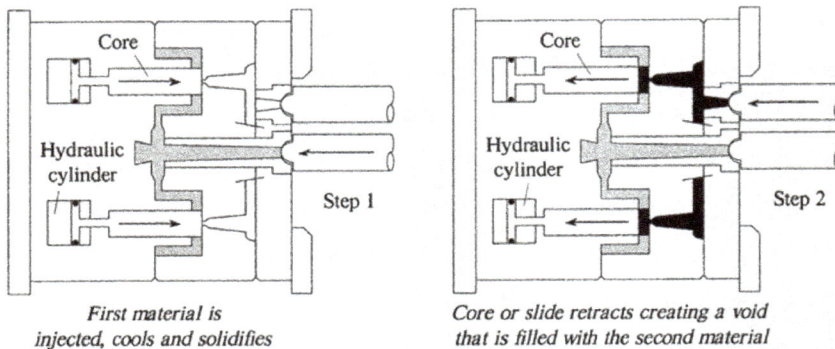

Figure 4.19 The two-shot molding process shown is known as the retractable core two-shot molding process. The primary material is molded (filled, packed and cooled) while the retractable core in the non-retracted position. Once the primary material has cooled, the retractable core is pulled back, and the secondary material is injected through a second runner system into the open space left by the now retracted core. The multi-material part can be ejected once the second material has solidified [2].

As stated, there are several variants of the two-shot molding process. While there are process variants, all two-shot molding process utilize an injection molding machine that has two, independently controlled injection units. These machines can either be purpose built for two-shot molding applications, or they can be conventional molding machines that are retrofitted with a second injection unit and an independent control. One example of a two-shot molding technology involves the use of an expandable mold cavity. Expandable cavity two-shot molding process are more commonly known as retractable core or core-back two-shot molding processes. As an example, the two-shot mold shown in Figure 4.19 has hydraulically activated cores that are retractable. The first step of the retractable core process is really conventional injection molding in that the parts are molded using the primary material with the cores in the non-retracted position. The primary parts are filled, packed and cooled in the same way as in the conventional injection molding process. Once the primary material has solidified, a core section within the mold cavity retracts (without the mold opening), and the secondary material is then injected (through a second runner system) into the open space left by the now retracted core. After filling, the secondary plastic material is then packed and cooled. Once the secondary material has solidified, the mold then opens and the multi-material part is ejected. Conceptually, this two-shot process is similar to plastic-over-plastic insert molding, however the two-shot process does offer a greater degree of automation and generally better interfacial adhesion, since the primary plastic part is still warm when the second (overmolded) material is injected onto it. In most cases, it is desirable for the two materials to exhibit good interfacial adhesion in order to achieve structural reliability for the multi-material part.

A more common two-shot molding technology involves the use of a rotating mold or rotating platen as depicted in Figure 4.20. The rotating mold process differs from the retractable core process shown above in that both injection units operate simultaneously, resulting in a cycle time advantage for the rotating mold two-shot process. With this two-shot process, the lighter-colored part shown in Figure 4.20 is molded conventionally using the primary injection unit and material. Once this part has been filled, packed and cooled, the mold will open. At that time, only the multi-material light and dark colored part in the lower cavity will be ejected. The mold then rotates and closes, automatically transferring the solidified light color part into the larger lower cavity in Figure 4.20. Once the smaller part has been transferred, both injection units are energized. The open cavity space surrounding the lower part will be filled (or overmolded) with the secondary material, while the primary material simultaneously fills the upper cavity. The fully automated process then repeats. This process is commonly used to produce multi-color, multi-material plastic parts, or rigid plastic parts that are overmolded with a thermoplastic elastomer. When two-shot molded parts are produced using materials of different stiffness, the more rigid material is usually the primary material due to its ability to resist the deformation forces associated with the overmolding process. The more flexible of the two materials is normally the overmolded or secondary material.

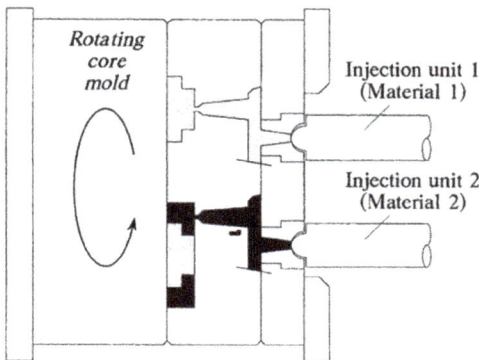

Figure 4.20 A two-shot molding process that utilizes a B-side rotating mold or rotating platen. Once the primary part has been filled, packed and cooled, the rotating mold transfers the solidified primary part is to a larger mold cavity, where the secondary material is used to fill (overmold) the empty space associated with the larger mold cavity. The primary and secondary molding processes occur simultaneously [2].

The two-shot molding process offers a number of advantages over a conventional plastic-plastic insert molding process. Some of these advantages are economic, especially in high production applications since two-shot molding processes are automated. Two-shot molding can also eliminate costly secondary assembly operations (such as welding) since the process is essentially an in-mold assembly operation. Two-shot molding processes also allow for improved positional registration, and improved interfacial bonding between the two materials. A key characteristic of two-shot molding is that the primary part surface temperature is still warm at the time when the secondary material is overmolded onto the primary material. It is generally accepted that a higher primary part surface temperature will improve the interfacial bond quality.

A few examples of common two-shot molding applications are shown below. These multi-material applications often involve materials that have different colors or optical properties, or materials that exhibit very different mechanical properties, or both. Two-shot molding processes have become so commonplace today that the process is often used for the production of one-time-use or disposable plastic items. However, it should be noted that multi-material plastic products (including two-shot molded parts) are more difficult to recycle, especially when the materials cannot be separated for individual material recovery [2].

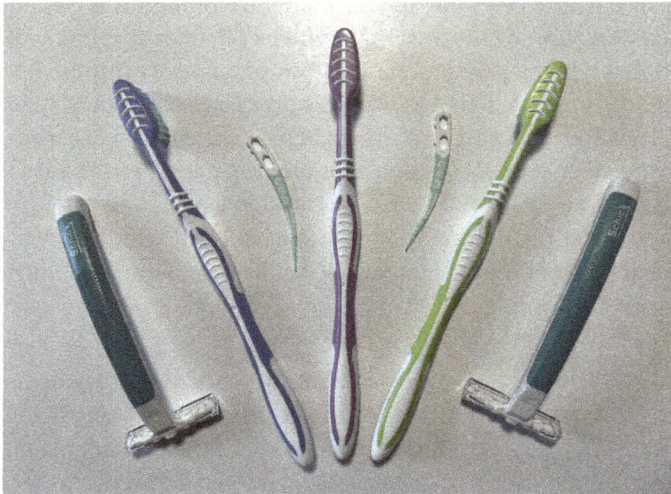

Figure 4.21 The items shown are all produced by the soft-touch two-shot molding process. In each case, the more rigid material is the primary material, and a softer, more flexible thermoplastic elastomer is the secondary or overmolded material.

Rigid plastic – flexible thermoplastic elastomer applications: Many two shot-molding applications utilize one relatively rigid primary plastic material, and a more flexible or elastomeric secondary plastic material. A rigid plastic part that incorporates an overmolded (integral) flexible gasket or seal is one example of such an application. Soft-touch applications such as those shown in Figure 4.21, are also a very common example of a hard-soft application. Soft-touch applications utilize a relatively soft overmolded thermoplastic elastomer as the secondary material. The softer material is used to provide improved grip or tactile feel to strategic areas of the molding (e. g., the handle). Soft-touch applications can be both aesthetically and ergonomically pleasing. Soft-touch two-shot parts are commonly used in applications that include; power tool housings, pens, cooking utensils and personnel care products. Note that with two-shot molding, it is common to gate the primary part in such a way that the primary part's gate vestige will be fully encapsulated (covered) by the softer overmolded secondary material.

Another key with two-shot molding is to ensure that the two materials being molded have a good degree of chemical compatibility with one another. For example, it would be advantageous to use a polyolefin-based thermoplastic elastomer when producing a soft-touch polyolefin or polypropylene two-shot part. On the other hand, it would be more appropriate to use a styrenic-based thermoplastic elastomer when producing a soft-touch two-shot molding where polystyrene is the primary material. It is always best to consult with your material supplier when selecting an overmolding material. It is also common to incorporate geometric features (holes, texture) on the primary part in order to provide for more surface area and/or undercuts that will help improve the interfacial bond integrity.

Rigid plastic – rigid plastic applications: The automobile tail-light lens assembly shown in Figure 4.22 is one of the original and still more common applications for the two-shot molding process (or the three-shot molding process). In this application, the plastic materials that make up the multi-material part are very similar (or the same) except the materials have different colors or optical properties (different pigments or dyes). Two-shot computer keys, such as those shown in Figure 4.18, are another example of two-shot parts that are produced with two rigid thermoplastic materials, each material having a different color.

Figure 4.22
The multi-material, multi-color automobile tail-light lens assembly is typically manufactured using a multi-shot molding process.

4.7 Insert Injection Molding Process

Some multi-material plastic parts are produced by the insert molding process. The process is most widely used for the production of plastic parts that incorporate metallic components, such as a metallic shaft hubs or electrical connector pins or lead frames. With insert molding, a portion or all of the insert is placed within the mold cavity prior to melt injection. The insert becomes permanently embedded in the overmolded plastic part as the material cools and shrinks around the insert. Non-metallic inserts, such as a previously molded thermoplastic part (plastic insert), can also be overmolded using the insert molding process. While plastic insert molding is possible, the two-shot molding process described in Section 4.6 is more commonly used for the production of all thermoplastic multi-material parts.

Insert molding is very widely used in thermoset molding processes, such as transfer molding, since thermosetting materials cannot be resoftened after molding. This makes it difficult or impossible to insert metal components into a thermoset molded part after molding. Insert molding is also common with thermoplastic injection molding processes, especially for inserts having a large or more complex geometry. However, small metallic inserts, such as a threaded brass insert (or a nut-sert) can be thermally or ultrasonically inserted into hollow bosses after a thermoplastic injection molded part has cooled. This is not possible with thermoset plastic parts since they cannot be re-softened after molding. The insert molding discussion that follows applies to thermoplastic parts and metallic inserts that have a geometry that does not lend itself to post-mold insertion.

The molding machinery used for insert molding is conventional in most respects. One exception is that of the molding machine clamp system. Most conventional injection molding machines have a horizontal clamp system. Insert molding can be accomplished using a molding machine with a horizontal clamp, but molding machines having a vertical clamp are generally preferred for insert molding. When a vertical clamp is used, the inserts can be manually or robotically placed into the lower mold half, and can be more reliably held in place by the force of gravity. Inserts placed in a horizontal clamp may be more likely to become dislodged due to machine vibration or clamp acceleration or deceleration. Regardless of the clamp system used, it is extremely important that the clamp is set up correctly, especially in terms of low-pressure mold protection. An improperly placed (or dislodged) metallic insert can cause severe and costly mold damage.

The mold clamping system used for insert molding may also include either a shuttle table or a rotary platen. As an example, many insert molds have one A side mold half and two B side mold halves. The inserts can be loaded (and subsequently unloaded) into one of the B mold halves while the second B mold half is being filled (i. e., overmolded), packed and cooled.

The metallic inserts that are placed into the injection mold must also be precisely located within the mold cavity. The insert must be supported such that it does not bend or distort when the overmolded thermoplastic is injected into the mold cavity. The rod-like metal insert shown in Figure 4.23, or the mold itself would also incorporate a feature such that axial registration of the insert is repeatable. In addition, both the A and B sides of the mold would have half-cylindrical segments at the mold parting line that firmly hold the metal insert in place when clamp force is applied. These depressions also serve as a shut-off surface to prevent molding flash. These mold depressions also help prevent the insert from shifting or repositioning when the melt is injected. It is important that insert-molded parts are gated in such a way that the forces associated with melt flow during mold filling act on the insert in a balanced manner. Unbalanced mold filling can cause the insert to bend or deflect during mold filling, particularly for thinner sheet metal inserts (e. g., electrical lead frames). Internal mold cavity pins or stand-offs are sometimes used to provide additional support for more flexible inserts to help minimize insert deformation.

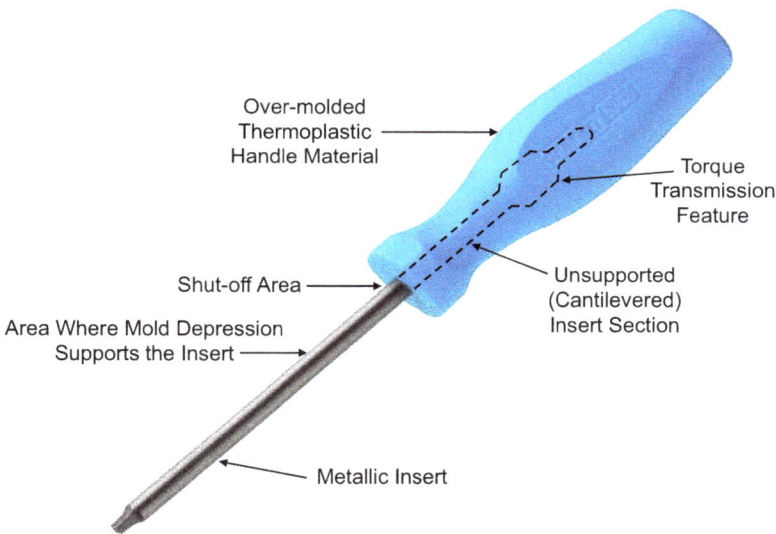

Over-molded
Thermoplastic
Handle Material

Torque
Transmission
Feature

Shut-off Area

Area Where Mold Depression
Supports the Insert

Unsupported
(Cantilevered)
Insert Section

Metallic Insert

Figure 4.23 A typical screw or nut driver is manufactured by (i) manually or robotically placing a rod-like steel insert into an injection mold cavity, (ii) closing the mold, and (iii) overmolding a thermoplastic material around the cantilever in-cavity portions of the insert.

Most insert molding applications require some degree of stress transfer between the metallic insert and the overmolded thermoplastic material. The surface characteristics of both the insert and the plastic material are important factors in terms of wetting, adhesion and stress transfer. Ideally, any type of surface contamination (e. g., oils) should be removed from the insert prior to overmolding. Insert surfaces are sometimes treated prior to molding in order to improve the interfacial adhesion between the metallic insert and the overmolded thermoplastic plastic. Metallic inserts are sometimes pre-heated to help improve wetting and to minimize the cooling of the plastic melt during mold filling. This is an extra step that can help improve the interfacial properties in critical applications, but the pre-heating process must be repeatable for consistent results. Inserts, such as the screw driver insert shown in Figure 4.23 may also be textured or geometrically modified to enhance tortional or pull-out strength by introducing mechanical undercuts as shown by the hidden lines in Figure 4.23.

Shrinkage stresses can also be a concern for insert-molded parts. The thermoplastic screw driver handle shown in Figure 4.23 will shrink as the part cools. The higher modulus of the metal insert will prevent the overmolded thermoplastic in the area around the insert from shrinking in the hoop direction. As a result, the thermoplastic handle will have (internal) shrinkage stresses in the hoop direction. These shrinkage stresses will help anchor the insert and provide torsional strength, but they can also lead to cracks or crazes, particularly if the part is exposed to an aggressive chemical. This can be a particular problem for high-shrink-rate plastics that have low break strains. It is also best if the inserts do not contain features (e. g., very sharp corners) that act as stress risers and further increase the likelihood of cracking.

4.8 Injection – Compression Molding Process

Another modified injection molding process is the injection-compression molding process (ICM). The ICM process is typically used in a number of demanding applications where the traditional injection molding process has some deficiencies. The process offers a number of unique advantages, particularly for very thin-wall plastic parts that require a low degree of internal stress, low warpage and superior pattern transfer or surface feature replication.

The ICM process has some characteristics of both the conventional injection molding process and the conventional compression molding process. The ICM process is in some ways analogous to some very early thermoplastic transfer molding or melt fed compression molding processes where a fixed volume of pre-plasticated thermoplastic melt is extruded or injected into an open and cooled compression mold. The compression mold is then closed to form and cool the part. The ICM process machinery is conventional in most ways, but the machine must have a clamp system and control system suitable for ICM. There are a number of ICM process variants, but the sequential ICM process, and semi-sequential ICM process are most common.

In the first stage of a conventional or traditional injection molding process, plasticated melt is injected into a fully closed mold. This first stage of the traditional molding process can become difficult for very thin cavities, particularly those having a high flow length to part thickness ratio that require very high injection pressures. While it is true that molding machine injection pressure capabilities have increased significantly over the years to satisfy the demand for thin-walling, very thin-wall parts may require even greater mold filling pressure. Higher melt flow rate material grades have also been developed for thin wall applications, but their use is not always appropriate. The ICM process is a molding process alternative for the production of thin-wall part geometries, particularly when flow lengths are long and low internal stress levels are desirable [2, 7].

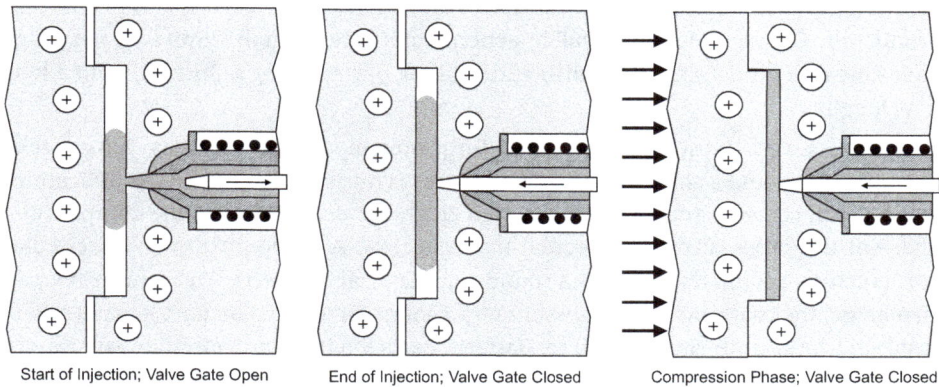

| Start of Injection; Valve Gate Open | End of Injection; Valve Gate Closed | Compression Phase; Valve Gate Closed |

Figure 4.24 The process sequence for a typical sequential injection compression molding process begins with the injection of a full shot into a partially open mold cavity (left). The mold clamp then closes (as it would in a compression molding processes) causing the melt to flow radially into the now thinner cavity. A valve gate or shut-off nozzle is used to prevent backflow of the melt into the hot runner or molding machine nozzle during the compression phase of the process.

The basic sequential ICM process is depicted schematically in Figure 4.24. The process begins with the injection of the conventionally plasticated melt into a partially open mold cavity. The cavity wall thickness at the time of melt injection is typically two times (plus or minus) the final cavity or part thickness. Immediately after mold filling, the molding machine clamp is then energized to compress the molten material in the mold, forcing material to flow towards the unfilled regions of the now thinner mold cavity. A valve gate or shut-off nozzle is used to prevent back flow of material into the hot runner or hot molding machine nozzle. The mold used with this process must be capable of containing the pressurized melt in all directions (including laterally). In the simplest case, this can be accomplished using some type of shear edge tooling as shown in Figure 4.24.

A benefit of the ICM process is that the mold cavity is relatively thick when the injection of the melt into the cavity occurs. Relatively thick mold cavities have a low pressure drop and can be filled very quickly. The flow length associated with the temporarily thicker mold cavity is also shorter than the mold cavity length. The mold then closes quickly in the compression phase of the ICM process causing the melt to flow to the extremities of the now thinner cavity. When compared with conventionally molded thin wall parts, ICM parts exhibit more uniform (and sometimes lower) mold shrinkage, lower residual stress (or lower birefringence in the case of optically transparent parts), low warpage, and good surface replication. These advantages are largely due to the fact mold cavity pressures with the ICM process tend to be more uniform over the cavity surface area than those associated with the conventional injection molding process. The ICM process can also be more versatile in terms of possible material melt viscosities (or melt flow rates) that can be used for thin-wall part production. Conventional thin-wall injection molding is normally limited to very high melt flow rate materials, due to difficulties filling parts having a thin wall and a long flow length.

ICM processes are most commonly used for producing plastic parts that have a relatively thin wall and a shallow draw. The process is most easily applied to applications when the parts are circular, since the melt flow that occurs during the compression phase of the process results in radial material flow. A sprue- or disc-gated circular part (such as a compact disc) has a round symmetrical geometry. The radial flow pattern associated with melt compression of a center-gated circular part geometry will result in a very uniform mold filling pattern. Parts having a non-circular asymmetrical geometry can be produced by ICM processes, but the gating scheme utilized should be as symmetrical as possible in terms of the mold filling flow pattern. Flow leaders can also be utilized to improve the flow pattern when molding asymmetrical parts.

One part defect that can occur, particularly with the sequential ICM process, is related to the flow hesitation that will occur between the end of the melt injection phase, and the beginning of the compression phase. The stop – start action or hesitation tends to create a dull surface appearance or halo mark on the surface of the plastic part (near the hesitating melt). The halo issue can be minimized or eliminated with a slight ICM process modification. With the semi-sequential ICM process, the compression phase of the process is initiated shortly before the injection phase is completed. The later portion of the injection phase and the compression phases occur simultaneously. This helps eliminate any flow hesitation and the related surface defect or halo. Both the sequential and semi-sequential ICM process require the use of injection molding machines that are ICM capable.

4.9 Metal (and Ceramic) Injection Molding Processes

Metal and ceramic injection molding processes are typically used to produce smaller-size, complex, tight-tolerance net shape metal or ceramic parts. Both metal and ceramic injection molding processes involve a series of complex processes that begin with material preparation and end with post mold finishing. The discussion here will focus primarily on the more popular metal injection molding processes, but the same principles generally apply to ceramic injection molding.

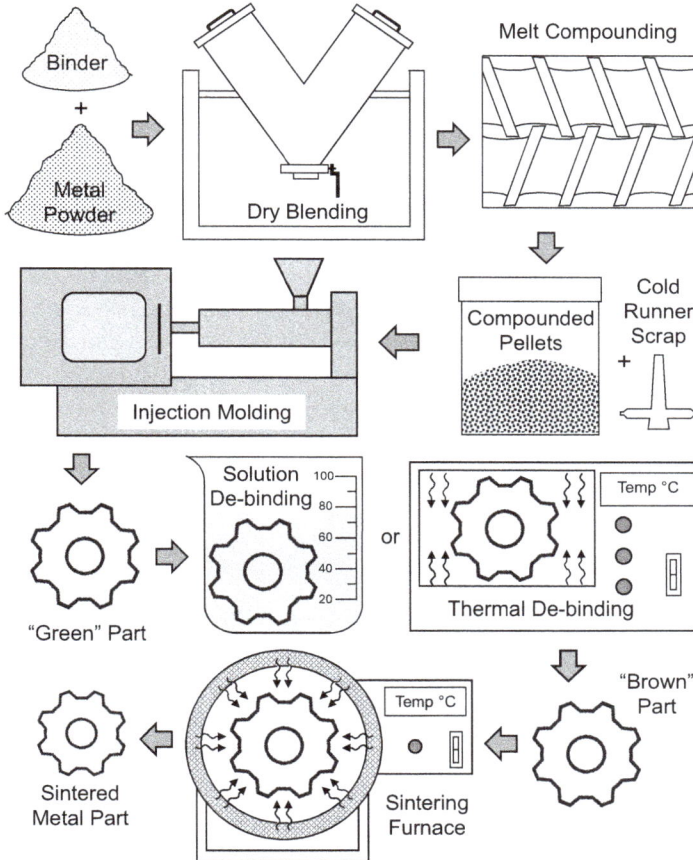

Figure 4.25 The metal injection molding process involves a large number of processing steps that include pre-molding steps, injection molding, and post mold processing that includes de-binding, sintering and post sinter finishing.

Compared to the conventional injection molding process that produces thermoplastic parts, the metal injection molding process is used to produce metallic parts that have greater strength, stiffness and greater temperature capability. The metal injection molding process is one of many ways to produce metal parts. Processes for the production of metal parts include; (i) conventional subtractive machining, (ii) investment casting, (iii) die casting, (iv) additive manufacturing or (v) metal injection molding. The die casting process is most commonly used for the production of small- to medium-size, high-volume non-ferrous metal parts. Die casting is commonly used for the production of aluminum, zinc, and magnesium parts. The metal injection molding process is more commonly used for producing metallic parts from a wider variety of higher melting point metals or metal alloys than those used for die casting. The metal injection molding process is more complex than die casting, but it can be used for the economical production of tight-tolerance, complex-geometry metal parts, from a wide variety of high-performance metal alloys [13–16]. The metal injection molding process is a capital-intensive process involving many processing steps that begin with formulation preparation, as depicted in Figure 4.25.

Raw materials: The raw materials required for the metal injection molding process are (i) a fine metal powder, and a polymeric or wax binder. The type of metallic powder utilized will be the primary factor in determining the mechanical properties, hardness and corrosion resistance that can be achieved for the finished metal part. The powdered metals that are most commonly used with this molding process are stainless steels, while high performance alloys based on titanium, tungsten and cobalt are used to a lesser extent. The metallic powder can be produced by conventional size reduction techniques (e. g., pulverizing) or using proprietary techniques that are capable of producing more spherical metal powders. The metal powders can have particle sizes in the range of 5–20 microns, and have a particle size distribution that will lead to both good flow behavior (during mold filling) and a good packing density (during sintering). The powdered metal grades used for metal injection molding should also be very pure (i. e., free from contamination).

The binder material is typically a low molecular weight formulation based on powdered polymer or wax flakes that will exhibit; (i) good flow properties to facilitate mold filling, and (ii) complete vaporization when the molded parts are sintered. Common thermoplastic binder materials include neat polyethylene, polypropylene and polyoxymethylene. Like most plastic materials, the binder may also contain smaller concentrations of other additives to enhance processability and final part properties.

Blending and compounding: The metal powder and the binder materials must be thoroughly mixed prior to injection molding. Some of the more commonly used metal injection molding formulations are available as pre-compounded and pelletized commercial grades from material suppliers. Other more specialized or proprietary metal injection molding formulations require custom compounding. This compounding can be accomplished using different types of batch or continuous mixing equipment. Because most injection molding machine plasticating units are *not* particularly good mixers, melt compounding is typically done before the formulation is injection molded. Twin-screw melt compounding extruders are commonly used to prepare metal or ceramic filled formulations. It should be noted that some of these formulations can be very abrasive due to their relatively high metal or ceramic powder concentrations; this means that the compounder mixing equipment elements should have good wear resistance. The powders for the formulation can be pre-blended or each powdered material can be metered into the process separately to minimize any material segregation or settling [15]. Twin-screw compounding extruders are also very good dispersive mixers, capable of breaking up any material agglomerates that may have formed over time. This melt compounding process produces well mixed, free flowing pellets, which are easier to handle and feed for the single screw injection molding process.

Injection molding: The injection molding process that is used to produce metal injection molded parts is more or less conventional. Mold runner systems and gating schemes are typically designed to ensure the mold filling pressures and flow pattern are optimized, while minimizing production scrap. Cold runner scrap can be blended with the virgin material feedstock in some metal injection molding applications provided the regrind is free from contamination. Special attention should also be given to the design of the injection mold ejection system, since the mechanical properties of what is known as the green (as molded) metal powder formulation are typically limited compared to those of a neat plastic material. A well-designed part ejection system should not overstress and damage the green parts during the part ejection process.

Thermal or solution de-binding: After molding, the green molded parts undergo some level of de-binding to remove most of the binder that was primarily added to impart the flow properties required for injection molding. Once the injection molding process is complete, the binder can be removed. This can be done using a thermal process, a solution (or solvent) process, or both. Catalytic de-binding is also a de-binding option for a binder material that can undergo catalytic depolymerization. Parts that have undergone the de-binding process are described as brown parts that now have a porous structure that is associated with the binder removal.

Thermal sintering: The brown (de-bound) injection molded parts are then sintered in a furnace like batch or continuous sintering chamber. Sintering is a high temperature process that is conducted in a controlled atmosphere (e. g., under vacuum or in an inert nitrogen or argon environment to prevent oxidation). The sintering temperatures must be high enough to achieve a high final part density that is typically 96–99% of the theoretically possible part density. The sintering time and sintering temperatures used are sometime high enough to cause partial melting of the metal, but conditions are generally optimized to allow for optimum mechanical property development. Brown metal parts also shrink significantly during the sintering process. The shrinkages associated with the sintering process can be as high as 15–25%.

Post sinter processing: After the sintering process is completed, the sintered parts may be finished in a variety of ways depending on the metal type and the final application. These post processes can include; conventional (finish) machining, surface polishing, heat treatment, pressing, plating etc.

Applications for metal injection molded parts range from medical devices (e. g., orthopedic implants or medical tools) to consumer products (e. g., golf club heads) to automotive components (e. g., turbo impellors or valve train parts). Ceramic injection molded parts are typically used in applications that require both complex geometries, outstanding mechanical performance, corrosion resistance and abrasion resistance.

References

[1]   Tosello, G., "Micro Injection Molding", Hanser Publishing, Munich (2018).

[2]   Malloy, R., "Plastics Part Design for Injection Molding", 2nd Ed., Hanser Publications, Munich (2011).

[3]   Technical Bulletin, "Foaming Agent Masterbatches", RTP Company, Winona, MN 55987 (2025).

[4]   Lee, J. et. al., "Study of the Foaming Mechanisms Associated with Gas Counter Pressure and Mold Opening using the Pressure Profiles", *Chemical Engineering Science*, Vol. 167, August (2017).

[5]   Li, S. et.al., "Influence of Relative Low Gas Counter Pressure on Melt Foaming Behavior and Surface Quality of Molded Parts in Microcellular Injection Molding Process", *Journal of Cellular Plastics*, **50** 5 (2014).

[6]   Avery, J., "Gas Assist Injection Molding", Hanser Publication, Munich, Germany (2001).

[7]   Osswald, T., Turng, T., and Gramann, P., "Injection Molding Handbook", Hanser Publications, Munich (2022).

[8]   Technical Bulletin, "Gas Assist Molding", Mack Molding, Arlington, VT 05250.

[9]   Hsu, J., "Predicting Gas Fingering Behaviors Using 3D GAIM Simulation Tool", Moldex3D, Farmington Hills, MI 48334, May 1 (2018).

[10] Su, H. et.al., "Reducing Ghost Marks in Injection Molded Plastic Parts Using External Gas Assisted Holding Pressure", *International Communications in Heat Transfer and Mass Transfer*, **66** August 15 (2015).

[11] Ham, S., "External Gas Molding", https://steveham.com, November 19 (2019).

[12] Loaldi, D. et.al, "Manufacturing Signatures of Injection Molding and Injection Compression Molding for Micro-Structured Polymer Fresnel Lens Production", *Micromachines*, 9 (2018).

[13] Acharya, D., "Metal Injection Molding (MIM): A Complete Guide", Pro-Lean Corporation, Shenzhen, China (2024).

[14] Technical Bulletin, "Seven Tips for Successful Metal Injection Molding with Powder Metallurgy", Redstone Manufacturing, Defiance, OH 43512, December 3 (2023).

[15] Leister, D., "Optimize Powder Injection Molding by Compounding the Perfect Feedstock", ThermoFisher Scientific, Karlsruhe, Germany (2024).

[16] Heaney, D., "Handbook of Metal Injection Molding", 2nd Ed., Woodhead Elsevier Publishing (2019).

Index